Coastal and Estuarine Studies

Coastal and Estuarine Studies

formerly Lecture Notes on Coastal and Estuarine Studies

32

F. Wulff J.G. Field K.H. Mann (Eds.)

Network Analysis in Marine Ecology

Methods and Applications

Springer-Verlag

Berlin Heidelberg New York London Paris Tokyo Hong Kong

Editors

F. Wulff
Institute of Marine Ecology, Askö Laboratory, University of Stockholm
10691 Stockholm, Sweden

J.G. Field
Marine Biology Research Institute
Zoology Department, University of Cape Town
Rondebosch 7700, South Africa

K.H. Mann
Department of Fisheries and Oceans
Habitat Ecology Division, Bedford Institute of Oceanography
Box 1006, Darthmouth, Nova Scotia, Canada B2Y 4A2

ISBN-13: 987-3-642-75019-9 e-ISBN-13: 987-3-642-75017-5
DOI: 10.1007/987-3-642-75017-5

Softcover reprint of the hardcover 1st edition 1989

2131/3140-543210 – Printed on acid-free paper

To obtain a diskette containing the software for the techniques used in this book, (for IBM or compatible computers on a MS DOS 360 kB diskette, for a Macintosh computer on a 3.5" diskette) send a money order or bank draft for US $ 10.00 payable to SCOR to:

Excecutive Secretary, SCOR
Department of Oceanography
Dalhousie University
Halifax, N. S. CANADA
B3H 4JI,

requesting Network Analysis Software.

CONTENTS

DEDICATION

PART I INTRODUCTION

Chapter 1 The need to analyze ecological networks 3
J. G. Field, F. Wulff & K. H. Mann

PART II METHODS AND THEORY

Chapter 2 A detailed guide to network analysis. 15
J. J. Kay, L. A. Graham & R. E. Ulanowicz.

Chapter 3 Construction of flow networks using inverse methods. 62
A. F. Vézina

Chapter 4 A generic simulation model for treating incomplete sets of data. 84
R. E. Ulanowicz

Chapter 5 Dealing with the "mixed units" problem in ecosystem network analysis. 90
R. Costanza & B. Hannon

PART III APPLICATIONS

Chapter 6 Network information indices with an estuarine model. 119
J. N. Kremer

Chapter 7 Network analysis of simulated succession after an upwelling event. 132
J. G. Field, C. L. Moloney & C. G. Attwood

Chapter 8 Derivation and analysis of flow networks for open ocean plankton systems. 159
H. W. Ducklow, M. J. R. Fasham & A. F. Vézina

Chapter 9 Network analysis of the North Inlet salt marsh ecosystem. 206
M. L. Asmus & H. N. McKellar, Jr

Chapter 10 Analysis of the flow network in an estuarine benthic community. 220
R. M. Warwick & P. J. Radford

Chapter 11 A comparative anatomy of the Baltic Sea and Chesapeake Bay ecosystems. 232
F. Wulff & R. E. Ulanowicz

PART IV DISCUSSION AND CONCLUSIONS

Chapter 12 Network analysis in marine ecology: an assessment. 259
K. H. Mann, J. G. Field & F. Wulff

INDEX 283

DEDICATION

This volume is dedicated to the memory of Dr David F. Smith who died suddenly on 19 May 1987 at the age of 52. From 1976 to 1982 he was a member of the working group " Mathematical Models in Biological Oceanography" established by the Scientific Committee on Oceanic Research (SCOR). By his knowledge, wit and wisdom he made a major contribution that will be long remembered with affection by his colleagues and friends. He made a further important contribution to a second working group,"Ecosystem Theory in Relation to Biological Oceanography" whose deliberations culminated in the present volume.

Those who worked with him mourn his early death but recognize that it was a privelege to have known him.

PART I

INTRODUCTION

Chapter 1

THE NEED TO ANALYZE ECOLOGICAL NETWORKS

John G. Field
Marine Biology Research Institute
University of Cape Town
7700 Rondebosch
South Africa

Fredrik Wulff
Askö Laboratory
University of Stockholm
S-10691 Stockholm
SWEDEN

Kenneth H. Mann
Bedford Institute of Oceanography
PO Box 1006
Dartmouth, NS
CANADA B2Y 4A2

BACKGROUND TO THIS VOLUME

This volume arises out of the activities of two working groups organized under the auspices of the Scientific Committee on Oceanic Research (SCOR). The first, SCOR Working Group 59 on "Mathematical Models in Biological Oceanography" met in 1977, 1979 and 1982, under the co-chairmanship of Kenneth Mann and Trevor Platt. A book with the same title (Platt, Mann & Ulanowicz 1981) resulted from these meetings. In a nutshell, the group achieved some consensus that simulation modeling was a useful activity, but that simulation models of ecosystems had strict limitations: they were not predictive because qualitative changes of state often occurred in ecosystems, and these could at that time usually not be predicted or modeled. The group therefore resolved to explore the properties of ecosystems other than those revealed by simulation models. A theme which developed was that it is necessary to have at least as much information on the fluxes among the components of marine ecosystems as on the biomasses. This led in 1982 to a symposium reviewing knowledge of the "Flows of Energy and Materials in Marine Ecosystems", organized in the NATO Advanced Research Institute series, the proceedings of which have been published (Fasham 1984). Here the theme persisted that there was a need to measure flows of energy and materials.

In 1982 Working Group 59 disbanded, and a new SCOR Working Group 73 was established under the same chairmanship and with the same nucleus of oceanographers to explore "Ecosystem Theory in relation to Biological Oceanography". Five topics were identified with potential to reveal ecosystem-level properties: *Thermodynamics, Statistical Mechanics, Input-Output Analysis, Information Theory,* and *Ataxonomic Aggregations*. Experts in each of these fields were invited to address the topic and discuss ecosystem theory with oceanographers at a colloquium held in Quebec in 1984. The colloquium was an outstanding success and the proceedings were published as a volume (Ulanowicz & Platt 1985).

There emerged from the topics and discussions at the colloquium, a recognition that input-ouput analysis, flow analysis, information networks and dynamic network analysis are a family of related techniques with some potential to provide additional insights into properties of whole ecosystems. This volume attempts to explore that potential.

NETWORK ANALYSIS IN OTHER DISCIPLINES

Economic *input-output analysis* was developed by Leontief (1951) to quantify the amount of raw materials and industrial services required to produce a quantity of consumer goods. This work traced outputs from components of an economic flow diagram or model (the demand for consumer goods) back to the sources of raw materials or inputs. Thus input-output analysis is the term normally applied to tracing flows backwards from the ouputs to the sources. Hannon (1973) was the first to apply Leontief's analysis to ecological systems. The backward form of analysis in ecology is usually termed *flow analysis*. In flow analysis, flows into each compartment are defined as linear functions of the total outflow from that compartment.

Augustinovics (1970) later worked out a forward form of economic analysis to trace the fate of system inputs through the economic flow diagram to their destinations. Finn (1976) developed a forward case analysis for ecological systems, and simultaneously introduced several measures of ecosystem function, including cycling, that aid in the ecological interpretation of ecosystem model results. The forward form of analysis is usually termed *compartmental analysis*. In compartmental analysis, outflows from each compartment are defined as linear functions of the total inflow into that compartment.

The term *flow analysis* has also been used in a more general way to include both forward and backward analyses, but we prefer the term *network analysis* for the general case to avoid ambiguity. Network analysis goes further than the original input-output analysis in that it includes analysis of cycles, through-flows, storages (Patten 1985) and information theory.

Network analysis has been applied in many disciplines. It forms a branch of statistics, is used in control theory, and is most widely applied to analyze electrical networks in which flow diagrams represent the flows of electric currents in circuit diagrams with resistors, transistors, condensors etc. as the components. Forms of network analysis have been developed to a very advanced level for the design of printed-circuit boards in electronics, and sophisticated software packages such as SPICE are used to develop simulation models from flow networks, bridging the gap between flow diagrams and simulation models. A theory of "network thermodynamics" has also been developed and applied to analyze the distribution of drugs through human body organs (Mikulecky 1985). "Network thermodynamics" may include non-linear relationships between flows in networks, but has not yet been applied to ecosystem flows.

In this volume we are only concerned with linear relationships of flows to donor and recipient compartments of ecosystems. We use the term network analysis to indicate that we are dealing with ecosystems which form networks of flows of energy, nitrogen, carbon, and many other elements. The term is also used to include the analysis of flows using information theory.

RELEVANCE OF NETWORK ANALYSIS TO ECOLOGY

Ecologists are used to the concepts of food chains and food webs, and to those of species richness and diversity. Network analysis provides tools for bringing these and many other concepts together by depicting ecosystems as a number of compartments interconnected by flows of energy or matter from one compartment to another. The compartments are defined by the ecologist. They may be species or they may be functional groups such as detritivores, a nutrient pool, particulate organic matter, etc. Thus each network analysis is based upon a model of the ecosystem being studied, at the level of detail decided upon by the ecologist. The analysis therefore depends upon the structure of flows among the system components and their magnitude as measured, estimated, or deduced by the researcher. The flows are of course rates per unit of time, and may vary greatly from second to second, but at the ecosystem level are more usually measured on a daily, monthly, seasonal or annual time scale. We believe that these flows contain a lot of information about the dynamic structure of a whole system and of how it functions, and introduce some of the terms and concepts of ecological network analysis below. Details of the methods and formulae are given by Kay, Graham & Ulanowicz (1989) in Part II of this volume.

A cycling index

A cycling index shows the proportion of the flow in a system that is recycled compared with the total flow. For example, it is well known that in tropical forests, nutrients such as nitrogen and phosphorus are tightly recycled. At any given time a large proportion of the nutrients present are bound up in the biomass of trees. Those nutrients that leave the trees as a result of grazing, leaf fall or leaching are rapidly taken up again. New inputs from the atmosphere are also taken up rapidly. The forest has a large cycling index for both nitrogen and phosphorus. Coral reefs, on the other hand, have a large cycling index for phosphorus but a smaller one for nitrogen, with extensive nitrogen fixing on the reef and considerable downstream export (Webb *et al.* 1975). It is therefore clear that the magnitude of the cycling index depends on the substance for which the flux is being modeled.

As suggested by the comparison between phosphorus and nitrogen in coral reefs, the amount of recycling depends upon the currency. Another commonly used currency is energy. Ecosystems are usually open with regard to energy flow, with little energy cycling and a small cycling index. On the other hand, if carbon is used as a currency in a model ecosystem, the carbon respired as CO_2 may be included in a CO_2 pool compartment, or it may be regarded as being exported from the system. Very different cycling indices will result, emphasizing that network analysis is very dependent upon the structure of the model chosen.

Average path length

The average path length measures the average number of steps a unit of flux must travel from entry into the system until it exits from the system. Thus it is commonly held that food chains in upwelling areas of the oceans are short, with the major primary production by large diatoms which are grazed by macrozooplankton or by anchovies or sardines. In warm open oceans food chains are much longer, with picoplankton playing a large role in primary production, and energy and nutrients being passed along from smaller to larger organisms in a series of steps ending in large top carnivores such as tuna. The concept of food chain length is easy to visualize. However, in reality, one is dealing with food webs, not food chains, and the situation is much more complex. The length of a food web is described by the average path length.

Diversity of flows in networks.

Flow diversity measures how many interactions there are in an ecosystem and is a therefore much more dynamic concept than species diversity. It is formulated from both the number and evenness of flows. For example, fish production in the North Sea is based on relatively few species of exploited fish, hence one would expect relatively few pathways of food leading to the total fish catch, compared to the fishery off Thailand (Longhurst & Pauly 1988) which is based on many species of exploited fish and presumably a great many flow pathways. The number of flows can be contrasted with the evenness of flows, or how equal the magnitudes of flows are to one another.

Indices of ecosystem growth and development.

Margalef (1968) has shown that succession occurs in marine ecosystems. For example, in temperate latitudes at the time of a spring bloom, diatoms are the main primary producers with a food chain leading to zooplankton and fish larvae. As nutrients are exhausted from the stable surface layer, flagellates take over the main role as primary producers and a much more complex food web develops. The plankton community evolves from an immature to a mature system through succession. Odum (1969) and Margalef (1968) have documented the many changes in biomass, production-to-respiration ratios, etc. that occur as a food web develops with succession. As these successional changes happen, the network of flows changes to a more complex network with richer connections. Ulanowicz (1980, 1986) has proposed to index the maturity of an ecosystem using three measures based on information theory.

Total system throughput (T) measures the size or growth of the system in terms of the flows through all its components, the more material flowing through the system, the bigger the *T*.

Ascendency (A) represents both the size and organization of the flows in a single measure, the product of *T* and the diversity of flows in a system. According to Ulanowicz, as an ecosystem matures and goes through a series of successional stages, the Ascendency of the system should increase. As a corollary, if a system is affected by disturbance or pollution, this should be reflected by change in the flows of energy or materials through the system, and Ascendency should change too.

Development Capacity (C) measures the potential of a particular network to develop, given its particular set of connections and total throughflow. *C* gives the highest possible value of *A* .

Some of the finer points of Ulanowicz's ideas on Ascendency are disputed by Kay (1984) who believes that two additional measures are necessary to characterize the evolution of ecosystems in succession.

Flow diversity (D) measures the diversity with which components of the system are used as food resources by other living components, in other words it measures the evenness of inputs to living components and reaches a maximum if all flows are equal in magnitude.

Flow specialization (S) measures the diversity of flows, averaged over all living components in the system; it is the average resource niche breadth. When *S* is large, most species or components are generalists and it appears to us that the term "Specialization" might better be applied to *1/S*. As a system matures, *D* should increase, and according to Kay (1984), *S* should decrease.

Tracing the origin or fate of a particular flow

Network analysis allows one to analyse the connections between any components in a food web.

Dependency coefficients describe what fraction of the total amount of material or energy leaving one compartment eventually enters another compartment, via both direct food chains and cycling. Thus we might be interested to find out how much of the diet of codfish is dependent on benthic meiofauna in the community. In this hypothetical example we might find that , say, 7% of cod diet in terms of carbon is contributed by meiofauna. Other fractions might be contributed by phytoplankton, zooplankton, benthic bacteria, benthic invertebrates and small fish, etc.

Contribution coefficients give the fractions a component contributed to the "diets" of other components. We might want to know the fate of bacterial production in the water column in a model of a pelagic system. Thus if the model currency were nitrogen, we might find that 80% of bacterial production ended up in the nitrogen pool, 85% contributed to flagellate diets, 40% to larger protozoan diets, 20% sank out of the euphotic zone and was exported, etc. Note that the sum of contribution coefficients may be more than 100% because direct, indirect and recycled flows are all included.

Comparing trophic positions in a network.

In a simple food chain one may easily assign animals to a trophic position: herbivore, carnivore, detrivore etc. The real situation is much more complex, because many organisms have mixed diets and the whole trophic level concept has been strongly criticized (Cousins 1980, 1985). Nevertheless, it may be of interest to know the trophic positions of living components in an ecosystem.

The *effective trophic position* is a weighted average of the number of steps a living component is from the primary producers or non-living components of the system. For example, a fish may get 90% of its diet from zooplankton and 10% from diatoms, placing it as 90% at position 3 (carnivore) and 10% at position 2 (herbivore). The average or effective trophic position of the fish would thus be 2.9. We might also be interested in comparing the postions of two similar species in different systems (Wulff & Ulanowicz 1989, Chapter 10, this volume).

A "*Lindeman Spine* " is an abstract food chain, derived from the network by collapsing it into a finite number of steps. This is analagous to the way a principal components analysis condenses a multi-species assemblage into a smaller number of principal components. A Lindeman spine shows the relative contributions of each component in the food web to each food chain step or category. By convention, autotrophs and non-living organic matter and nutrients are allocated to trophic category 1. Those components in the model that are assigned to particular trophic categories can then be extracted, for example trophic category 3 may consist of 95% carnivore activity and 5% detritivore activity.

The *trophic efficiency* of a food web can be calculated from a Lindeman spine by comparing the inflow to a trophic category with its outflow to the next category.

THE PROBLEM OF INCOMPLETE DATA SETS

There are never enough measurements of the rates of processes in an ecosystem to specify all the flows of all the elements among compartments. It is therefore necessary to restrict oneself to one or very few currencies (often energy, carbon or nitrogen) and estimate, deduce, or calculate by difference those flows that it is not practical to measure.

A priori and inverse methods

One way of estimating flows is to assume that over a considerable period, say one year, flows balance and the average biomasses are constant so that the system is in steady state over the period. By carefully choosing realistic values for various flow rates and efficiencies one can derive missing flows. This is called the *a priori method* (Ducklow *et al.* 1989, Chapter 8 this volume). Another approach is to use a *simulation model* of the processes in the system to estimate flows when the model runs to steady state (Ulanowicz 1989, Chapter 4, this volume). A third approach is to use *inverse methods* (Vézina & Platt 1988; Vézina 1989, Chapter 3, this volume; Ducklow *et al.* 1989, Chapter 8, this volume). *Inverse methods* are based on the available flow data, biomass data, and realistic ranges within which each parameter value should lie. Solutions for the parameter values are then calculated numerically according to specified criteria. The *inverse* and *a priori methods* are compared by Ducklow *et al.* (1989, Chapter 8, this volume).

The problem of mixed units

In ecosystems one is dealing with flows of many elements and with energy, often bound together in complex organic compounds. Ecologists arbitrarily choose to work with energy as a unit of currency, or with carbon, nitrogen or phosphorus because it would be impossible to simultaneously budget for all the energy and elements flowing through ecosystems. The chosen element or energy is used as a tracer of the overall rates of functioning of the ecosystem. However, at different stages and in different parts of an ecosystem the rates of functioning may be retarded by bottlenecks in the supply of different particular nutrients (Liebigs Law of the Minimum). For example, the rate of supply of nitrogen is likely to limit phytoplankton production in the surface waters at temperate latitudes in summer, whereas after an upwelling event the nitrogen supply rate would be much less important to the rates of ecosystem processes because it is already abundant.

The question is, how can we express the limitation of nitrogen supply in, say, an energy flow diagram. The dilemma is often expressed in the saying "You cannot add apples and oranges". Costanza & Hannon (1989; Chapter 5, this volume) haver worked out a technique for dealing with the flows of different commodities in the biosphere. By analogy with the prices in an economic model they calculate energy intensities for various commodities, thus deriving a single currency that can be used for network analysis.

A SCOR WORKSHOP AND ITS FOLLOW-UP.

Since the Advanced Research Institute on flows of energy and materials (Fasham 1984) a number of measurements had been made, and data sets assembled, of flows among components of marine ecosystems. Thus it seemed timely for SCOR Working Group 73 on "Ecological Theory in relation to Biological Oceanography" to organize a workshop at which the various related techniques would be brought together to be used on a common computer with data in the same format, so that different data could be analyzed with the same techniques, and different techniques compared on the same sets of data.

A preliminary trial with Robert Ulanowicz, John Field and Fredrik Wulff, was hosted by Michael Fasham at the Institute for Oceanographic Sciences, Wormley, England early in November 1985. The workshop itself took place in December 1986 at the University of California, Santa Cruz, hosted by Mary Silver assisted by Tony Michaels. Participants were: Robert Costanza (USA), Hugh Ducklow (USA), Michael Fasham (UK), John Field (South Africa -convenor), Lee Graham (USA), James Kay (Canada), James Kremer (USA), Tom Malone (USA), Kenneth Mann (Canada), Hank McKellar (USA), Donald Mickulecky (USA), Trevor Platt (Canada), Dave Smith (Australia), Jo Smitz (Belgium), Alain Vézina (Canada), Richard Warwick (UK), Fredrik Wulff (Sweden), and Robert Ulanowicz (USA).

At the end of the workshop overlapping techniques were identified, and groups and individuals agreed upon tasks to be undertaken in preparation for the present volume. These included an agreed convention for terminology, preparing computer software for distribution, and analysis of various data sets using the conventions and software developed as a result of the workshop.

REFERENCES

Augustinovics, M. 1970. Methods of international and intertemporal comparison of structure, pp. 249-269 in: A. P. Carter & A. Brody (eds), *Contributions to Input-Output Analysis. Vol. I*. North Holland, Amsterdam.

Costanza, R & B. Hannon. 1989. Dealing with the "mixed units" problem in ecosystem network analysis. In: Wulff, F., J. G. Field & K. H. Mann (eds), *Network Analysis in Marine Ecology: Methods and Applications*. Lecture Notes on Coastal and Estuarine Studies. Springer-Verlag, New York.

Cousins, S. H. 1980. A trophic continuum derived from plant structure, animal size and a detritus cascade. J. Theor. Biol. 82: 607-618.

Cousins, S. H. 1985. The trophic continuum in marine ecosystems: structure and equations for a predictive model. In: Ulanowicz, R.E. & T. Platt (eds), *Ecosystem Theory for Biological Oceanography*. Can. Bull. Fish. Aquat. Sci. 213: 76-93.

Ducklow, H., M. J. R. Fasham & A. Vézina. 1989. Derivation and analysis of flow networks for open ocean plankton systems. In: Wulff, F., J. G. Field & K. H. Mann (eds), *Network Analysis in Marine Ecology: Methods and Applications*. Lecture Notes on Coastal and Estuarine Studies. Springer-Verlag, New York.

Fasham, M. J. R. (Ed). 1984. *Flows of Energy and Materials in Marine Ecosystems: Theory and Practice*. Plenum, New York, 733 p.

Finn, J. T. 1976. Measure of ecosystem structure and function derived from the analysis of flows. J. Theor. Biol. 56: 363-380.

Hannon, B. 1973. The structure of ecosystems. J. Theor. Biol. 41:535-546.

Kay, J. J., L. A. Graham, & R. E. Ulanowicz. 1989. A detailed guide to network analysis. In: Wulff, F., J.G. Field & K. H. Mann (eds), *Network Analysis in Marine Ecology: Methods and Applications*. Lecture Notes on Coastal and Estuarine Studies, Springer-Verlag, New York.

Kay, J. J. 1984. Self-organization in living systems. Unpublished PhD thesis, University of Waterloo, Waterloo, Ontario.

Leontief, W. 1951. *The structure of the American Economy, 1919-1939*, 2nd ed. Oxford Univ. Press, New York, 264 p.

Longhurst, A. R. & D. Pauly . 1988. *Ecology of Tropical Oceans*. Academic Press, New York, 407 p.

Margalef, R. 1968. *Perspectives in Ecological Theory*. University of Chicago Press, Chicago, 111 p.

Mikulecky, D.C. 1985. Network thermodynamics in biology and ecology: an introduction. In: Ulanowicz R.E. & T. Platt (eds), *Ecosystem Theory for Biological Oceanography*. Can. Bull. Fish. Aquat. Sci. 213: 163-175.

Odum, E. P. 1969. The strategy of ecosystem development. Science 164: 262-270.

Patten, B. C. 1985. Energy cycling, length of food chains, and direct versus indirect effects in ecosystems. In: Ulanowicz, R. E. & T. Platt (eds), *Ecosystem Theory for Biological Oceanography*. Can. Bull. Fish. Aquat. Sci. 213: 119-138.

Ulanowicz, R. E. 1980. An hypothesis on the development of natural communities. J. Theor. Biol. 85: 223-245.

Platt, T., K. H. Mann & R. E. Ulanowicz. (eds). 1981. *Mathematical Models in Biological Oceanography*. The UNESCO Press, Paris, 156 p.

Ulanowicz, R. E. 1984. Community measures of marine food networks and their possible applications. In: Fasham, M. J. R. (Ed.), *Flows of Energy and Materials in Marine Ecosystems*. Plenum Publishing Corp., New York, p. 23-47.

Ulanowicz, R. E. 1986. *Growth and Development: Ecosystems Phenomenology*. Springer-Verlag, New York, 203 p.

Ulanowicz, R. E. 1989. A generic simulation model for treating incomplete sets of data. In: Wulff, F., J. G. Field & K. H. mann (Eds), *Network Analysis in Marine Ecology: Methods and Applications*. Lecture Notes on Coastal and Estuarine Studies. Springer-Verlag, New York.

Ulanowicz R. E. & T. Platt. (Eds). 1985. *Ecosystem Theory for Biological Oceanography*. Can. Bull. Fish. Aquat. Sci. 213, 260 p.

Vézina, A. F. 1989. Construction of flow networks using inverse methods. In: Wulff, F., J. G. Field & K. H. Mann (eds), *Network Analysis in Marine Ecology: Methods and Applications*. Lecture Notes on Coastal and Estuarine Studies. Springer-Verlag, New York.

Vézina, A. F. & T. Platt. 1988. Food web dynamics in the ocean. I. Best estimates of flow networks using inverse methods. Mar. Ecol. Prog. Ser. 42: 269-287.

Webb, K. L., W. D. DuPaul, W. Wiebe, W. Sottile & R. E. Johannes. 1975. Enewetak (Eniwetok) Atoll: Aspects of the nitrogen cycle on a coral reef. Limnol. Oceanogr. 20: 198-210.

Wulff, F. & R. E. Ulanowicz. 1989. A comparative anatomy of the Baltic Sea and Chesapeake Bay ecosystems. In: Wulff, F., J. G. Field & K. H. Mann (eds), *Network Analysis in Marine Ecology: Methods and Applications*. Lecture Notes on Coastal and Estuarine Studies. Springer-Verlag, New York.

PART II

METHODS

Chapter 2

A DETAILED GUIDE TO NETWORK ANALYSIS

James J. Kay
Department of Environment and Resource Studies,
University of Waterloo
Waterloo, Ontario
N2L 361 Canada

Lee A. Graham
Department of Entomology,
Texas A&M University,
College Station,
Texas 77843
USA

Robert E. Ulanowicz
Chesapeake Biological Laboratory
University of Maryland
Solomons, Maryland 20688-0038
USA

INTRODUCTION

This chapter will introduce a number of the more frequently used network analysis techniques, many of which trace their origins from Leontief's (1936) economic input-output analysis and Shannon's information theory (1948). We begin by presenting a standard format for recording all of the flow data related to the ecosystem in a network representation. Having constructed the representation, the direct and indirect relationships that exist among the different components can then be examined in depth using a suite of network analyses that have been developed through the years by a number of researchers. These analyses consist of calculating a number of measures which synthesise some part of the information about flows of energy or materials through an ecosystem. These measures and analytical techniques cover from the microscopic level, the level of a component, right up to the macroscopic level, the level of the whole ecosystem. The measures are presented in a gradation from micro to macroscopic. The set of measures described herein provides a very rich description of an ecosystem; a description which looks at the system from many different perspectives. Clearly anyone applying these measures will have need for only those which reflect the perspectives of interest to the researcher .

All of the analytical techniques and measures discussed in this chapter have been written up as a set of computer programs (NETWRK, STRUCTURE, ENVIRON and others). This software is available from SCOR (see elsewhere in this book, Wulff *et al.* 1989) or from the authors. Details of how to use the programs are included with the software. Also, a list of the subheadings in this chapter and the programs which do the corresponding analysis is also included. Finally, reference to the programs has been made in the text where relevant.

FLOW NETWORK REPRESENTATION OF ECOSYSTEMS

This section explains how a network representation can be used to describe the flows of mass and energy in an ecosystem. The process of building such a representation serves two purposes. It brings together all of the available data related to mass and energy flows in the ecosystem, and it identifies what information is missing. A standard data format is presented allowing for easy calculation and comparison of different ecosystems is presented.

The term "network" describes a collection of elements called *nodes*, pairs of which are joined to one another by a collection of elements called *edges*. Each node represents a compartment (biotic or abiotic) of the ecological system and is referred to by a numerically-referenced name, such as x_1 or x_6, which can also be used to reference the compartment's standing stock. The edges that connect the nodes are usually directed, the direction indicating that matter and energy flow from the initial to the terminal node. Directed edges are called *arcs*. Arcs are named using the numerical identifiers of the nodes they connect. For example, the arc connecting components x_1 and x_6 is referenced as f_{16}.

Each arc in an ecological flow network can have an associated value. This value represents the magnitude of flow that occurs from the initial to the terminal node of the arc in a given unit of time. A network of this sort is said to be weighted. Living systems also exchange matter and energy with their environment. Thus, ecological networks have at least some arcs that originate or terminate outside of the system. These are called open networks. All of the networks discussed in this book are open, weighted networks.

The flow diagram in Figure 2.1 illustrates a generalized scheme for diagramming ecological flow networks. Values are associated with either standing stocks or one of four classes of flow: (1) inputs from outside the system, (2) flows between components, (3) exports to other systems, and (4) respiration losses. Note that all flow values are presented in terms of units of mass or energy per unit area or volume per unit time, while standing stocks are measured as units of mass or energy per unit area or volume.

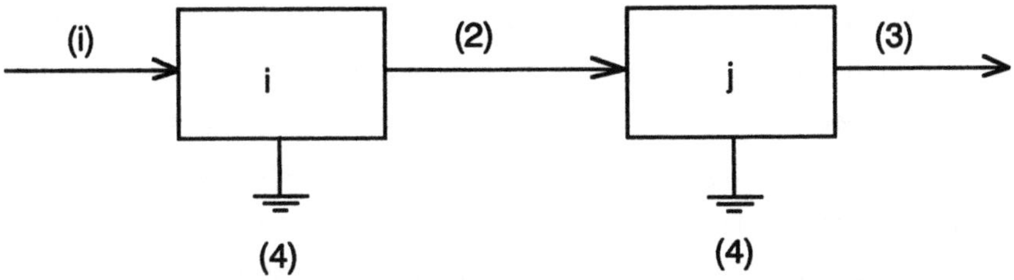

Fig 2.1. The four classes of ecological flows: (1) exogenous inputs, (2) intercompartmental exchanges, (3) exports of useable medium, and (4) dissipation of unuseable medium.

Figure 2.2a depicts a simple weighted network representing the five components of the Cone Spring ecosystem described by Tilly (1968) and quantified by Williams and Crouthamel (unpub. ms.). Flows of energy (kcal/m^2/yr) are indicated by the values that appear on the arcs, while standing stocks (kcal/m^2) are indicated inside the boxes that represent the nodes. There are two inputs that originate outside of the system. The arcs that terminate outside the system represent exports of energy in a form that can be used by other systems. The special ground symbols represent energy that is dissipated through respiration. This energy is lost from the system and is unusable by any system at the same scale (Ulanowicz 1986a).

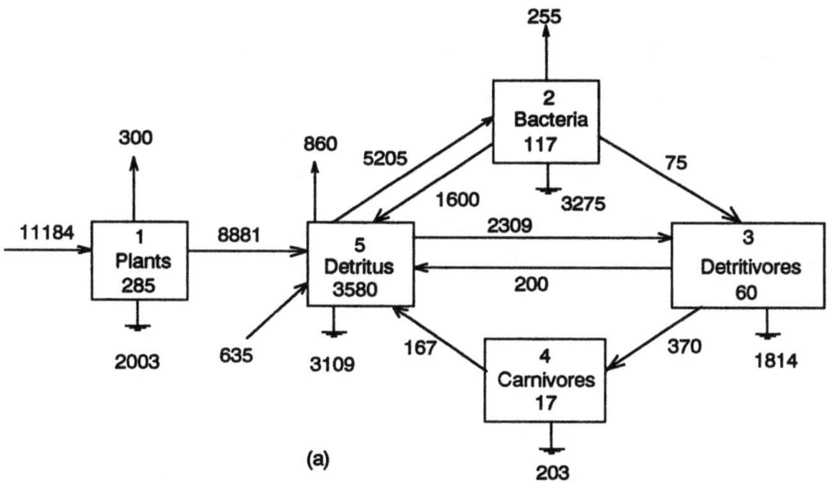

Fig 2.2a. Energy flows (kcal/m^2/yr) and densities (kcal/m^2) among the 5 components of the Cone Springs network.

Diagrammatical representations of networks are common, but can be confusing and cumbersome when used to describe large systems. A mathematical network representation which describes an ecosystem's structure and function is necessary. In this book a standardized method is used for representing ecological flow networks using a single matrix (flows between compartments) and four vectors (standing stocks, inputs, exports, and respirations). (See Figure 2.2b for an example, the Cone Spring model). This convension must be adopted by the reader in constructing networks for analysis by the software associated with this book.

$$\underset{\mathbf{F}}{\begin{bmatrix} 0 & 0 & 0 & 0 & 8881 \\ 0 & 0 & 75 & 0 & 1600 \\ 0 & 0 & 0 & 370 & 200 \\ 0 & 0 & 0 & 0 & 167 \\ 0 & 5205 & 2309 & 0 & 0 \end{bmatrix}} \quad \underset{\mathbf{Z}}{\begin{bmatrix} 11184 \\ 0 \\ 0 \\ 0 \\ 635 \end{bmatrix}} \quad \underset{\mathbf{E}}{\begin{bmatrix} 300 \\ 255 \\ 0 \\ 0 \\ 860 \end{bmatrix}} \quad \underset{\mathbf{R}}{\begin{bmatrix} 2003 \\ 3275 \\ 1814 \\ 203 \\ 3109 \end{bmatrix}}$$

Fig 2.2b. The standardized method for representing ecological flow networks using a single matrix of flows between compartments (F) and three vectors representing inputs **Z**, exports **E**, and respiration, **R**.

The number of compartments n in the Cone Spring network is 5, 4 of which are biotic and one of which is abiotic. The vector **x** contains elements $x_i \ldots x_n$ that indicate the magnitudes of the standing stocks of each compartment. In order to reap the full benefit of the analysis programs and software associated with this book, we recommend that the flow network be conceptualized and quantified so that, when ordering the nodes (assigning positions $x_1 \ldots x_n$), the biotic components are numbered before the abiotic ones. In the Cone Spring example, component 5 (abiotic detritus) is assigned the last of the 5 possible numbers. The software program, REORDER, can be used in those cases where improperly ordered network data already exists (see software documentation for details).

The vector **z** is used to store flows that originate from outside the system (inputs). Flows between system components are stored in the $n \times n$ exchange matrix **F**. The orientation of the matrix is such that flow travels from row to column components of the matrix. We adopt the From Row - To Column orientation throughout this book, but note that many of the analyses we will describe (those developed by Patten, Finn, Barber, and Bosserman) often use the To Row - From Column format. One should always be careful to note matrix orientation when examining the literature relating to specific analyses.

Losses from the compartments are denoted by the elements of the export vector **e** and the respiration vector **r**. The export and respiration vectors are sometimes summed to produce a total loss vector **y**.

Most network analysis techniques were originally formulated for steady state systems. This means that, for each compartment, the sum of all its inputs must equal the sum of all its outputs. Most analyses can handle non-steady state systems, but care must be taken in interpreting the results. The discontinuities and nonlinearities that exist in the interactions occurring in nature cannot always be represented by using a simple matrix description of the ecosystem.

THEORETICAL BACKGROUND

Leontief (1936, 1963) developed economic input-output analysis as a means of quantifying the amount of raw materials and industrial services required to produce a quantity of consumer goods. His work introduced the backward case analysis, concerned with tracing output or demand back to inputs. Augustinovics (1969) later developed the forward case economic analysis that traces the fate of system inputs through the system to outputs.

Hannon (1973) was the first to apply Leontief's analysis to ecological systems. Hannon was interested in determining the direct and indirect contributions of the system to a unit of output, and therefore modified Leontief's analysis to solve the equations differently. His success spawned a flurry of efforts among systems ecologists. Finn (1976) later developed a forward case analysis for analyzing ecological networks. In addition he introduced several measures of ecosystem function, including cycling, that aid in the ecological interpretation of the results of input-output analysis. Ulanowicz (1986a) has documented these analytical techniques in textbook fashion.

It is important to note that the network perspective provided by flow analysis is the foundation on which all of the analyses in this book are built

Input-Output Flow Analysis

The basics of flow analysis are quite simple (see also Costanza & Hannon 1989, this volume). Given the system information provided in the **F** matrix and **z**, **e**, and **r** vectors, one can construct a production matrix, which serves as the basic computational unit for flow analysis. The basic structure of the production matrix is given in Figure 2.3a, with the Cone Spring network presented as an example in Figure 2.3b. The important thing to realize about the production matrix is that the **F22** partition is the **F** matrix, the **F12** partition holds the system inputs (**z**), while the latter partition contains the **e** and **r** vectors (perhaps summed as the total loss vector **y**).

F11	F12
F21	F22

0	11180	0	0	0	635
2300	0	0	0	0	8881
3525	0	0	75	0	1600
1810	0	0	0	370	200
203	0	0	0	0	167
3970	0	5205	2309	0	0

Fig 2.3. The basic structure of the production matrix (top) and the Cone Spring network as an example (below).

We can use the production matrix to compute the throughflow, T_i, for each compartment *i*. This value is defined as the rate of energy or material flow through compartment *i*. It is calculated as the sum of the inflows to compartment *i* (obtained by summing the *i*th column of the production matrix) or the sum of the outflows (calculating the row sums) from *i*. In a steady state system, the two methods for calculating T_i produce the same result. For a dynamic system, throughflow is modified to include changes in storage as inflow or outflow. A positive state derivative (**xd**$_+$) is considered a loss from the system pool of mobile energy, while a negative state derivative (**xd**$_-$) is viewed as a gain to the pool (Patten *et al.* 1976, Finn 1977). The equations for throughflow then become either, for outflows:

$$T_i = \sum_{j=1}^{n} f_{ij} + e_i + r_i + xd_{+i} \qquad (1)$$

or, for inflows:

$$T_i = \sum_{j=1}^{n} f_{ij} + z_i + x_{d\text{-}i} \qquad (2)$$

Total system throughflow T (also called total system throughput) is defined as sum of all compartmental throughflows. It measures the total amount of material or energy flowing in the system, multiplied by the number of compartments that the flow is passing through. T is one of the total system properties discussed later in this chapter.

Fractional Coefficient and Structure Matrices

We now consider separately the *backward* (Leontief/Hannon) and *forward* case (Augustinovics/Finn) flow analyses, beginning with the backward case input or inflow analysis (also called creaon analysis in Patten *et al.* 1976). Inflow analysis traces system output back through the network to system input. As a first step, the fractional inflow matrix **G'** is generated for a network by dividing each f_{ij} element of the exchange matrix by T_j, the throughflow entering compartment j. The elements of this matrix indicate the fraction of compartmental throughflow that originates from a another node in the network. Ulanowicz (1986a) calls the elements of **G'** partial feeding coefficients, since each value represents the proportion of a component's "diet" that is derived from a particular "feeding" interaction.

The next step involves calculating the Leontief inverse (**N'**) matrix. We will follow Hannon's practice and call **N'** the input structure matrix. **N'** and its output structure analog **N"** are often referred to as a fundamental matrix (Kemeny and Snell 1960) or a transitive closure matrix (Patten *et al.* 1976). The elements of the structure matrices show the direct and indirect contributions of each interaction to system output. Remember that once matter and energy enter a system, they may visit many compartments along a large number of possible pathways before exiting. The input structure matrix is obtained by subtracting **G'** from the identify matrix **I** (a matrix that is all zeros except for the diagonal elements, which have values of 1), and then inverting the result using matrix algebra:

$$\mathbf{N'} = (\mathbf{I} - \mathbf{G'})^{-1} \qquad (3)$$

The elements of **N'** can be interpreted as (1) n'_{ij} is the amount of throughflow from i required to produce a unit of output in j, or (2) n'_{ij} is the expected number of times energy or matter that ended up in compartment j has passed through compartment i. The variances (stored in the **U'** matrix) associated with these expectations can also be calculated using the equations presented in Patten and Matis (1982). These variances are useful indicators of the complexity of network structure and flow distribution.

The forward-looking output or outflow analysis (also called genon analysis) is concerned with tracing the fate of inputs through the system to outputs. Because ecologists want to know what drives living

Table 2.1. Matrices produced by Input/output analysis of the Cone Spring network. Flows travels from rows (i) to column (j) compartments in this format.

The fractional inflow matrix; **G'**

	1	2	3	4	5
1	.000	.000	.000	.000	.773
2	.000	.000	.032	.000	.139
3	.000	.000	.000	1.000	.017
4	.000	.000	.000	.000	.014
5	.000	1.000	.969	.000	.000

The fractions of the total inputs to j that flow directly from i.

The fractional outflow matrix; **G"**

	1	2	3	4	5
1	.000	.000	.000	.000	.794
2	.000	.000	.014	.000	.307
3	.000	.000	.000	.155	.084
4	.000	.000	.000	.000	.451
5	.000	.453	.201	.000	.000

The fractions of the total outputs from j that flow directly from i.

$(I\text{-}G')^{-1}$ inverse matrix (also known as **N'**)

	1	2	3	4	5
1	1.000	.933	.933	.933	.933
2	.000	1.169	.201	.201	.169
3	.000	.039	1.039	1.039	.039
4	.000	.018	.018	1.018	.018
5	.000	1.207	1.207	1.207	1.207

Represent the fractions of donor throughflow (direct and indirect) needed to produce one unit of recipient outflow.

$(I\text{-}G'')^{-1}$ inverse matrix (also known as **N"**)

	1	2	3	4	5
1	1.000	.434	.199	.031	.958
2	.000	1.169	.092	.143	.374
3	.000	.084	1.039	.161	.186
4	.000	.247	.113	1.018	.545
5	.000	.547	.251	.039	1.207

Represent the fractions of recipient throughflow generated by one unit of inflow to the donor.

Table 2.1. Continued

U' variance matrix (of N')

	1	2	3	4	5
1	.000	.062	.0623	.062	.062
2	.000	.198	.229	.229	.198
3	.000	.040	.040	.040	.040
4	.000	.018	.179	.018	.018
5	.000	.249	.249	.249	.249

U" variance matrix (of N")

	1	2	3	4	5
1	.000	.393	.175	.031	.436
2	.000	.198	.091	.015	.389
3	.000	.106	.040	.141	.228
4	.000	.269	.109	.018	.473
5	.000	.433	.207	.039	.249

systems, and since outputs in ecological systems are generally dominated by respiration losses, outflow analysis has been applied more frequently in ecology. It has also been refined and redefined to produce new insights into the intricacies of ecological organization. The analysis begins with the calculation of the fractional outflow matrix G" (together, G" and G' are referred to as fractional or technical coefficient matrices). Rather than normalizing the f_{ij} by the throughflow T_j entering j, we normalize by the throughflow T_i leaving the donor compartment i. Now, the elements of the fractional coefficient matrix indicate the fraction of throughflow that is generated as a result of the demand from another node. The elements of the fractional outflow matrix are also called partial host coefficients by Ulanowicz (1986a).

The Augustinovics inverse matrix N" is obtained in a manner similar to the Leontief inverse:

$$N'' = (I - G'')^{-1} \tag{4}$$

(Due to our desire to keep matrix orientation in the from Row-to Column format, the Augustinovics inverse is seen here in its transposed form.). N" is the output structure matrix for the network. Each element can be interpreted as (1) the output y_j in compartment j generated by a unit throughflow T_i in compartment i, or (2) the expected number of times a unit of matter or energy originating in i will pass through compartment j before leaving the system. U" is the matrix of variances associated with the elements of G".The fractional coefficient matrices (G', G") and the structure (N', N",) and variance (U', U") matrices for the Cone Spring ecosystem are presented in Table 2.1.

We will later see how the structure matrices can be used to determine the origins and fates of any given flow in a network, including specific techniques for elucidating the underlying trophic structure. These matrices also serve as the foundation for the development of a number of information measures, many of which can be used to measure growth, development, specialization, and diversity in ecological systems.

Attributes of Individual Compartments

Network analyses produce results that permit the description of an ecological system in terms of: 1) attributes of the components, 2) relationships between the components, and 3) total system properties. In what follows, this section focuses on attributes of components, the next three sections focus on between compartment relationships, and the following three sections focus on total system properties. One of the goals of this work is to identify what ecological insights are gained from thoroughly analyzing a network and how well these insights carry over in comparisons of similar systems and very different ecosystems.

In characterizing individual compartments, the estimation of compartmental throughflows (T_i) is useful for revealing the relative rates at which the flow medium enters and leaves each component. Where standing stocks are also available, the turnover rate for each compartment can be calculated by dividing compartmental throughflow by the component's standing stock. Turnover rates are measured in units of inverse time and indicate how fast matter and energy are exiting from the compartment (because turnover rates are based on losses from compartments, they are presented as negative numbers). Fast compartments have high turnover rates compared to slow ones. The inverse of the turnover rate is turnover time, which is the time required for the equivalent amount the original standing stock to pass through the compartment. In the Cone Spring network, detritus (x_5) turns over every 113 days -a much slower rate than the living plant biomass (17 days) that supplies the detritus pool and the bacteria (8 days), detritus feeders (9 days), and carnivores (17 days) that depend on it for nourishment.

Finn (1976) and Patten *et al.* (1976) introduced a number of flow analysis measures, some of which characterize properties of individual system components. By calculating the rows sums of the **N"** matrix, we obtain the outflow path length vector **ply**. If ply_i equals 1, the outflow from i is directly connected to compartment i. A value greater than 1 is obtained when indirect links (longer paths and/or cycles) help produce the outflow from i.

The column sums of **N'** produce the inflow path length (**plz**) vector. Each plz_j measures the penetration of exogenous and endogenous inputs to compartment j into the network.

The outflow path lengths for the Cone Spring model are 1.0 for plants, 4.37 for bacteria, 3.4 for detritus feeders, 4.4 for carnivores, and 3.37 for detritus. The inflow path lengths are 2.6 (plants), 1.6 (bacteria), 1.5 (detritus feeders), 1.9 (carnivores), and 2.06 (detritus). The inflow path length values for the animal components (including bacteria) of the ecosystem are derived entirely from endogenous inputs. For the compartments receiving exogenous inputs, the solar inputs the plants receive travel deeper into the network structure than do inputs to the detritus pool.

Finn is also responsible for introducing the first measures for estimating the amount of cycling in ecological flow networks. The diagonal elements of the structure matrices **N"** and **N'** represent the number of visits that a unit of flow starting or ending in a given compartment makes to that same compartment. The number of straight through visits is always 1. Therefore, the cycled portion of throughflow in the *i*th compartment is calculated as:

$$c_{ii} = [(n_{ii}-1)/n_{ii}] / T_i \ . \qquad (5)$$

The recycling efficiency is therefore expressed as:

$$r_{ei} = (n_{ii}-1) / n_{ii} \, . \qquad (6)$$

The compartmental cycling efficiencies for the Cone Spring compartments are 0.79, 0.32, 0.24, 0.45, and 0.65 respectively.

BILATERAL RELATIONSHIPS

Whereas the suite of measures discussed in the last section focus on the status of single compartments as they are imbedded in the whole network, much useful information exists concerning the bilateral relationships between any two nodes of interest where the entire rest of the network serves as the conduit between them.

Input and Output Flow Environs

An input environ is the set of fractional flows generated in the system by one unit of inflow into one of the compartments receiving input from outside the system. There are as many input environs as there are compartments receiving imports. Similarly, an output environ is the set of fractional flows which occur among compartments when one unit of flows leaves one of the exporting compartments. There are as many output environs as there are compartments which export useful flows. In matrix terminology, the fractional coefficient and structure matrices for the ecosystem can be used to trace the origins or fate of any input, output, or internal flow represented in the network. Let us examine the backward case first. Because the elements of **G'** indicate the fraction of the recipient component's throughput that is generated from each donor compartment's throughflow, it is possible to estimate the intrasystem flow required to sustain the output from the recipient compartment. One must first diagonalize the column of the **G'** matrix that corresponds to the output of interest (for example, to look at the output from the bacteria in the Cone Spring system, we would use column 2 to form a matrix whose diagonal elements are the elements of column 2 and whose other elements are 0), then multiply it by the input structure matrix **N'**. The resultant matrix $\mathbf{T'_i}$ is the unit input environ matrix for outflow y_i. The elements of the matrix are generally used to construct a diagram that indicates the location and magnitude of flow required to generate a unit of output within or from compartment *i*. Since all compartments in ecological systems lose medium to their environment, there will be one diagram for each compartment. The unit input environs for the Cone Spring ecosystem are show in Figure 2.4.

The unit outflow environs $\mathbf{T''_i}$ are constructed in a similar manner as the unit inflow environs. This time, the elements of the resultant matrix indicate the fate of a unit input into compartment *i*. One can

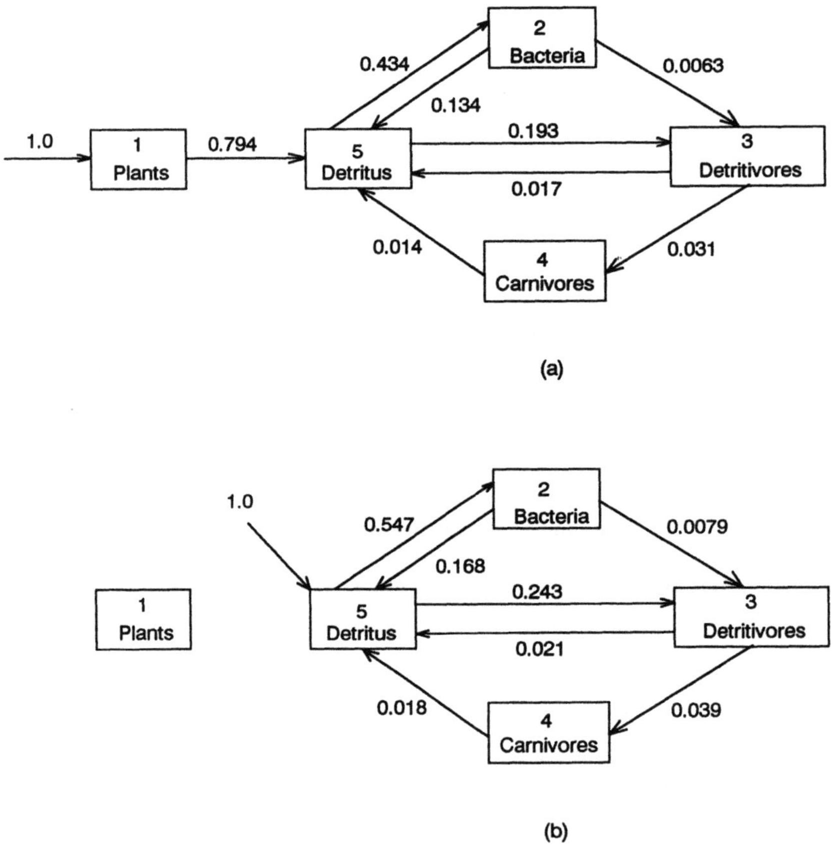

Fig 2.4. The unit input environs for the Cone Spring network.The relative amounts of internal exchanges generated by a unit input to a) Plants, b) Detritus.

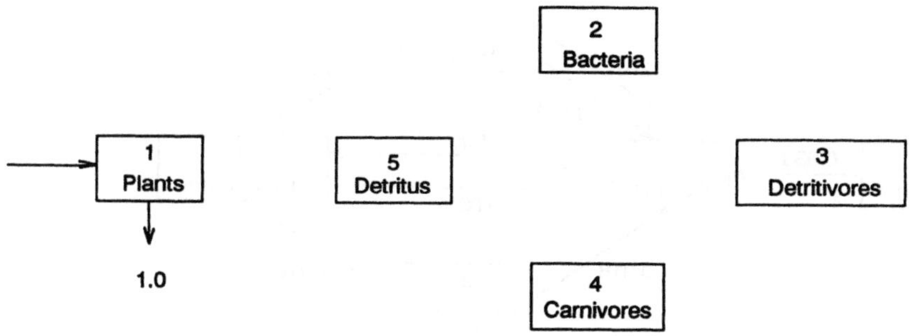

(a)

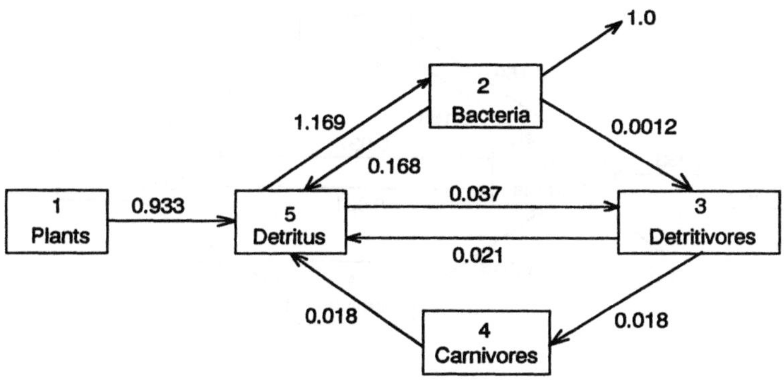

(b)

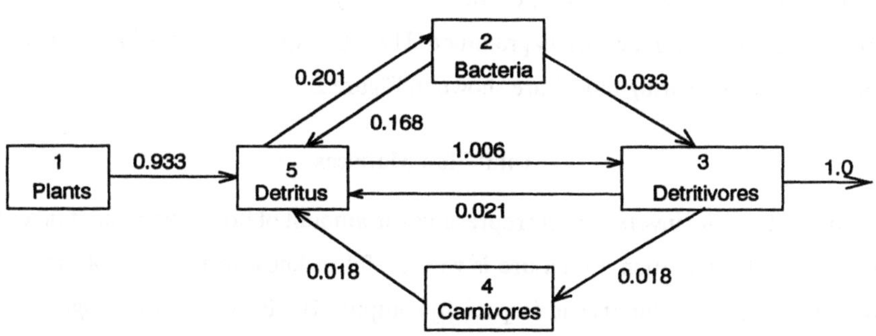

(c)

Fig 2.5a-e. The unit outflow environs for the plant and detrital components of the Cone Spring network. Each figure depicts the amounts of internal flows necessary to sustain the single unit output shown. Flow units are relative to outputs.

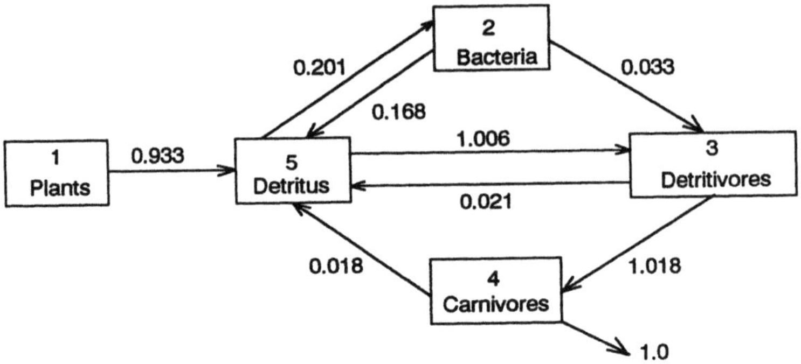

(d)

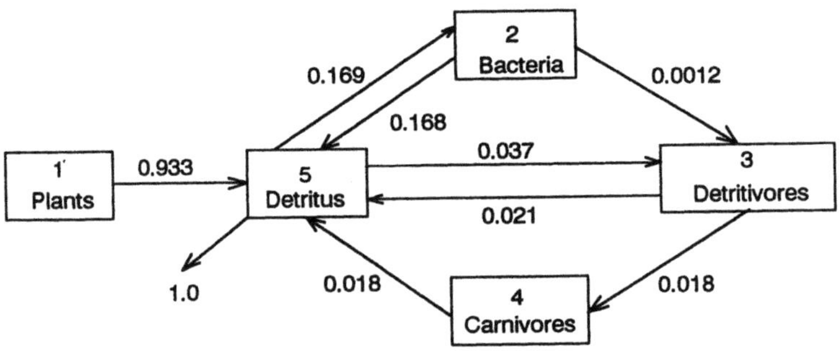

(e)

Fig 2.5 a-e. (continued).

construct unit input environs for all compartments in the system, but only the ones for compartments receiving external inputs are generally diagrammed. The unit outflow environs for the plant and detrital components of the Cone Spring model are shown in Figure 2.5.

Total Flow Matrices

Each value in the Leontief inverse matrix represents the amount of flow from i that it takes to generate one unit of output from j. Such a measure is very useful to know in the field of economics, where emphasis is upon exogenous inputs and (especially) outputs. But in ecology the exogenous outputs are dominated by respiration, and the question of how much each compartment contributes, directly and indirectly, to the sustenance of another, is of more interest. Conversely, one might wish to know the degree to which a particular species depends upon another for its material existence.

Szyrmer and Ulanowicz (1987) show how these total intermediate flows can be calculated from the components of the Leontief and Augustinovics inverses. Without going into the mathematical details here, the total flow from compartment i to j (excluding that which has been recycled back through i) is:

$$z_{ij} = [(n'_{ij} - d_{ij}) / n'_{jj}]\, T_j \qquad (7)$$

where z''_{ij} is the total flow from i which reaches j (without recycling through i) over all pathways of all lengths, n'_{ij} are the elements of the Leontief inverse, $d_{ij} = 0$ when $i \neq j$ and $d_{ij} = 1$ when $i = j$, and T_j is the amount flowing through compartment j.

Looking backwards, it is possible to say how much of the medium arriving at j passed through i at some point in its journey through the system:

$$z'_{ij} = [(n''_{ij} - d_{ij}) / n''_{ii}]\, T_i \qquad (8)$$

where z'_{ij} is the flow reaching j that at some earlier time passed through i and n''_{ij} is the ijth element of the Augustinovics inverse.

In general, because of cycling $z'_{ij} \neq z''_{ij}$. In the Cone Spring example, of the 1675 kcal/m^2/yr leaving the bacteria, $z''_{24} = 73.02$ of those units eventually reach the carnivores; whereas of the 370 units consumed by the carnivores, $z'_{24} = 63.54$ of them passed through the bacteria along the way.

Perhaps the total flow information is most useful in its normalized form. Thus, the quotient z''_{ij} / T_i represents the fraction of the total output of i which reaches j and z'_{ij} / T_j, the fraction of j's total consumption which passed through i. In the output from the program NETWRK, the former quotients are called "total contribution coefficients", while the latter are termed "total dependency coefficients." The dependency coefficients are especially useful diagnostic indices because they represent the extended diet of the recipient compartment j. If one reads down the jth column of the normalized flow matrix, **G'**, one sees the fractions of the actual diet of the jth species that come from the various other members of the community. Often the entries in the column are sparse, as most species derive food from no more than about 4 prey items. However, the corresponding column of the dependency coefficient matrix is usually filled with many non-zero values that portray the history of the same food. Occasionally, the extended diets reveal important differences not apparent in the rations of direct prey (Baird & Ulanowicz 1989). A Chesapeake Bay example would be the striped bass (*Monrone saxatillis*) and the bluefish (*Pomatomus saltatrix*). Both appear as pelagic carnivores. Their extended diets, however, reveal significant niche separation. The striped bass is a recipient of energy and material that was passed up the pelagic grazing chain, while the medium reaching the bluefish had arrived mostly via the benthic community. This disparity in the extended diets provided a partial explanation of why the pesticide Kepone (a contaminant of the James River sediments) was appearing in high concentrations in the flesh of the bluefish, but was not prominent in the striped bass.

Investigators are troubled at first by the fact that the extended diet columns usually sum to a number greater than unity (i.e., they do not represent well-normalized probabilities). Their anxiety eases, however, once they realize that a given particle or quantum of energy usually passes through several compartments before reaching the predator in question. It is only natural, therefore, that the dependency coefficients represent a systematic multiple counting scheme. In fact, the sum of any column of this matrix is closely related to the average trophic status of the corresponding species (see section on trophic analysis below).

Markovian Transition Probabilities

Earlier, we introduced the fractional coefficient matrices **G'** and **G"**, which partition a compartment's input and output throughflow relative to its origins or destination within the ecological network. Another interpretation treats the fractional coefficients as discrete-time finite state Markov transition probabilities (Kemeny & Snell 1960, Barber 1977a,b). A Markov process is the simplest type of stochastic process described in statistics. The condition of the ecosystem is represented as a finite set of states representing the biomasses in the compartments at one time and the flows between them. The condition of the system at any time is assumed to depend only upon the condition immediately beforehand and is independent of all time prior to that. Thus the condition of the system at one instant is assumed to contain enought information to predict the next condition or transition (Legrendre and Legendre 1983). It is called a first order process, since it is only going back one step in time to predict the next condition. Higher powers of **G'** and **G"** (for example, $\mathbf{G}''^{1}$, $\mathbf{G}''^{2}$,...$\mathbf{G}''^{3}$) represent the *k*th order distribution of a unit of medium in each compartment to all compartments following *k* transitions.

One of the obvious network characteristics that has largely been ignored in previous discussions is the importance of standing stocks in determining system structure and function. In fact, only in our discussion of turnover rate and turnover time have standing stocks come into play. Barber (1977a, b) was the first to cast forward and reverse case flow analysis in terms of a Markovian model that incorporates storage as well as flow probabilities. Environ analysis (Matis & Patten 1981, Patten & Matis 1982) was later introduced as a formal methodology for computing storage partitions in addition to the usual flow partitions produced by traditional inflow and outflow analyses.

Environ analysis is founded on the development of linear donor and recipient controlled models of system dynamics. Any network (such as the Cone Spring network we use in our examples) can be described by a set of difference or differential equations. For a static linear system, this is achieved by making each flow a function of its donor compartment, $f_{ij} = a_{ij} x_i$. The linear coefficients a_{ij} can be stored in a matrix **A"** so that

$$\mathbf{x} = \mathbf{A''x} + \mathbf{z}. \qquad (9)$$

where **x** is the vector of standing stocks and **z** the vector of imports into the system.

The diagonal elements of **A"** are the turnover rates (expressed as negative numbers) for each compartment. Alternatively, one can normalize each flow by the standing stock of its recipient compartment. In this case, the change in standing stock is expressed using the following equation:

$$\mathbf{x} = \mathbf{A'x} + \mathbf{y}. \qquad (10)$$

where **y** is the vector of total exports. **A"** and **A'** (Table 2.2) are the forward and reverse linear system matrices for the ecological network.

Table 2.2. The forward (**A"**) and reverse (**A'**) linear system matrices for the Cone Spring network.

A" matrix

	1	2	3	4	5
1	-39.240	.000	.000	.000	31.160
2	.000	-44.640	.643	.000	13.720
3	.000	.000	-39.730	6.167	3.333
4	.000	.000	.000	-21.760	9.824
5	.000	1.454	.645	.000	-3.208

A' matrix

	1	2	3	4	5
1	-39.240	.000	.000	.000	.000
2	.000	-44.640	.000	.000	44.640
3	.000	1.250	-39.730	.000	38.480
4	.000	.000	21.760	-21.760	.000
5	2.481	.447	.056	.047	-3.208

Knowing **A"** and **A'**, we can compute the discrete time Markov transition matrices **P"** and **P'** as follows:

$$\mathbf{P''} = (\mathbf{I} + h\mathbf{A''}) \qquad (11)$$

and

$$\mathbf{P'} = (\mathbf{I} + h(-\mathbf{A'})) \qquad (12)$$

where **I** is the identity matrix and h is a scalar selected to guarantee that each element along the diagonal of the transition matrices is positive.

In terms of time, h represents the number of discrete points within one unit of time t. This makes sense when one remembers that the inverse of the turnover rate is turnover time, meaning that h must be small enough that it captures the dynamics of the fastest component (where t is given in whatever real time units used in formulating the flow rates for the model). The value of h must therefore be less than or equal to the value of fastest compartment's turnover time. An h of 1 means that 1 time unit (days, weeks, etc.) is sufficient for recording the turnover of standing stock of each compartment. An h of 0.25 days indicates that one must sample at least 4 discrete points within a time unit to catch the dynamics of the faster components. Each Markovian transition would correspond to 0.25 days.

An element p''_{ij} of **P"** indicates the probability that an initial unit of medium will flow from i to j during t/h units of time, while an element p'_{ij} of **P'** indicates the probability that the unit of matter or energy now in j traveled from i during one transition interval. The probabilities on the diagonal (p''_{ii} and p'_{ii}) then indicate the probability that the medium originated or remained in i during one transition. Table 2.3 shows the Markov transition matrices for the Cone Spring ecosystem.

Table 2.3. Forward (**P"**) and reverse (**P'**) Markov transition matrices for the Cone Spring network.

P" matrix

	1	2	3	4	5	6
1	.387	.000	.000	.000	.487	.126
2	.000	.303	.010	.000	.214	.473
3	.000	.000	.379	.096	.052	.472
4	.000	.000	.000	.659	.154	.187
5	.000	.023	.010	.000	.949	.017
6	.946	.000	.000	.000	.054	.000

P' matrix

	1	2	3	4	5	6
1	.387	.000	.000	.000	.000	.613
2	.000	.303	.000	.000	.698	.000
3	.000	.019	.379	.000	.601	.000
4	.000	.000	.340	.659	.000	.000
5	.039	.007	.001	.001	.949	.003
6	.195	.299	.154	.017	.335	.000

This approach essentially relates the flow and standing stock properties of the system. However, as Barber (1978b) points out, a high resolution perspective of an ecosystem might view ecological resources as particulate in nature. Since all of the particles that comprise a component's standing stock presumably possess the same potential behavior, it follows that a model of resource flow could follow the movement of one such particle through the network, where the distribution of all particles is represented in terms of the random variables of a Markovian transition matrix, and the flow of particles among components is described as a stochastic process. The meaningful application of such a model to an ecological network requires that one exercise care in aggregating components. For example, it would be incorrect to lump a fast component with a slow one, since the probabilities for retention of medium would vary significantly between the two.

Markovian Flow Partitions

The higher powers of the **P"** and **P'** matrices can be used to trace the fates or origins of a unit medium of flow through the network over k transitions in the same manner as that described earlier for the **G"** and **G'** matrices. **P"** and **P'** are time-referenced by *h*, so the probabilities now represent probabilities of transfer over *t/h* units of real time. The Leontief and Augustinovics inverses (and associated variances **U"** and **U'**) can be computed for each of the Markovian probability matrices using the following equations:

$$\mathbf{Q''} = (\mathbf{I} - \mathbf{P''})^{-1} \qquad (13)$$

and

$$\mathbf{Q'} = (\mathbf{I} - \mathbf{P'})^{-1} \qquad (14)$$

The elements of **Q"** and **Q'** indicate, in the first case, the number of times material in *i* will visit *j* before leaving the system, and in the second case, the expected number of times substance in *i* has visited *j* since entering the system. When the model is parameterized so that *h* approaches infinity (corresponding to continuous time), the **P"** and **P'** matrices are identical to the **G"** and **G'** matrices, so that **N"** and **N'** are equivalent to **Q"** and **Q'** (Patten & Matis 1982). Patten (1985) has developed an extended environ analysis (not presented in this volume) that is a microscopic analysis for tracing the position of particles in the network through time.

A pair of Markovian unit input and output environs ($\mathbf{E'}_i$ and $\mathbf{E''}_i$) can be computed for each system input and output by multiplying the linear coefficient matrices **A'** and **A"** by a diagonalized column of their negative inverses (see Patten & Matis 1982). These environs are storage-referenced rather than throughflow-referenced (as are the environs $\mathbf{T''}_i$ and $\mathbf{T'}_i$). The elements of the output environ matrix $\mathbf{E''}_i$ are interpreted as the flow from *i* to *j* that is generated from a unit input to *i*. The values in the input environ matrix $\mathbf{E'}_i$ give the quantities of flow needed to generate a unit of outflow from i. Standing stock is now treated as a source of flow for this set of environs.

Origin and Destination Probabilities

When one computes the row or column sums over all environs $\mathbf{E'}_i$ and $\mathbf{E''}_i$, the end result is an expression of the origin (**W'**) and destination (**W"**) probabilities for the ecosystem resources. These matrices are shown for the Cone Spring model in Table 2.4. Each entry w'_{ij} indicates the probability that medium in i had entered the system via the inflow to j. The elements of **W"** give the probability that a unit of medium entering the system as z_i will exit via the outflow from j. In Cone Spring, we can see most of the exogenous energy entering from both primary production and detrital inputs will exit via the detritus and bacteria compartments. The origin probabilities show that the overwhelming majority of energy in all compartments can be traced back to the sun.

Storage Partitions and Residence Times

The most significant analytical contribution of discrete time environ analysis is its ability to estimate the mean residence times and storage partitions for network medium. The mean residence time matrices **M'** and **M"** (Table 2.5) are calculated as $-(\mathbf{A'})^{-1}$ and $-(\mathbf{A''})^{-1}$, respectively. When corrected for the Markov time step h (as are all the matrices computed using the ENVIRON program), the values stored in these matrices indicate the mean time that a particle now in compartment i has resided in j since entering the system (**M'**) or the average time that resources in i will spend in j before leaving the system (**M"**). When referenced back to system input (**z**) or output (**y**), one can estimate what portion of each compartment's standing stock was generated by each system input (stored in matrix **X"**) and what part of each standing stock will leave the system via each outflow (**X'**) . For example, for the Cone Spring ecosystem (Table 2.6), all of the plant biomass is generated via the input to x_i, but the other compartments owe some of their standing stock to external inputs to the detritus pool. On the other hand, energy that began as plant biomass may ultimately leave through any of the system's outflows. However, an almost equal amount of the energy bound in the detrital biomass will ultimately exit through bacterial metabolism as that directly leaving the detritus compartment.

TROPHIC ANALYSIS

Anyone who has worked in the field of ecology for even a short while is usually impressed by the complexity of the feeding relationships within an ecosystem. The feeding "web" is often quite convoluted. Given this, one might well ask of what use is it to continue to speak of herbivores and carnivores as if the feeding relationships always could be sequenced in a linear fashion? If one focuses on any species beyond the primary producers, then one is often faced with a situation where there exists

Table 2.4. Matrices of destination (W'') and origin (W') probabilities for the Cone Spring network.

W'' Matrix

	1	2	3	4	5
1	.206	.295	.151	.017	.331
2	.000	.000	.000	.000	.000
3	.000	.000	.000	.000	.000
4	.000	.000	.000	.000	.000
5	.000	.371	.191	.021	.417

W' Matrix

	1	2	3	4	5
1	1.000	.000	.000	.000	.000
2	.933	.000	.000	.000	.067
3	.933	.000	.000	.000	.067
4	.933	.000	.000	.000	.067
5	.933	.000	.000	.000	.067

Table 2.5. Mean residence time matrices for Cone Spring ecystem resources

M' Matrix

	1	2	3	4	5
1	.025	.000	.000	.000	.000
2	.024	.026	.001	.001	.376
3	.024	.004	.026	.001	.376
4	.024	.004	.026	.047	.376
5	.024	.301	.001	.001	.376

The mean time that a particle now in compartment i has resided in j since entering the system.

M'' Matrix

	1	2	3	4	5
1	.025	.009	.005	.001	.299
2	.000	.026	.002	.001	.117
3	.000	.002	.026	.007	.058
4	.000	.006	.285	.047	.169
5	.000	.012	.006	.002	.376

The average number of times that a particle in i will cycle to j before leaving the system.

Table 2.6. Storage partitions using environ analysis for the Cone Spring network. Variances in parentheses.

Inflow Z_1 partitioned among compartments X_i

X 1	285.000	(81225.01)
X 2	108.819	(51919.48)
X 3	55.996	(29603.41)
X 4	15.866	(16339.81)
X 5	3340.548	(.169E+08)

Inflow Z_5 partitioned among compartments X_i

X 1	.000	(.000)
X 2	7.781	(198.309)
X 3	4.004	(116.878)
X 4	1.134	(66.068)
X 5	238.852	(57050.450)

Outflow Y_1 partitioned among compartments X_i

X 1	58.686	(3444.160)
X 2	.000	(.000)
X 3	.000	(.000)
X 4	.000	(.000)
X 5	.000	(.000)

Outflow Y_2 partitioned among compartments X_i

X 1	83.951	(8055.764)
X 2	92.469	(8550.570)
X 3	3.426	(620.549)
X 4	2.846	(931.387)
X 5	1327.794	(1763035.000)

Outflow Y_3 partitioned among compartments X_i

X 1	43.141	(2127.315)
X 2	8.160	(708.929)
X 3	47.415	(2248.191)
X 4	1.463	(245.954)
X 5	682.328	(465571.111)

Outflow Y_4 partitioned among compartments X_i

X 1	4.827	(26.641)
X 2	.913	(8.878)
X 3	5.306	(28.155)
X 4	9.491	(90.073)
X 5	76.357	(5830.469)

Outflow Y_5 partitioned among compartments X_i

X 1	94.392	(10184.020)
X 2	15.057	(2904.263)
X 3	3.852	(784.492)
X 4	3.200	(1177.450)
X 5	1492.921	(2228813.000)

a multitude of feeding pathways of different lengths that can be traced back from the animal of interest to the primary producers. As Cousins (1985) observes, "A hawk feeds at five trophic levels."

One strength of network analysis techniques is that they offer the investigator a systematic and mechanical way of keeping track of the myriad of feeding pathways upon which any given population within the network depends. Thus, it might have happened that the hawk in Cousin's network of feeding relationships received its sustenance along some 75 or more separate pathways leading back to the plants. If necessary, those pathways could be enumerated using the backtracking methods presented in the next section. However, if one also has a weighted network description of the ecosystem, then the simpler input/output methods portrayed earlier will suffice to determine how much of the flow reaching the hawk arrives after traversing pathways of different integer lengths. Suppose that an analysis reveals that 5% of the hawk's diet is plants, 35% arrives after two feeding transfers, 40% after 3, 15% after 4 and 5% after 5 exchanges. Then the hawk functions on the average trophic level 3.8 (= .05 x 2 + .35 x 3 + .40 x 4 + .15 x 5 + .05 x 6). A unique number has been assigned to the trophic status of the hawk. This value can be compared with the corresponding trophic positions of other members of the community for purposes of ranking. More interestingly, any change in this value as the external conditions or community composition change could be used to help assess how well the hawk is adapting to the new conditions. (A lowered trophic status has been hypothesized to be indicative of stress on the population in question.).

Given the partitioning of flows over pathways of various lengths, one could estimate the average trophic status for each compartment in the network. Also this information is useful in constructing a picture of the trophic status of the entire community. Lindeman (1942), for example, envisioned a trophic pyramid, or concatenation of trophic relations wherein the amounts transferred to each higher level would become progressively smaller. The same matrix methods we used to portray the predator's diet, makes it possible to apportion the activity of, say, the hawk to a series of abstract trophic levels 2 thru 6 in such a way that the matter and energy of the system is conserved. The resulting hypothetical trophic chain would yield a profile of the efficiencies of trophic transfer that could serve as a diagnostic of how well the community is functioning. For example, well-developed communities could be expected to have longer trophic chains with more material and energy reaching the higher levels. Stressed communities, on the other hand, are hypothesized to possess short, but intense chains of transfers.

The pivotal question remains how to calculate the trophic apportionments. A basic method for doing this was described in Ulanowicz & Kemp (1979) and is sketched below:

Recall that the elements of the input structure matrix, g'_{ij} represent the fraction of the total input to j that flows directly from i. The utility of the matrix **G'** is that its successive powers quantify how much flows from i to j along all pathways of length equal to the power to which **G'** is raised. For example, the ijth element of $[\mathbf{G'}]^3$ (the matrix **G'** multiplied twice by itself) is precisely the fraction of total input to j which flowed from i along all pathways of length 3 connecting i to j. Hence, if one begins with a knowledge of how much each population acts as a primary producer, one can use the powers of **G'** to see how that primary production reaches the organisms at higher trophic levels after each succeeding transfer.

If z_i represents the external input to compartment i and T_i the total flow through i, then the quotient z_i / T_i will be taken to quantify the degree to which i acts as a primary producer (or supplier) of medium to the network. Consider a row vector with elements $L_i = z_i / T_i$. It follows that the product **LG'** will also be a row vector whose elements describe how much arrives at each node after exactly one internal transfer, that is, how much each population feeds at the second trophic level. In general, $\mathbf{L}[\mathbf{G'}]^{m-1}$ will be a row vector giving the fraction of its total consumption which each population receives at the mth trophic level.

If the feeding network contains no directed cycles (a simple cycle is observed when the initial and terminal nodes of a sequence of arcs are identical, but none of the intermediate nodes are repeated), the sequence of row vectors just described will truncate (yield a row vector of all zeros) after at most n-1 steps, where n is the number of nodes in the feeding web. It then becomes possible to construct an n x n matrix **L** whose ith row is identical to $\mathbf{L}[\mathbf{G'}]^{i-1}$. This matrix is called the Lindeman trophic transformation matrix. The ith row of **L** describes how much of the feeding activity of each population takes place at the ith trophic level. It usually happens that only the first m rows ($m < n$) are non-zero, hence the aggregation of n species into m levels. The jth column of **L** shows the fractions of the total feeding activity of population j that occur at each trophic level. In a network without cycles these columns will always sum to unity, indicating that all activity has been accounted for.

The concept of trophic level is best confined to living, feeding organisms, so that the creation of **L** will involve only the living species. It is for this reason that in the standard data format used in this book the living members of the community should appear first in the compartment list. This allows the trophic aggregation routine in NETWRK to operate initially only on the living food web. Pimm (1982) relates how directed cycles are rare among the livings members of an ecosystem and are usually of very small magnitude if they do occur. Hence, in the event that an ecosystem network does possess a few small cycles in the food web, these usually can be removed (see the section on cycle identification below) without significantly impacting the subsequent trophic analysis.

The entire ecosystem is generally larger than just the living components, and most networks of ecosystems flows include one or more abiotic compartments such as detritus or inorganic nutrient pools. It is through these non-living elements that the preponderance of material or energy cycling in the ecosystem occurs. Although the method of trophic partitioning does not involve the abiotic elements, the trophic aggregation would be incomplete if it did not in some way incorporate the recycled quantities. Following an assumption commonly made during the course of the IBP program, Ulanowicz (1989) assigns all the non-living components a trophic status of one. This has the effect of equating herbivory and detritivory in the trophic sense (all of which appeals to the intuition, because it is sometimes difficult to separate living from dead plant tissue). To complete the mass balance, the Lindeman transformation matrix can be expanded by d (the number of non-living members in the community) rows and columns that are all zeros except for the terminal d x d submatrix,which would contain a single one somewhere in each column (usually in the last entry).

The expanded Lindeman transformation matrix is then used to complete the trophic analysis. In the Cone Spring example, there are 4 living populations, as shown in Figure 2.2a. The corresponding Lindeman transformation matrix looks like the following:

Effective Trophic level	Plants	Bacteria	Detritivores	Carnivores
1	1	0	0	0
2	0	1	0.969	0
3	0	0	0.031	0.969
4	0	0	0	0.031

(Here the number of trophic levels just happens to equal the number of living species.) To find the effective trophic level of each of the species, one simply multiplies all the members in each row of the Lindeman matrix by the number of that row and then adds down the column. As a result, one finds that the plants are at level 1.0, the bacteria at 2.0, the detritivores at 2.03 and the carnivores at 3.03. The detritus has been assigned to level 1.0.

The trophic "aggregation" of the Cone Spring network (keeping the autotrophs and the detritus separated for the time being) is depicted in Figure 2.6. The trophic levels are designated by Roman numerals to avoid confusion with the species designations. The exports and respirations of the trophic levels are calculated by premultiplying the actual losses by the Lindeman matrix, i.e., **Le** and **Lr**, where **e** and **r** are the column vectors of exports and respirations, respectively. (These results are called the canonical exports and canonical respirations in the output from NETWRK.) In the Cone Spring example there are no exports from levels III and IV.

To calculate the interlevel transfers and recycles one employs the transformed flow matrix, **LFLT** in ways that need not be detailed here (Ulanowicz, 1989). The focal structure in Figure 2.6 is the "grazing chain," or the transfers from each level to the next higher up. (The Cone Spring network is peculiar in that it evinces no herbivorous grazing. The usual circumstance is for a network to possess an uninterrupted grazing chain.) In general each level returns a certain amount to the detrital pool, much of which is fed back into the grazing chain as detritivory (the 7514 kcal/m^2/yr shown flowing from D to level II.) The exogenous inputs to, exports and respirations from, and circulation within the detrital pool (none) all appear separately on Figure 2.6.

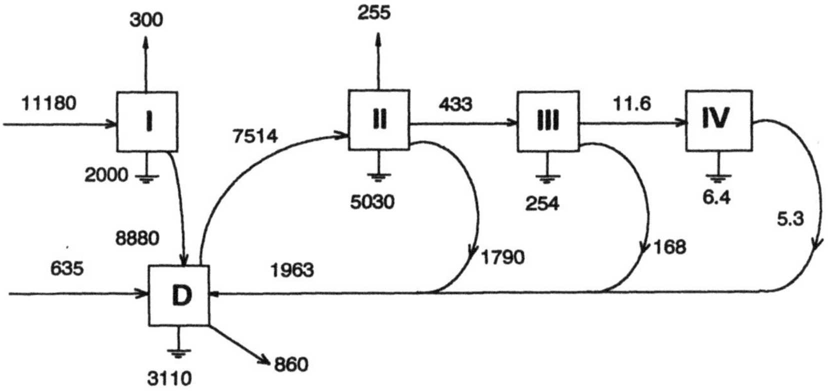

Fig 2.6. The trophic aggregation of the Cone Spring network with the autotrophs and detritus separated. The trophic levels are designed by Roman numerals, D represents the detritus pool.

If the detrital pool is merged with the autotrophs, one sees the classical trophic pyramid as portrayed in Figure 2.7. Here the interlevel transfers form a "Lindeman spine" of ever decreasing flows falling from 13,778 units into trophic level I to a scant 11.6 into trophic level IV. The uniform decrease allows one to calculate the trophic efficiency of each level as the amount it transfers to the next higher level divided by the amount it received from the one below it. In Cone Spring these efficiencies are uniformly decreasing. Experience with a variety of ecological networks shows that there is a tendency for trophic efficiencies to decrease at higher levels. The trend is not without exceptions, however, and it appears that each exception can be traced back through the Lindeman transformation matrix to an interesting peculiarity of the network being analyzed (Baird & Ulanowicz 1989).

Despite the obvious complexity of most observed foodwebs, it is apparent that network analysis can be employed to reveal an underlying canonical trophic structure that is useful for interpreting how a particular ecosystem is functioning.

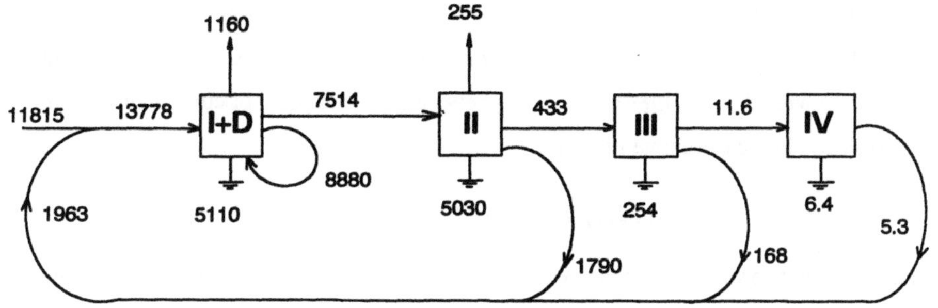

Fig 2.7. The trophic aggregation of the Cone Springs network with the autotrophs merged with the detrital pool (I + D).

CYCLE ANALYSIS

Positive feedback is a critical phenomenon in determining overall system structure. Positive feedback occurs in ecosystems when flows cycle among compartments. Compartments engaged in positive cycling will have an increase in output relative to those compartments which do not. This could result in a strong competitive advantage over non cycling compartments during growth phases. As well, positive feedback may also be a stabilizing factor in systems with a time lag. Cycles in ecosystems are an important factor contributing to their autonomous behaviour (Ulanowicz 1983, 1986a). Furthermore, it has been argued (Kay 1984) that the organization of flow in ecosystems reflects a tendency to degrade the exergy content of the flowing energy while conserving materials flowing within the system. Exergy degradation is related to straight-through flows, while material conservation is related to the cycles in the ecosystem. These are reasons for studying the cycling in ecosystem flows and in particular for separating the flow network into two components: the cycles and the straight-through flows.

The process of separating cycles from straight-through flows has two distinct aspects. The first involves identifying which components are connected together to form a cycle. This is a straightforward topological problem whose solution can be found in most elementary graph theory texts. The second aspect involves partitioning the flows among compartments into those associated with cycles and those associated with straight-through flows. This is a problem which has not been dealt with extensively and which still requires further investigation.

What follows is a brief description of the process of identifying cycles. More detailed descriptions can be found in Ulanowicz (1983, 1986a). The description of how NETWRK accomplishes this task can be found in Ulanowicz & Kay (1986). The flows between various pairs of compartments are defined by non-zero entries in the exchange matrix.

Topological Cycles and their Enumeration

The identification of the structure of the cycle, that is, what is topologically connected to what to make up the cycle, begins with a depth-first search (See Ch. 6 of Horowitz & Sahni 1984). The purpose of the search is to identify the cycle arcs. Cycle arcs are connections between compartments whose direction is from a higher compartment number to a lower compartmental number. In the Cone Spring example the arc from detritus to bacteria (from 5 to 3) is a cycle arc. If one searches for cycles starting only from those compartments which have cycle arcs into them, one is assured of finding all the cycles. This significantly cuts down on the time for searching out and enumerating all the cycles.

The next step is a backtracking search. The idea of backtracking search is to start with one compartment and see if it is connected to the next higher numbered compartment (i.e., is 3 connected to 4?) If it is connected, then we see if this next compartment (4) is connected to the lowest numbered compartment not already visited whose number is as least as great as the starting compartment number (i.e., is 4 connected to 5?). The process continues until the starting compartment is reached or until we find a compartment which is not connected to any compartment whose number is higher than the starting compartment number and which has not been visited. In the former case a list of the visited compartments constitutes a simple cycle. In the latter case we back up one compartment in our list and see if it is connected to the second next lowest numbered compartment not already visited (i.e., is 3 connected to 5, given that 3 is connected to 4, and 4 is not connected to any compartment whose number is greater than 3, nor to 3) and so on.

This is an algorithm which is best described by pictures rather than words. The idea is to treat each compartment as a dot. The dots are numbered. Compartments i and j are connected if the ijth element of the exchange matrix is non-zero. One tries to find a path, along the arcs connecting the dots, which starts and finishes with the same dot and which contains no other dot more than once. The trick is that from any dot you may only try to follow a path either to the lowest numbered dot (whose number is greater than the original dot) which you have not already tried, or to the original dot.

For the Cone Spring example there are 5 single cycles (See Figure 2.8). For the Crystal River ecosystem of 17 compartments (see Ulanowicz 1986a) there are 119 cycles. The number of possible cycles goes up as approximately the factorial of the number of compartments. Luckily, most ecosystems are very sparse, that is, very few of all possible connections are actually made.

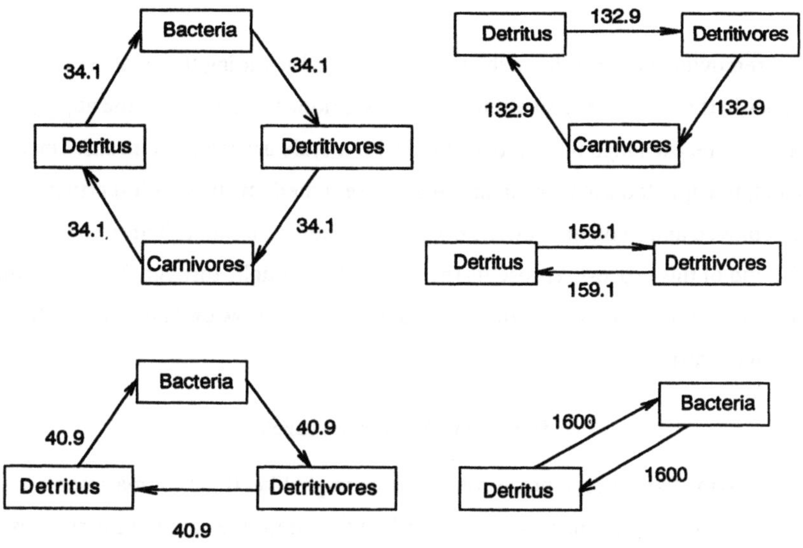

Fig 2.8. The five simple, directed cycles of energy flow in the Cone Spring network weighted according to circuit completion probabilities.

Functional Cycles and their Removal

Once the structure of the cycles has been identified, the next step is to separate the flow network into 1) arcs associated with cycles and 2) arcs associated with straight-through flow. This process begins by identifying the critical arc in each cycle. In each cycle, there is one arc which has the smallest flow through it. If this arc is removed and the value of the flow through it is subtracted from the flow in each of the other arcs in the cycle, then the cycle is effectively removed from the flow network. There is no other arc in the cycle whose flow rate could be set to zero (thus breaking the cycle) without some flows in some cycle arcs becoming negative. Thus, the critical arc is the cycle arc with the smallest flow through it.

It is most likely that an arc is part of more than one cycle. The collection of all the cycles which share the same critical arc is called a nexus. When a critical arc is removed from the network, all the cycles in the nexus are broken. The portion of the flow through the critical arc which is subtracted from each of the arcs in each of the cycles, is determined by calculating the circuit probabilities. This is done by multiplying the g'_{ij} for each cycle arc to obtain a weight for the whole cycle. The weight for each cycle is then summed for all the cycles in the nexus to give the nexus weight. The portion of the flow through the critical arc which is subtracted from each arc of a particular cycle is that cycle's weight divided by the nexus weight.

The process of removing the cycles from the network begins by finding the smallest critical arc and its nexus. The cycles in the nexus are then eliminated. This is done by subtracting the appropriate amount from each entry in the exchange matrix associated with a cycle arc from the nexus. Once a nexus has been eliminated, the updated exchange matrix is again searched for the smallest critical arc. This arc's nexus is then removed and so on until all the cycles have been eliminated. The exchange matrix which remains at the end of this process represents the network of straight-through flows. The matrix which represents the cycles is found by subtracting the straight-through flow exchange matrix from the whole network exchange matrix.

Issues for Further Research

The algorithm used in this process is not as robust as one might like. If the compartments are numbered differently, the resulting cycles might be slightly different. Where the value of the smallest critical arc is shared by more than one arc, the order in which these arcs are removed is arbitrary. (The order in which arcs are removed affects the overall partitioning of cycles and straight-through flow.). There is a problem when a cycle has more than one arc with the value of the critical arc (multiple critical arcs). It is not clear if these problems can be resolved in a satisfactory way. However, Ulanowicz and Kay have run a number of networks through the algorithm and found that these problems affect a very small percentage of the cycles.

How one decides on a critical arc and the partitioning of its flow is a subject of debate. By subtracting the total value of the smallest arc one is assuming that all the flow through it is part of a cycle and none contributes to straight-through flow. Using the circuit probabilities to assign the flow to the various cycles in a nexus is only one of several possible schemes. However, the schemes used in the partitioning algorithm are the most intuitively satisfying ones proposed so far.

Finally, partitioning the network into two components (cycles and straight-through flows) is a reductionist analysis which assumes superposition. It could be argued that the synergistic effects in the network make it irreducible. Exercises, such as this book, which test the analytical techniques suggested in this chapter on a broad spectrum of ecosystems, will ultimately determine the utility of this analysis.

TOTAL SYSTEM PROPERTIES: MEASURES OF FLOW

The order in which analyses are being discussed reflects an increase in the degree to which the associated variables characterize the total system. This progression culminates now in the consideration of how to quantify attributes of the whole system. That is, how might one encapsulate a property of all of the bilateral relationships into a single number? Such a number would resemble a state variable in thermodynamics. It would describe the overall status of the system, but not always its exact configuration. Many configurations can give rise to the same value of the system attribute. Thus, one should not expect whole system properties to specify exactly what is occurring at, say, the predator-prey level. The prediction capability of a whole-system index as to what is occurring at finer scales is minimal at best.

What is gained, however, is the potential for weaving several of these global variables into a coherent picture of ecosystem dynamics at the macroscopic level, the possible discovery of a new phenomenological principle that describes ecodynamics in the same sense that the laws of thermodynamics describe the evolution of macroscopic physical systems. Of course, it may eventually turn out that some of the variables to be described will turn out to be epiphenomenological, and do not quantify truly emergent properties of ecosystems. Such possibility notwithstanding, the verdict on this issue is still far from being written, and experimentation with these quantities is today one of the most exciting and potentially fruitful endeavors in all of science. Given the tools presented here, the reader is invited to apply these ideas to his/her ecosystem of interest and to join in the excitement of the search.

Total System Throughflow

One of the first questions one might ask about an ecosystem is how big it is. Size is generally reckoned in terms of how much material or energy an object possesses. Thus, one could begin by calculating the total system biomass, or the total energy content of all the ecosystem parts.

While these values are certainly germane to the concept of system size, they are by no means its exclusive determinants. For example, it is of distinct secondary interest to an economist how much money resides within a particular economic community. What appears to matter more to the general welfare is how much this capital actually circulates. Hannon (1973) suggests that ecosystems bear more resemblance in this regard to economic communities than to physical systems. The activity of a system is often the matter of most immediate concern, as evidenced by the intense interest in such activity as the gross national product.

Hannon suggested that the aggregate of all the transfers, otherwise known as the total system throughflow or throughput, should be the primary measure of system size. The equations used to calculate *T* depict it as the sum of all internal and exogenous inputs to system components, or the sum of all outputs (endogenous flows, exports, and respirations) of all the compartments. There remains some disagreement on exactly how to calculate *T* (Hannon 1973, Finn 1976). The most inclusive sum (of all external and internal flows) is now gaining favor, i.e., the sum of both of the input and output representations of *T* described earlier. This all-inclusive form of the total system throughflow is the same whether viewed from the input or the output perspective and remains well-defined, even if the system is not in balance. We therefore recommend its use as calculated in the programs NETWRK, ECOSYS and ENVIRON.

The non-inclusive value for *T* for the Cone Spring network is 30,626 kcal/m^2/yr. The all-inclusive value for total system throughput is 42,445 kcal/m^2/yr.

It should be noted that *T* increases with the number of subdivisions into which the system is divided and is therefore dependent upon the degree and manner of aggregation decided upon by the investigator. A special case occurs when compartments are aggregated by the program AGG which pools compartments specified by the reseacher. When two or more compartments are pooled, the flows between the previously separated compartments are modelled as loops leaving and re-entering the new pooled compartment. Under these circumstances (with "self loops") *T* remains the same when compartments are aggregated (see Warwick and Radford 1989, this volume, for examples).

Finn Cycling Index

Of the activity represented by *T*, some represents medium flowing through the system, whereas the rest of the activity consists of medium being cycled within the community. As discussed in the previous section, the exact proportions of these two components is ecologically significant. Odum (1969), for example, cites an increased amount of cycling as an important attribute of more mature ecosystems. (There are indications [Ulanowicz 1985, Schneider 1988], that the immediate response of a system to a eutrophying stress is to increase its proportion of internal cycling. This is interpreted as a sign of a healthy system coping with stress.) Most agree that the proportion of recycle in a network is a significant component of its overall status.

The reader will recall that the diagonal elements of the total flow describe exactly the amount of flow which leaves compartment *i* and eventually makes its way back to *i*. That is, it represents how much of the throughput of *i* is engaged in cycling. It follows that the sum of the diagonal elements quantifies the total amount of system activity devoted to cycling as:

$$T_{cyc} = \sum_{i=1}^{n} z'_{ii} = \sum_{i=1}^{n} z''_{ii} \qquad (15)$$

where T_{cyc} is that part of T which represents cycled flow. Accordingly, the fraction which this amount comprises of the total system throughput is called the Finn cycling index, CI, where

$$CI = T_{cyc} / T. \qquad (16)$$

In Cone Spring $T_{cyc} = 2816$ kcal/m^2/yr, giving a Finn cycling index of 0.0663.

Average Path Length

The cycling index gives the investigator an idea of how retentive a system is. There is another measure of system retention called the average path length, which combines the attribute of recycle with that of trophic length. The average path length (*APL*), as its names implies, is the average number of transfers an arbitrary quantum of medium will experience on its trip through (and around) the network. If Z is the sum of all the exogenous inputs from other systems, then $T - Z$ is the amount of flow occurring after the medium has entered the system. It follows that $APL = (T - Z) / Z$ will yield the average number of times this input is transferred by the system. In Cone Spring this quantity comes to 2.59. One would expect *APL* to rise under normal succession and to decrease in the face of stress.

TOTAL SYSTEM PROPERTIES: MEASURES OF STRUCTURE

The measures examined in this section characterize the structural self-organization of an ecosystem. By structural organization is meant the way in which the different compartments are interconnected. These interconnections are determined by who is eating what. To be more precise, a compartment can be treated as a source or as a consumer of resources. The structural organization of an ecosystem is determined by the way in which the consumers make use of the other compartments as resources. The set of sources used by a consumer is its resource niche. Characterizing the resource niche of the compartments describes the structural self-organization of an ecosystem.

A description of the resource niche must characterize each of the individual sources and the total set of sources used by the consumers. The individual sources are characterized by the quantity and quality of resources extracted from them. The set of sources is characterized by the relative sustenance provided by each source to each consumer. The use of the set of sources, from the perspective of the whole ecosystem, must also be characterized.

Mathematically, the flows in the ecosystem can be represented as follows: If F is the matrix of resource flow between compartments, the total output from a compartment (outj) is given by:

$$out_j = \sum_{i=1}^{n} f_{ji} \tag{17}$$

Similarly, the total input to a compartment (in_i) is given by:

$$in_i = \sum_{j=1}^{n} f_{ji} \tag{18}$$

Then the proportion of the total outputs from all compartments which flows from i is given by:

$$P_i\,[X_s] = out_i \,/ \sum_{j=1}^{n} out_j \tag{19}$$

and the proportion of the total flow into compartment j which comes from compartment i is given by:

$$P_i\,[X_s / X_{cj}] = f_{ij} \,/\, in_j\ . \tag{20}$$

Similarly, the proportion of the total inputs to all compartments which flow into i is given by:

$$P_i[X_c] = in_i \,/ \sum_{j=1}^{n} in_j \tag{21}$$

X_s is associated with output from a compartment (i.e. it acts as a source) and $P_i[X_s]$ can be taken as the probability of a quantum of resource coming from compartment i. Similarly X_c is associated with input to a compartment (i.e. acts as a consumer) and $P_i[X_c]$ can be taken as the probability of a quantum of resource flowing into compartment i. The probability distribution $P_i[X_s]$ describes the compartments in the system as sources and the probability distribution $P_i[X_c]$ describes compartments as consumers.

It should be noted that only flows into living compartments are considered. It is not sensible to talk about the resource needs of a non-living compartment. Non-living compartments act as sources and sinks in ecosystems but not as consumers. These definitions also apply to the general non-steady state situation.

Measures of resource niche and structural organization (S & D)

Using these definitions and the information theory measures described in the appendix of this chapter, two measures of resource niche and structural organization can be defined:

$$D = H[X_s] \text{ or } -\sum_{i=1}^{n} P_i[X_s] \log P_i[X_s] \tag{22}$$

and

$$S = H[X_s / X_c] \tag{23}$$

or

$$S = -\sum_{i=1}^{n} P_i[X_c] \sum_{j=1}^{n} P_j[X_s / X_{ci}] \log P_j[X_s / X_{ci}]; \tag{24}$$

where $O < S < D < \log n$.

D measures the diversity with which compartments are used as sources by other living compartments. As the utilization of compartments as sources becomes more evenly distributed over all the compartments acting as sources, *D* increases. *D*, in a sense, indicates the degree to which the potentially available sources of resources are utilized in the particular ecosystem being examined.

S measures the diversity of sources utilized by each living compartment, averaged over all living, consuming compartments. When *S* is large, the living compartments are on average generalists, that is they use many sources of food evenly. When *S* is small, the compartments are on average specialists. They have a small number of sources, one of which is used heavily. In effect, *S* measures the average amount of choice in picking sources exercised by each living compartment during the time the ecosystem is observed. This is equivalent to measuring the average resource niche breadth. Note that this interpretation is particularly meaningful when the compartments correspond to individual species.

If the flow rates do reflect the proportion of available resource coming from each source, then *S* measures the average relative amount of sustenance provided to each living compartment by each source. That is, in an inverse sense, *S* indicates the average dependence of any one living compartment on the available sources. (The larger *S*, the less the compartments depend on any one source.). *S* measures the diversity of ecological interaction between the compartments which occurs through the food web.

In summary, D measures the diversity with which the compartment's outputs contribute to the total resource flows in the system. S measures the average resource niche breadth and, hence, indicates how specialized the compartments are. Together, S and D characterize the resource niche, that is how sources are being utilized by the compartments.

The structural organization of an ecosystem reflects the food consumption patters of the species which make up the compartments. The pressures which shape structural organization (e. g. positive feedback) act to modify the patterns of source utilization by the individual compartments. These patterns can be monitored using S. These pressures will also affect the overall utilization of sources in the ecosystem. This can be monitored by D. Thus, the structural self-organization of the ecosystem can be monitored using D and S.

Applications of S & D

Following are some examples to illustrate the utility of S and D. Figure 2.9 shows three hypothetical examples. The compartments labelled X are the compartments as sources and those labelled Y are the same compartments as consumers. Assume that the same amount of resource flows through each link. In system (a), each compartment is as specialized as possible ($S = 0$) and there is no diversity of source use ($D = 0$). In system (b), the compartments are still specialized, but the use of sources is as diversified as possible ($D = \log 3$). In system (c), the sources are used as evenly as possible ($D = \log 3$) and the one consuming compartment is a generalist, eating from each source equally ($S = \log 3$).

For the Cone Spring example, $S = 0.0419$ and $D = 0.2409$. These values alone do not provide us with much information. S and D are useful for relative comparisons of similar systems. They are not particularly useful as absolute measures of a system state in isolation from other states. When comparing values of S and D it should be remembered that they are logarithmic functions, small changes in value reflect large changes in the distribution of flows in the system.

Two examples explored later in this book, show the utility of S and D. One is a warm core ring system (Ducklow *et al.* 1989). Table 2.7 shows the values of S and D for spring, summer and fall. The diversity (D) with which the available sources are used remains constant. However, S changes quite dramatically (recalling that S and D are logarithmic). Initially, the average resource niche is somewhat diverse. In summer the resource niche breadth becomes quite broad (generalists) but by fall has become quite narrow (specialized). That D remains fairly constant means that the change in resource niche breadth does not reflect a change by consumers to a single source.

Table 2.7. Temporal variation of D and S.

	MAY	AUG	OCT
D	1.2	1.2	1.2
S	0.75	1.0	0.5

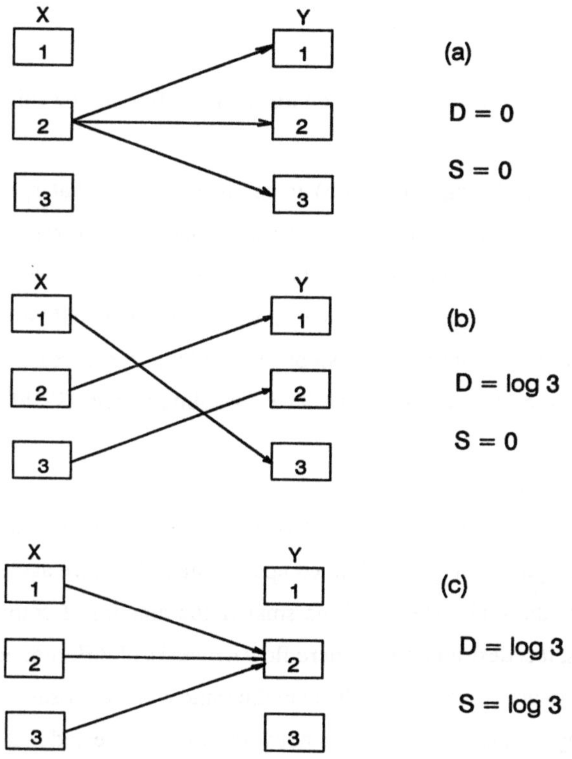

Fig 2.9. Three hypothetical examples of flow organization to illustate the utility of S and D.

An example from an upwelling (Field *et al.* 1989) is shown in Figure 2.10. In this example, *S* and *D* increase together during the initial stages of succession after the bloom peaks. This reflects the increased availability of resources. As the system moves from the beginning of the final stage of succession to its culmination, *S* falls off rapidly as the food web is dominated by specialists while *D* increases slightly. As the system dies out *S* continues to fall off sharply while *D* declines gradually. Further examples and discussion of the use of *S* and *D* can be found in Kay (1984).

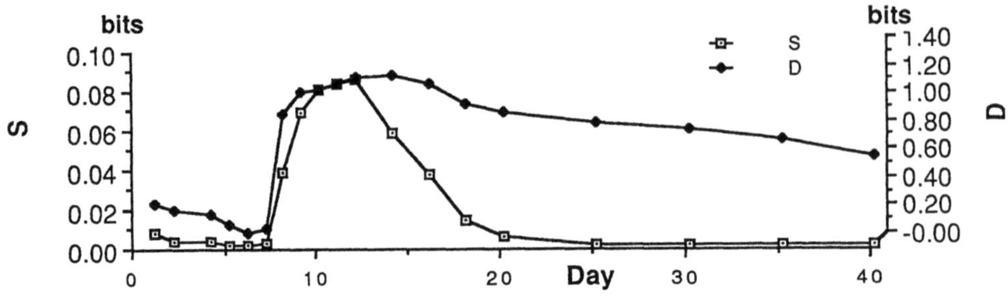

Fig 2.10. The change in S and D with time of a plankton system after an upwelling event.

As an ecosystem develops (i.e., during succession) it has been argued that species will become specialists. Specialization is linked to a more efficient use of the available resources. It would be expected that *S* would decrease, i.e., the resource niche would narrow with ecosystem development. It would be expected that, on the average, the sources in an ecosystem would be used more evenly, since this would normally result in a more total use of the available resources. Hence, it would be expected that *D* would increase with development. The upwelling system (Field *et al.* 1989) is an example of this behaviour.

The problem of aggregation in ecosystem modelling has been troublesome for a long time. There is always a trade-off in the aggregation of species into compartments. The more aggregated the system is, the less it tells you about the real system and the smaller the amount of empirical information required. In some situations, it is desirable to compare flow networks of different ecosystems. This is only possible if the ecosystems are compartmentalized in the same way. One issue is how to aggregate compartments without losing the essence of the flow network. It was suggested that *S* and *D* could be used in this context as optimizing functions. One would aggregate compartments in a way which minimized the changes in *S* and *D*. In this way the degree of connectivity within the flow network would be preserved.

Ascendency and Related Measures

The foregoing indices *S* and *D* are intended by their originator (Kay 1984) to quantify the topological attributes of ecosystem networks. They represent a continuation of the efforts of Rutledge *et al.* (1976) to use information theory to measure attributes of network structure. Ulanowicz (1980) had also recognized the strength of the Rutledge approach for attaching a number to intensive attributes of ecosystems networks. Ulanowicz argued that ecosystem networks also have extensive properties, and that a full quantitative description of the successional process must include both extensive and intensive factors.

Accordingly, Ulanowicz took the average mutual information of a flow network and scaled this property by the popular measure of system activity - the total system throughflow T. The resultant product was called the network ascendency and defined as:

$$A = T \sum_{i=1}^{n} \sum_{j=1}^{n} (f_{ij} / T) \log (f_{ij} T / T_i T_j) . \tag{25}$$

The reader will notice that A is the product of size (T with the dimensions of flow of medium) times "organization" (the summed terms with the dimensions of information, e.g., bits).

That the average mutual information is related to organization follows from the notion that within an organized network there exists less ambiguity about the pathways over which medium flows. That is, an organized network is well articulated. An example of increasing network articulation is given in Figure 2.11. All three hypothetical configurations have identical total system throughputs of 96 units. Figure 2.11a is wholly unarticulated. At any of the four nodes there is maximal uncertainty about where a quantum will next flow. In Figure 2.11b there is only half as much uncertainty, for only a single choice between two recipients is necessary. Configuration b is better articulated than a. Finally, Figure 2.11c is maximally articulated. If a quantum is at any node, there is no uncertainty about where it will flow next.

Ascendency was tailored to encapsulate numerous phenomenological observations into a single quantitative statement. Odum (1969) had listed 24 separate attributes of more "mature" ecosystems. These criteria can be grouped into four generic categories: (a) more speciation, (b) finer specialization, (c) longer retention and (d) more cycling. All other things being equal, an improvement in any one of these generic properties of a network serves to increase its ascendency.

Ascendency can become a quantitative tool in diagnosing ecosystem change at the level of the whole system. Maturing ecosystems should, on the average, increase in ascendency. Early in system development the rise in ascendency will be due mostly to an increase in aggregate activity, T, while later on the continuing rise will mostly be caused by the finer articulation resulting from competition between parallel pathways. A decrease in ascendency is almost always indicative of a stress on the system. A small class of perturbations,however, can actually serve to increase the overall ascendency.

A sudden increase in exogenous inputs can actually increase the ascendency by disproportionately raising the total throughflow T, while at the same time decreasing the mutual information factor. Such a set of circumstances is usually called eutrophication and leaves its characteristic signature on the rise in ascendency (Ulanowicz 1986b).

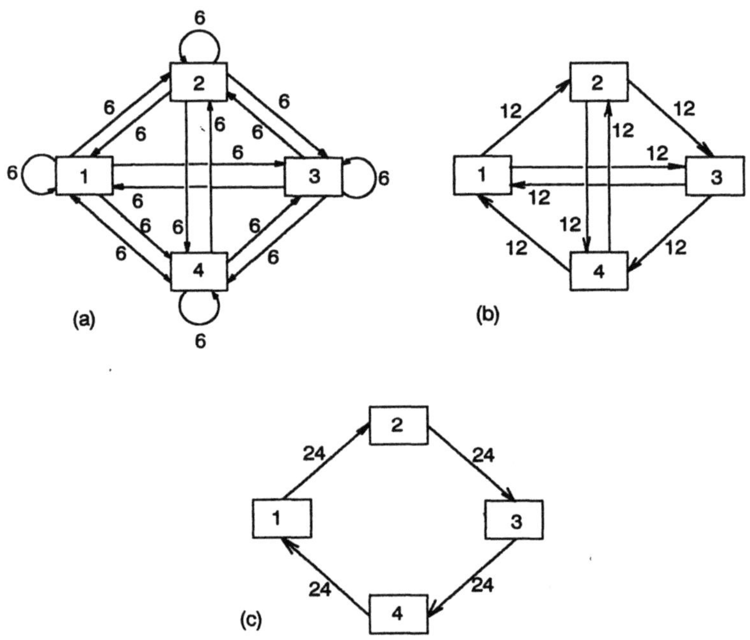

Fig 2.11. Three artificial, closed networks having the same total system throughputs but differing in their degrees of articulation. (a) The maximally connected but minimally articulated configuration. (b) An intermediate level of articulation. (c) The maximally articulated configuration.

There must be limits on the increasing ascendency, as no system can grow without bounds. To better narrate these limits in quantitative terms, one may define five other scaled information indices for any arbitrary network. One upper limit to the mutual information of the network is the entropy of flows defined earlier. One may scale this entropy by T to define a quantity known as the Development Capacity,

$$C = -T \sum_{i=1}^{n} (T_i / T) \log (T_i / T) \quad . \tag{26}$$

In Figure 2.11c the maximally articulated network became identically equal to its upper limit, C. That was a hypothetical and unreal situation; however, and in any actual network there will always remain a positive difference $C - A$, which is referred to as the system's overhead.

Part of the overhead is generated by the exogenous transfers. It is possible to identify three separate components attributable to the inputs, exports and dissipations, respectively. A fourth component represents the residual uncertainty associated with the presence of multiple or parallel pathways among the network components, and it is termed the redundancy.

If one excludes from the capacity and the ascendency those terms generated by the exogenous transfers, one is left with what may be termed the internal capacity and the internal overhead. These measures may be of interest if one is focusing on endogenous system behavior. For algebraic reasons not important here, the internal overhead (the difference between the internal capacity and the internal ascendency) contains terms generated by the exports and respirations. These components are called the tribute and dissipation, respectively. They, along with the aforementioned redundancy, comprise the internal overhead.

The value of the development capacity of the Cone Spring network, along with its four components is:

Development Capacity	=	93,172	kcal-bits m^2/yr
Ascendency	=	56,725	"
Overhead on Inputs	=	2,652	"
Overhead on Exports	=	1,920	"
Overhead on Respiration	=	21,364	"
Redundancy	=	10,511	"

The reader will notice that the ascendency and the four overhead terms all sum to give the capacity. The corresponding decomposition of the internal capacity looks like:

Internal Capacity	=	71,372	kcal-bits m^2/yr
Internal Ascendency	=	29,332	"
Tribute	=	2,971	"
Dissipation	=	28,558	"
Redundancy	=	10,510	"

A more complete exposition of the meaning of the individual overhead terms and the algebraic details involved in the decomposition of the capacity are all given in Ulanowicz (1986b).

Ascendency, *S* and *D*: The Differences

Ascendency (*A*), *S*, and *D* are all measures of organization. *A* is meant as a measure of total ecosystem development. *S* and *D* are meant as measures of structural organization. *S* measures the average (over all compartments) diversity of source use by each compartment and *D* measures the diversity of outflow use over the whole ecosystem. *A* is meant as an ecosystem level measure (as is *D*), and not as a compartment level measure (which *S* is). This is the reason Kay uses two measures while Ulanowicz uses one.

A, S, and D are related mathematically as follows. A is the product of the total throughput (T), and the mutual information (I). Looking at equation 4 in the appendix, $I = D - S$. Thus,

$$A = T\ (D\text{-}S). \tag{27}$$

It should be noted that S and D are logarithmic measures. A is the product of a linear function and a log function. Thus, A will be more sensitive to system changes which affect the value of T relative to those changes which affect S and D. It is possible that significant changes in S and D will not be reflected by significant changes in A. However, changes in total throughput (T) are not measured by S and D as they are with A.

A, S, and D are meant to be used in different situations and are therefore different even though they are mathematically related. See Mann *et al.* (1989) for arguments as to when to use which of these measures of ecosystem development.

CONCLUDING COMMENTS

In this chapter an overview of commonly used ecosystem network measures was provided. These measures are all static and provide a snapshot of the state of the ecosystem flow network. The measures tell us about the compartments, their relationships and overall system status. All of the measures describe only the flow in an ecosystem and not the driving forces behind the flows. The picture of how an ecosystem food web functions will always be incomplete until both the flows and driving forces are known and network thermodynamic models of the ecosystems can be built.

The realities of available data limit the utility of all the measures presented. The authors of this chapter found that the limit on the utility of these measures is the lack of data of the detail (level of aggregation, temporal and spatial refinement) necessary to make the measures useful for real world applications. It is our hope that by making the measures more accessible to field ecologists, through computer software like those associated with this book, that further collection of the data needed for the application of the measures will be stimulated. Only if this happens will theoreticians be able to refine the measures so as to give environmental managers the tools they need to monitor the health of our ecosystems.

REFERENCES

Aczel, J., & Z. Daroczy 1975. *On Measures of Information and Their Characterizations*, Academic Press, New York.

Augustinovics, M. 1981. Methods of international and intertemporal comparison of structur,. in A. P. Carter and A. Brody (Ed.) *Contributions to Input-Output Analysis*, Vol. 1, North Holland, Amsterdam.

Baird, D. & R. E. Ulanowicz. 1989. The seasonal dynamics of the Chesapeake Bay ecosystem, Ecol. Monogr (in press)

Barber, M. C. 1978a. A retrospective Markovian model of ecosystem resource flow, Ecol. Mod. 5: 125-135.

Barber, M. C. 1978b. A Markovian model for ecosystem flow analysis, Ecol. Mod. 5:193-206.

Barber, M. C. 1979. A note concerning time parameterization of Markovian models of ecosystem flow analysis, Ecol. Mod. 6: 323-328.

Bosserman, R. W. 1981. Sensitivity techniques for examination of input-output flow analyses, in W.J. Mitsch & J.M. Klopatek (eds), Energy and Ecological Modeling, Elsevier Publishers, Amsterdam.

Costanza, R. & B. Hannon 1989. Dealing with the "mixed units" problem in ecosystem network analysis. In: Wulff, F., J.G. Field & K.H. Mann (eds), *Network Analysis in Marine Ecology: Methods and Applications*, Lecture Notes on Coastal and Estuarine Studies, Springer-Verlag, New York.

Cousins, S. 1985. Ecologists build pyramids again, New Scientist 107: 50-54.

Ducklow, H. W, M. J. R. Fasham & A. F. Vézina 1989. Derivation and analysis of flow networks for open oceanic plankton systems, in: Wulff, F., J. G. Field & K. H. Mann (eds), *Network Analysis in Marine Ecology: Methods and Applications,* Lecture Notes on Coastal and Estuarine Studies, Springer-Verlag, New York.

Field, J. G., C. L. Moloney & C. G. Attwood 1989. Network analysis of simulated succession after an upwelling event, in: Wulff, F., J. G. Field & K. H. Mann (eds), *Network Analysis in Marine Ecology: Methods and Applications,* Lecture Notes on Coastal and Estuarine Studies, Springer-Verlag, New York.

Finn, J. T. 1976. Measures of ecosystem structure and function derived from analysis of flows, J. Theor. Biol. 56: 363-380.

Finn, J. T. 1977. Flow analysis: A method for tracing flows through ecosystem models., PhD Thesis, Department of Zoology, University of Georgia, Athens, GA.

Hannon, B. 1973. The structure of ecosystems, J. Theor. Biol. 41: 535-546.

Horowitz E., & S. Sahni 1984. *Fundamentals of Data Structures in Pascal*, Computer Science Press.

Kay, J. J. 1984. *Self-organization in living systems,* Ph.D.Thesis, Department of Systems Design Engineering, University of Waterloo, Waterloo, Ontario, Canada.

Kemeny, J. G. & J. L. Snell 1960. *Finite Markov Chains*, D. Von Nostrand Co, Princeton, NJ.

Legendre, L & P. Legendre 1983. *Numerical Ecology*, Elsevier, Amsterdam, 419 pp.

Leontief, W.W. 1936. Quantitative input-output relations in the economic system of the United States, Rev. Econ. Stat. 18: 105-125.

Leontief, W.W. 1963. *The structure of the American Economy, 1919-1939*, 2nd edition. Oxford University Press, New York.

Lindeman, R. L. 1942. The trophic-dynamic aspect of ecology, Ecology 23: 399-418.

Mann, K. H., J. G. Field & F. Wulff 1989. Network analysis in marine ecology: an assessment, in: Wulff, F., J. G. Field & K. H. Mann (eds), *Network Analysis in Marine Ecology: Methods and Applications.* Lecture Notes on Coastal and Estuarine Studies, Springer-Verlag, New York.

Matis, J. H. and B. C. Patten 1981. Environ analysis of linear compartmental systems: the static, time invariant case, Bull. Int. Stat. Inst. 48: 527-565.

Odum, E. P. 1969. The strategy of ecosystem development, Science 164: 262-270.

Patten, B. C., R. W. Bosserman, J. T. Finn, & W. G. Cale 1976. Propagation of cause in ecosystems, pp. 457-579, in: B. C. Patten (ed), *Systems Analysis and Simulation in Ecology, Vol 4*, Academic Press, New York.

Patten, B. C. & J. H. Matis. 1982. The water environs of the Okefenokee Swamp: an application of static linear environ analysis, Ecol. Mod. 16: 1-50.

Patten, B. C. 1985. Energy cycling, length of food chains, and direct versus indirect effects in ecosystems, pp. 119-138, in: R. E. Ulanowicz and T. Platt (eds) *Ecosystem Theory for Biological Oceanography*, Fisheries & Oceans Canada .

Pimm, S. L. 1982. *Food Webs*, Chapman and Hall, London.

Rutledge, R.W., B.L. Basorre, & R. J. Mulholland 1976. Ecological stability: An information theory viewpoint, J. Theor. Biol. 57:355-571.

Schneider, E. D. 1988. Entropy, ecosystems and evolution, To appear in: B.Weber and D. Depew (eds): *Entropy, Information and Evolution*, MIT Press, Cambridge, MA.

Shannon, C. 1948. The mathematical theory of communication, Bell System Technical Journal, Reprinted 1949: *The Mathematical Theory of Communication*, University of Illinois Press, Urbana, Illinois.

Szyrmer, J. & R. E. Ulanowicz. 1986. Total flow in ecosystems, Ecol. Mod 35:123-136.

Tilly, L. J. 1968. The structure and dynamics of Cone Springs, Ecol. Monogr. 38:169-197

Ulanowicz, R. E. 1983. Identifying the structure of cycling in ecosystems, Math.Biosci. 65: 219-237.

Ulanowicz, R. E. 1984. Community measures of marine food networks and their possible applications, pp 23-47, in: Fasham, M. J. R. (ed.): *Flows of energy and materials in marine ecosystems,* Plenum Publishing Corp, New York.

Ulanowicz, R. E. 1986b. A phenomenological perspective of ecological development, pp 73-81, in: T. M. Poston & R.Purdy (eds.) *Aquatic Toxicology and Environmental Fate, Vol 9*. ASTM 921, Am. Soc. for Testing and Materials, Philadelphia, PA.

Ulanowicz, R.E. 1986a. *Growth and Development: Ecoystems Phenomenology*, Springer-Verlag, New York.

Ulanowicz, R. E. 1989. Ecosystem trophic foundations: Lindeman exonerata, To appear in: B. C. Patten & S.E. Jorgensen (eds.), *Progress in Systems Ecology*, Prentice-Hall, New York.

Ulanowicz, R. E. & W. M. Kemp 1979. Towards canonical trophic aggregations. Amer. Nat. 114: 871-883.

Ulanowicz, R. E., & J. J. Kay 1986. NETWRK 3 : A package of computer algorithms to analyze ecological flow networks, UMCEES Ref No. 82-7C CBL, University of Maryland, Solomons, MD.

Warwick, R. M. & P. J. Radford 1989. Analysis of the flow networks in an estuarine benthic community, in: Wulff, F., J. G. Field & K. H. Mann (eds), *Network Analysis in Marine Ecology: Methods and Applications,* Lecture Notes on Coastal and Estuarine Studies, Springer-Verlag, New York.

Wulff, F., J. G. Field & K. H. Mann 1989 (eds). *Network Analysis in Marine Ecology: Methods and applications,* Lecture Notes in Coastal and Estuarine Studies, Springer-Verlag, New York.

Yaglom, A. M., & I. M. Yaglom 1960. *Probability and Information* (Russian). Fitzmatgiz, Moscow.

APPENDIX 2A

Review of Information Theory

Consider an experiment A, [1] which consists of observing which of N possible outcomes occurs. Suppose that the probability of the ith outcome being observed is $P_i[A]$. Then the average uncertainty an observer has about which event (i.e., which outcome of the experiment) will occur is given by:

$$H[A] = \sum_{i=1}^{n} -P_i[A] \log P_i[A] \,;\quad 0 \leq H[A] \leq \log N \qquad (1)$$

H[A] also measures the information gained, on average, from observing which event occurs. It increases as the number of possible events (N) increases. It also increases as the likelihoods of the occurrences of the events become more similar, that is, as the differences between the $P_i[A]$ decrease. Maximum uncertainty occurs when each event appears equally likely ($P_i[A] = 1/N$). In this case H[A] = log N. The uncertainty (H[A]) decreases as one or a few events become more likely (larger $P_i[A]$) relative to the other events. The function H[A] is known as the **Shannon Entropy**. It has certain well known properties such as Symmetry, Normality, Expansibility, Decisivity, Stròng Additivity, Recursivity, Maximality, and Sub-Additivity. (See Aczel & Daroczy, 1975.)

Now consider a second experiment B which consists of observing which of M possible outcomes occurs. The probability of the *j*th outcome is $P_j[B]$. The possibility of experiment A influencing experiment B, or B influencing A, can be explored using the conditional probabilities $P_i[A/B_j]$, $P_j[B/A_i]$ [2]. These indicate the probability of an outcome of one experiment occurring given that the outcome of the other experiment is known. Let us assume that the probability distributions are complete. Then the observer's average uncertainty about experiment B, given that he already knows the outcome of A is:

$$H[B/A] = \sum_{i=1}^{n} -P_i[A] \sum_{j=1}^{m} P_j[B/A_i] \log P_j[B/A_i] \,;\; 0 \leq H[B/A] \leq H[B] \qquad (2)$$

This is the **Conditional Entropy**. It also measures the average information the observer gains from observing event B given he already knows which event A occurred. The inequality states that the observer's maximum conditional uncertainty about B is the same as the uncertainty (Shannon entropy) he would have if he had no a priori information. The minimum conditional uncertainty (H[B/A] = 0) occurs when the outcome of A completely predicts the outcome of B.

The composite event A+B has a complete probability distribution associated with it, such that the probability of ai and bj being the outcome of the two experiments is $P_{ij}[A+B] = P_j[B/A_i]P_i[A] = P_i[A/B_j]P_j[B]$. The observer's average **Uncertainty About The Composite Event** is given by (1).

$$H[A,B] = \sum_{i=1}^{n} \sum_{j=1}^{m} -P_i[A]\, P_j[B/A_i] \log(P_i[A]\, P_j[B/A_i]).$$

Using (1) and (2) it can be shown that:

$$H[A,B] = H[A] + H[B/A]\,;\; 0 \leq H[A,B] \leq H[A] + H[B] \qquad (3)$$

The inequality states that the maximum average uncertainty about the composite event occurs when the outcomes of experiment A and B are independent. Note that Bayes Theorem implies that H[A,B] is symmetric, that is H[A,B] = H[B,A].

The **Mutual Information** is defined as:

$$I[A,B] = H[A] + H[B] - H[A,B]$$

This is symmetric. Using equation (3) and inequality (2) leads to:

$$I[A,B] = H[B] - H[B/A];\; 0 \leq I[A,B] \leq H[B] \qquad (4)$$

I[A,B] measures the reduction in the observer's average uncertainty about event B due to his knowing which event A occurred (or vice versa). In other words it measures the average information gained about the outcome of experiment B from performing experiment A. It is a maximum when H[B/A] = 0, that is the outcome of A completely predicts the outcome of B. It is a minimum when H[B/A] = H[B], the outcome of A tells the observer nothing about the outcome of B. It is not an entropy measure in the strict mathematical sense. It increases as the average conditional uncertainty decreases.

The final measure to be discussed is called the redundancy. The redundancy (R) is defined as:

$$R = \frac{E((\max I[A,B] - I[A,B])}{(\max I[A,B]))} \qquad R = 0 \leq R \leq 1 \qquad (5)$$

(maxI[A,B] = the maximum value I[A,B] can take on for the given experiments A and B.)

The redundancy is a measure of how useful it is to perform experiment A if the experimenter's objective is to decrease his uncertainty about experiment B. Performing experiment A is said to be redundant if it does not provide any additional information about experiment B. In this case the redundancy is a maximum and I[A,B] = 0. In the other extreme the redundancy is a minimum when performing experiment A gives maximum information, that is tells the experimenter the outcome, a priori, of experiment B. Performing experiment A is not redundant relative to determining the outcome of B and I[A,B] = maxI[A,B] = H[B].

Some confusion has occurred over this measure. This is due to incorrectly interpreting R as a measure of the redundancy of performing experiment B after having performed experiment A. The reason for this interpretation, even though it is incorrect mathematically, is that it seems reasonable given the English language use of the word redundancy. However, the name "redundancy" for R makes English language sense in the context of network structures. For further reading Yaglom & Yaglom (1960) is recommended.

Endnote

[1] Experiment and event are used interchangeably.

[2] $P_i[A/B_j]$ = probability that outcome A_i occurs given that B_j has occurred.

Chapter 3

CONSTRUCTION OF FLOW NETWORKS

USING INVERSE METHODS

Alain F. Vézina
Département d'Océanographie
Université du Québec à Rimouski
310, Allée des Ursulines, Rimouski
Québec, Canada G5L 3A1

INTRODUCTION

The planktonic food web is intimately involved in the oceanic biogeochemical cycles that impinge on such important issues as the regulation of the world climate in the face of rising atmospheric CO_2 levels (US Global Ocean Flux Study 1986) and the behaviour of anthropogenic contaminants in the ocean. Changes in the structure and dynamics of planktonic food webs may to a large extent determine the residence time of elements or contaminants in the upper ocean (Frost 1984, Goldman & Caron 1985, Vézina & Platt 1987). It is thus imperative that we develop quantitative theories of the spatio-temporal variability in food web structure and dynamics in the sea. A first step in that direction would involve comparative analyses of the flows of energy and materials in the food webs of different marine environments (Leggett *et al.* 1985).

The basic object of comparison in that regard is the flow network, which summarises our knowledge of the flows of materials among compartments of the system. Flow networks are inferred from heterogeneous, and most often incomplete, sets of observations. At present, there is no standard methodology for deriving flow networks from the available information. This process is left to the individual practitioner, with the result that variations among networks can arise as much from differences in perspective among investigators as from "real" differences among systems. Even if the task of constructing the networks for different systems is left to a single investigator, he is forced, due to the deficiencies in the data, to make a number of subjective decisions which may account as much as the data for the differences among networks. Standard methodologies to produce flow networks must be applied if networks are to be compared or used as testing grounds for general hypotheses.

I will show here how inverse methods, borrowed from the physical sciences (Parker 1977, Wunsch 1978), can be used to generate flow networks from the available information. The inverse formalism ensures internal consistency, allows for uncertainties and redundancies in the data, and forces the flow networks to meet explicit operational requirements to make inter-site comparisons possible. The particular procedure I describe here invokes the principle of parsimony and derives the simplest possible flow network that is compatible with the data; that is, the network excludes any structure that is not required to satisfy basic mass balance and biological constraints. I will illustrate the techniques with the observations gathered by Holligan *et al.* (1984a,b) in the western approaches to the English Channel.

METHODOLOGY

The inverse problem in food web research

Generally, the flow networks for particular food webs cannot be reconstructed completely from the available data alone. This is particularly true for ocean systems where the breadth and intensity of the sampling coverage is severely constrained by logistic limitations. A similar situation is encountered, for example, in physical oceanography where the flow field is recovered from the spatial distribution of tracers (Wunsch 1978), and in geophysics where the structure of the earth is deduced from seismic wave propagation (Wiggins 1972). Investigators in these fields face the fundamental limitation that the number of independent observations they can make is usually far less than the number of parameters they need to know to describe ocean circulation or earth structure (Parker 1977, Wunsch 1978, Wunsch & Minster 1982). Inverse methods were developed to deal formally with this problem. It is reasonable then to assume that inverse methods can be applied profitably to the problem of constructing flow networks for marine food webs.

The inversion procedure starts with a model of the system, in this case a compartmental model of the food web. Normally, the model is used in numerical experiments to project the behavior of the state variables (i.e. standing stocks) from known parameter values and initial conditions. The inverse method is simply the reverse of that procedure: observations of the state variables are used to estimate the parameters. In this sense, inverse methods have strong affinities with regression analysis (Draper & Smith 1980) and the fitting of compartmental models to tracer kinetic data (Berman *et al.* 1962, Smith 1974). The crucial difference is in the low ratio of observations to unknowns in the inverse problem, leading to indeterminacy in the solution.

In what follows, I will describe how I applied inverse methods to reconstruct plausible flow networks for marine ecosystems. The subject of inverse methods is very rich, and many variations and extensions are possible. Readers interested in more details are referred to the extensive literature on the subject (Lanczos 1961, Jackson 1972, Wiggins 1972, Wunsch 1978, Wunsch & Minster 1982, Bolin *et al.* 1983, Enting 1985). Inversions can be performed on mainframe computers using general purpose programs such as LSEI (Least-Squares with Equality and Inequality constraints, available from Argonne National Laboratory) or by combining the appropriate subroutines from public domain libraries (IMSL, SLATEC). To my knowledge, general purpose programs for data inversion on microcomputers do not as yet exist, probably because small machines are not fast enough to comfortably handle the large number of computations involved.

Setting up the inverse problem

The principle underlying the application of inverse methods to flow network construction is that, given some data on food web structure and flows, simple mass balance rules and biological realities constrain the unmeasured flows in a particular food web. All that is needed then is to translate these constraints into mathematical form, substitute the empirical data for the variables in the resulting equations, and apply the inverse machinery to solve for the unknown flows.

The mass balance rules are expressed into difference equations that are functions of state variables (x_i) and rate parameters (r_j):

$$f(x_i, r_j) = \Delta x_i / \Delta t \qquad (1)$$
$$i = 1,...,m$$
$$j = 1,...,n$$

In the linear case, the system of difference equations that describes the food web fluxes can be expressed in compact matrix notation:

$$\mathbf{Ar} = \mathbf{b} \qquad (2)$$

where **A** is an m x n matrix that contains the state variables, **r** is an n x 1 vector of the unknown r_j's, and **b** is an m x 1 vector of the $\Delta x_i/\Delta t$ terms. Prior estimates of primary production and other flows can be used in the inversion in one of two ways. Measurements of individual fluxes can be lumped with the $\Delta x_i/\Delta t$ terms on the right-hand side of the affected mass balance equations and the corresponding r_j terms dropped from the equations. Alternatively, additional equations that specify the value of the flow (Q),

$$f(x_i, r_j) = Q \qquad (3)$$

can be appended to the mass balance equations. This is the only method available when measurements refer to a compounding of several flows. This method is also best for dealing with uncertainties in the observational data. Let m refer to the number of equations in the expanded set. Then **b** represents the flows and the standing stock changes that are fixed by observation or assumption, and will be termed here the dynamic vector. **A** represents the average state of the food web over the temporal and spatial scale of observation and is designated here as the state matrix. Thus, Eq. 2 is a statement of the relationship between the state of the food web and its known or assumed dynamics and will be called the continuity equation.

In addition to conservation of mass, other constraints that express a priori knowledge of the food web processes can be introduced to further constrain the unknown flows. This is particularly important here since it is most likely that the number of flows to be estimated greatly exceeds the number of independent mass balance conditions that can be posed; that is, $m << n$ in Eq. 2 above. Under those conditions, the solution to Eq. 2 will certainly be biologically meaningless. These additional constraints may specify certain bioenergetic realities, such that living organisms are not perfectly efficient in turning energy and materials into biomass, or pose bounds on the rate of biological processes, for example, lower limits on respiration or upper limits on the ingestion rate of consumers. Determining the bounds in this way is a subjective process. However, investigators are more likely to agree on such general criteria than on specific values for the rate and efficiency of biological processes.

Most useful constraints can be described as linear inequalities of the form:

$$\Sigma g_{ij} \cdot r_j \geq h_i \qquad (4)$$
$$i = 1,\ldots,n_c$$
$$j = 1,\ldots,n$$

where g_{ij} are the coefficients of the inequality and h_i is the lower bound condition that the inequality must meet. The inversion procedure can only solve for lower bound inequalities; an upper bound condition can be changed to a lower bound by multiplying the inequality through by -1. Expressing the set of n_c inequalities in matrix form, we have:

$$\mathbf{Gr} \geq \mathbf{h} \qquad (5)$$

where **G** is the n_c x n matrix of coefficients and **h** is the n_c x 1 vector of inequality conditions. Eq. 5 is termed here the constraint equation and x is the n x 1 vector of unknown rate parameters, r_j's.

The procedure that solves for the vector **r** which satisfies both sets of equations follows.

The inverse solution

The solution process relies on the principle of parsimony: it generates the simplest flow network that satisfies both the mass conservation and biological constraints. It is the network the flows are most evenly spread out among the possible pathways in the food web. This does not imply that food webs are organized on the principle of maximum simplicity. The aim is to produce a baseline flow network that can be compared within and across food webs. At this time, the operational requirement that the flow structure be as simple as possible seems no worse than any other, and certainly better than none.

The constrained inversion procedure works in two steps. First, the simplest solution to the continuity equation is found. The inequality constraints are then applied to this initial solution to yield a new solution that satisfies both the mass conservation and the additional biological constraints. The second solution is more complex than the first, but remains the simplest among the biologically realistic flow networks.

If **A** is square ($m = n$), the solution to the continuity equation is given by:

$$\mathbf{A}^{-1}\mathbf{b} = \mathbf{r} \tag{6}$$

In the inverse problem, however, $m < n$ (fewer equations than unknowns), and $\mathbf{A}^{-1}$ does not exist. At most m unknowns can be solved for explicitly. In practice, redundancies and errors in the data reduce the effective number of equations available to solve for **r**. Therefore, the index k ($k \leq m$) is used to designate the number of solvable unknowns.

Nevertheless, a particular solution to the inverse problem can be found if the continuity equation is reparametrized to express the n r_j's in terms of k unknowns. To do this, the n-dimensional space spanned by the original parameters (the columns of **A** or the parameter space) must be reduced to a k-dimensional space. This is exactly analogous to data reduction techniques, such as principal component analysis, where the information contained in multiple observations of a large number of variates (dimensions) is re-expressed into a much smaller number of transformed variables that are linear combinations of the original data. At the same time, since usually $k < m$, the m dimensions spanned by the mass balance equations (the rows of **A** or the observation space) must be reduced to k dimensions. By factoring **A** into k-dimensional subspaces of its observation and parameter spaces, Eq. 2 can be solved explicitly for k unknowns and, knowing the relationship between those unknowns and the original parameters, a particular solution to the continuity equation can be reconstructed. This is achieved through the Singular Value Decomposition (SVD) of **A**:

$$\mathbf{A} = \mathbf{U}\mathbf{L}\mathbf{V}^{T} \tag{7}$$

where **U** is an m x k matrix containing k independent linear combinations of the rows of **A** that span the observation space, **V** is an n x k matrix of linear combinations of the columns of **A** that span the parameter space, and **L** is a diagonal k x k matrix of scaling factors (l_i, $i = 1,\ldots,k$). Technically, the SVD is the solution to a coupled eigenvalue problem (Lanczos 1961); thus the columns of **U** and **V** are commonly referred to as the observation and parameter eigenvectors of **A** and the l_i terms as the singular values. The singular values are ranked in decreasing order along the diagonal of **L** and indicate the proportion of the total information contained in **A** that is accounted for by the *i*th column of **U** and **V**, in direct analogy to the eigenvalues in principal component analysis that denote the proportion of the total variance in the original data that is accounted for by the associated principal component. With the SVD, the continuity equation can be rewritten as:

$$\mathbf{U}\mathbf{L}\mathbf{V}^{\mathrm{T}}\mathbf{r} = \mathbf{b} \tag{8}$$

which can be solved immediately for **r**:

$$\mathbf{r} = \mathbf{V}\mathbf{L}^{-1}\mathbf{U}^{\mathrm{T}}\mathbf{b} \tag{9}$$

The formal role of the SVD as a reparametrization of the original model is now exposed. Let,

$$\mathbf{c} = \mathbf{L}^{-1}\mathbf{U}^{\mathrm{T}}\mathbf{b} \tag{10}$$

where **c** is a *k* x 1 vector of unknowns that can be solved explicitly as the weighted projection of the dynamic vector on the *k*-dimensional version of the observation space. Simplifying (9) using (10):

$$\mathbf{r} = \mathbf{V}\mathbf{c} \tag{11}$$

This equation is a statement of the relationship between the *k* solvable unknowns and the *n* parameters that we wish to estimate.

That the particular solution **r** is not unique can now be seen in a formal way. **V** could be made square by adding *n-k* column vectors that are mutually independent and independent of the original *k* columns. This new **V** would be an equivalent representation of the parameter space of **A**. Thus, appending arbitrary values to **c** to increase its length to *n* would result in a new solution that also satisfies the continuity equation. The *n-k* column vectors needed to square **V** fill the null space of **A**, the part of the *n*-dimensional parameter space that has no correspondence with the *k*-dimensional observation space. If these additional vectors are collected into a *n* x *n-k* matrix **V'**, a new solution ($\mathbf{r}_n$) to the continuity equation can be computed as:

$$\mathbf{r}_n = \mathbf{V}\mathbf{c} + \mathbf{V}'\mathbf{d} \tag{12a}$$

or,

$$\mathbf{r}_n = \mathbf{r} + \mathbf{V}'\mathbf{d} \tag{12b}$$

where **d** is a *n-k* x 1 vector of arbitrary constants. An infinity of solutions could be generated just by changing the components of **d**.

It is also clear that **r** will be the simplest among all non-trivial solutions to the continuity equation. The simplicity of a solution, as described earlier (absolute values of the solution components and the differences among components as small as possible), is geometrically the distance between the point it defines in the n-dimensional parameter space and the origin (0) representing the trivial solution. Algebraically, this distance is measured as the norm of **r** ($\| \mathbf{r} \|$), where:

$$\| \mathbf{r} \| = \Sigma r_j^2 \qquad (13)$$

It is evident that the addition of null space terms (**V'd**) can only increase $\| \mathbf{r} \|$, hence the complexity of the solution.

We now have a basis for the derivation of constrained solutions. Recall that the additional constraints are specified through the constraint equation,

$$\mathbf{G r_n} \geq \mathbf{h} \qquad (5)$$

Since **r** is fixed as the solution to the continuity equation, the constraint equation should be re-expressed in terms of the free parameters **d**. Substituting for $\mathbf{r_n}$ in Eq. 5 we get,

$$\mathbf{G(r + V'd)} \geq \mathbf{h} \qquad (14a)$$

$$\mathbf{GV'd} \geq \mathbf{h} - \mathbf{Gr} \qquad (14b)$$

Before proceeding, **V'**, which contains the n-k independent linear combinations that span the null space of **A**, must be constructed. Given that there is no information on the null space, it can be filled with columns of random numbers that are mutually independent and independent of the original k columns. I present without proof (see next section) the method of Fu (1981) to achieve this. **V'** is constructed by orthogonalizing the following matrix **M**:

$$\mathbf{M} = \mathbf{(I\text{-}VV^T)P} \qquad (15)$$

where **P** is a n x $(n\text{-}k)$ matrix of random numbers. The modified Gram-Schmidt method (Nash 1979) can be used for the orthogonalization. Let **G' = GV'** and **h' = h - Gr**. The result is a new set of linear inequality constraints on **d**:

$$\mathbf{G'd} \geq \mathbf{h'} \qquad (14)$$

Since, unnecessary complexity must not be introduced into $\mathbf{r_n}$, the solution for **d** must be the simplest possible that satisfies the inequalities, or:

Minimize $\| \mathbf{d} \|$

Subject to $\mathbf{G'd} \geq \mathbf{h'}$

This is a quadratic programming problem that can be solved using a least distance algorithm (Lawson & Hanson 1974). The vector **d** can now be substituted into Eq. 12 to recover the constrained solution $\mathbf{r_n}$.

Before I go any further, I should establish that the solution $\mathbf{r_n}$ is only a possible solution, not necessarily the correct solution. The non-uniqueness of the solution is a necessary consequence of the lack of information on food web dynamics in the ocean. Flow networks derived from ad hoc considerations are not necessarily correct either. With inverse methods, internal consistency is assured. Also, any number of solutions that result from different a priori assumptions can be computed. The most important advantage, however, is that the "inverse" flow networks for different scenarios or food webs are directly comparable: they are the simplest networks compatible with the site-specific data, conservation of mass laws, and a priori biological constraints.

Errors and resolution

I mentioned earlier that the inverse method is analogous to more traditional curve-fitting techniques in using observational data to solve for the parameters of a model. The analogy goes further in that the inverse solution, as the solution of a regression problem for example, cannot be perfect due to both errors in the data and inadequacies in the model representation of the food web. In other words, the solution cannot be expected to reproduce the mass balance ($\Delta x_i/\Delta_t$) and measured flow *(Q)* terms exactly. Therefore, the problem of estimating food web fluxes from observational data is really the least-squares problem of finding the solution $\mathbf{r}_n$ that

minimizes the residual error,	$\| \mathbf{A}\mathbf{r}_n - \mathbf{b} \|$
and the estimated rates,	$\| \mathbf{r}_n \|$
subject to the estimated flows equaling measured flows,	$\mathbf{A}\mathbf{r}_n \cong \mathbf{b}$
and the constraints on minimum and maximum flows,	$\mathbf{G}\mathbf{r}_n \geq \mathbf{h}$

In contrast to a normal regression problem, however, the solution to the inverse problem as expressed in Eqs. 2 and 5 is not necessarily inconsistent. With more unknowns than data, it is certainly possible to find a solution that fits each mass balance condition exactly, just as a curved line defined by 3 parameters can be made to pass exactly through 2 points. To fit the observational errors, the exact inverse solution becomes very complex (high $\| \mathbf{r}_n \|$) and includes many unrealistic features. In the context of the inverse problem, therefore, the least-squares solution is really a trade-off between the error left by the solution ($\| \mathbf{A}\mathbf{r}_n - \mathbf{b} \|$) and its complexity ($\| \mathbf{r}_n \|$).

There are a number of ways one could proceed to find the most appropriate least-squares solution (Wunsch 1978). The cut-off method (Wiggins 1972) was used here. This method exploits the fact that the magnitude of the residual error ($\| \mathbf{A}\mathbf{r}_n - \mathbf{b} \|$) and the complexity of the solution ($\| \mathbf{r}_n \|$) depend on k, the number of dimensions used to represent the observation and parameter spaces of **A** (Fig. 3.1). A disproportionate increase in the complexity of the solution for k close to m is diagnostic of the effect of errors on the inverse solution. On the other hand, rapid increases in the residual norm as k is lowered indicate that information required to approximate a closed mass flow budget is ignored. k is selected to discount the effect of uncertainties and redundancies in the observational data while preserving the mass balance as much as possible.

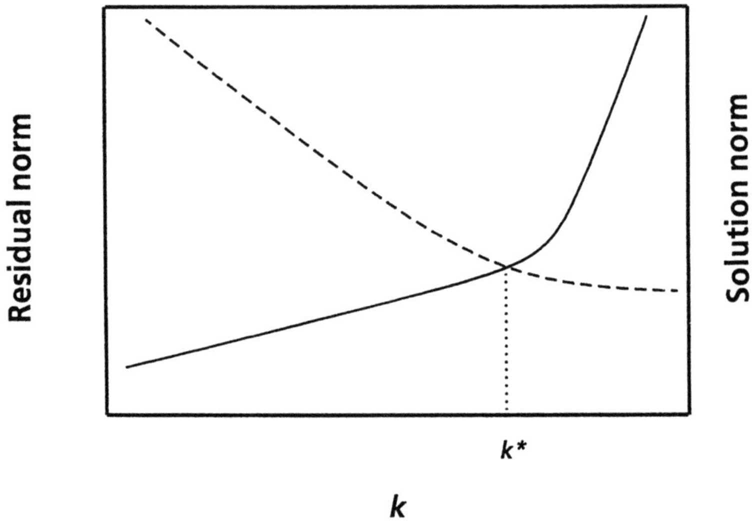

Figure 3.1. Schematic representation of the relationship between the complexity (solution norm, solid line) and the precision (residual norm, dashed line) of the solutions to a least-squares inverse problem and k, the number of linear combinations of the data used in computing the solutions. k^* corresponds to the solution that best compromises between complexity and precision (see text for more details).

The residuals left by the least-squares solution ($\mathbf{e} = \mathbf{Ar_n\text{-}b}$) can be useful in assessing a particular solution or comparing alternative solutions for the same data set. I assume that the errors should be randomly and independently distributed with uniform variance (σ^2), i.e. $\mathbf{e} = \mathbf{I}\sigma^2$ (unweighted least-squares). Obviously, the error may not be uniform over all compartments and it is possible to use prior estimates of the variance in the data to control the error structure of the solution (Wiggins 1972, Wunsch 1978, Wunsch & Minster 1982). Unfortunately, the internal variance of comprehensive food web data is poorly documented, and I will be confined here to assuming uniform variance.

Another feature of inverse solutions is that not all the information used is equally useful. With respect to the empirical observations, it is possible to compute indices of the weight carried by each mass balance condition in the final solution. This information is contained in the resolution matrix of the observations or $\mathbf{UU^T}$ (Jackson 1972, Wiggins 1972, Wunsch 1978). Basically, $\mathbf{UU^T}$ expresses the relationship between the solution obtained from using all m terms in the SVD of $\mathbf{A}$ ($\mathbf{r_m}$) and the actual least-squares solution:

$$\mathbf{r} = \mathbf{U}\mathbf{U}^T\mathbf{r_m} \tag{16}$$

When $k=m$, $\mathbf{UU}^T$ is the identity matrix, i.e. the equations contribute fully and equally to the solution. For inconsistent equations, the diagonal elements vary from 0 to 1 and indicate the degree to which the associated equation adds independent information to the solution. Equations with a zero resolution are redundant and could be dropped without changing the solution. Similarly, the solutions will be less sensitive to observations with low or mid-range resolution than to those with resolutions near 1. Also, the observations with low resolution will leave disproportionately large residuals that will tend to skew the distribution of errors away from normality. These resolutions are a quick guide to the varying information content of the mass balance constraints.

There will also be considerable variation in the degree to which the flows are constrained by the mass balance conditions. In evaluating the solution, it would be important to distinguish between the flows that are well resolved by the mass balance constraints and those that are very dependent on the additional biological constraints. This information is contained in the matrix $\mathbf{VV}^T$, which is the function that maps the mass balance conditions into the parameter space to obtain the particular solution. This matrix is identity when the number of independent equations equals the number of unknowns, i.e. when the particular solution is also the unique solution (the rationale behind the formula of Fu (1981) to compute the null space vectors (Eq. 15) is now clear: $\mathbf{I}\text{-}\mathbf{VV}^T$ represents the part of the parameter space that is not fixed by the mass balance and is used to scale columns of random numbers that are then orthogonalized to fill the null space). Thus, the diagonal elements are the parameter resolutions that vary between 0 and 1 . Parameters with high resolutions (approx. 1) will change little after application of the biological constraints. The lower the resolution the stronger the effect of the additional constraints on the parameter. Poorly resolved fluxes for which there are no, or only vague, prior expectations are clearly the weakest links in the solution for the flow network.

On the surface, resolutions appear useful in the design of observational programs since they rank the usefulness of observations and *a priori* information (Wunsch 1978). I caution against this interpretation, however, because resolutions are roughly proportional to the numerical magnitude of the observations that characterize the part of the model they represent (Wunsch & Minster 1982). In the case of parameter resolutions, for example, this is equivalent to saying that fluxes that involve compartments with large standing stocks are less dependent on *a priori* information than fluxes that involve compartments with small standing stocks. Different resolutions would arise from imposing different weights on the data. From the standpoint of inverse methods, all information is equally useful. The resolutions are introduced only as diagnostics of numerical problems that must be considered in interpreting the solution.

AN EXAMPLE

The data set

To illustrate the method, I will use the data of Holligan *et al.* (1984a,b) that describes the food web of sta. E5 (49°05'N, 6°37'W), located in stratified, relatively chlorophyll-poor waters west of the Ushant tidal front in the English Channel. Table 3.1 gives a summary of the data for the surface waters (0-24 m) of sta. E5. The rates of biomass change in each food web compartment were not observed and there is some evidence that the food web was not in steady state at the time of sampling. The inverse solution is offered here as an example only and should not be taken too literally. Details on how the original data was adapted for the inversion are available elsewhere (Vézina and Platt, 1988).

Table 3.1. Food web data for sta. E5, English Channel. Units of estimates are mg m^{-2} for standing stocks and mg $m^{-2}d^{-1}$ for fluxes. Adapted from Holligan *et al.* (1984a,b).

Variable	Estimate
Physical environment	
Temperature (°C)	15
NO_3-N concentration	40
NO_3-N gradient (mg m^{-4})	0.08
Standing stocks	
Particulate organic C	7356
Particulate organic N	850
Autotroph C	420
Bacterial C	142
Protozoan C	118
Microzooplankton C	286
Microzooplankton N	65
Mesozooplankton C	682
Mesozooplankton N	144
DOC	34656
Fluxes	
Gross primary production	720
Net primary production	480
Sedimentation[1]	216

[1] Assumed at 45% of NPP (Betzer *et al.* 1984).

The mass balance conditions

The data was incorporated into linear mass balance equations (Eqs. 1 and 2). Separate equations were written for C and N balance in each compartment. C:N ratios were used to estimate unmeasured N standing stocks and flows from the C data (Vézina and Platt, 1988). The food web system was assumed to be in steady state; that is, all $\Delta x_i/\Delta t$ terms are zero. The most complex pattern in the connecting flows that can be realistically assumed for an oceanic food web (Fig. 3.2) is used in writing the equations. The inverse solution will remove (zero) any flow that is not required to conserve mass and satisfy basic biological constraints. Adding the equations specifying the primary production and sedimentation flows gives 22 mass balance conditions. The food web model is discussed in more detail in Vézina and Platt 1988.

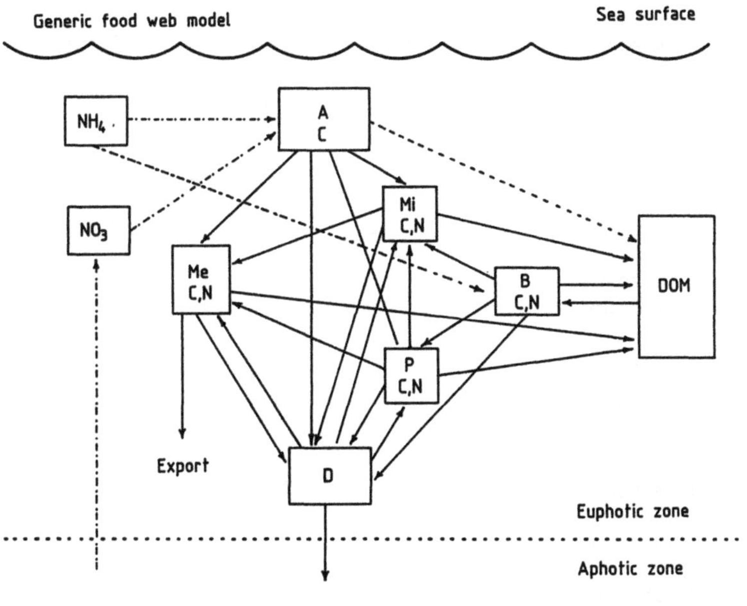

Figure 3.2. Generic model of planktonic food webs in stratified waters. The compartments are: autotrophs (A), bacteria (B), protozoans (P), microzooplankton (Mi), mesozooplankton (Me), detritus (D), dissolved organic matter (DOM), nitrate (NO_3) and ammonium (NH_4). The letters C and N inside the boxes for the living compartments designate the flows that regenerate inorganic forms of those elements, i.e. respiration and excretion. Solid arrows are the inter-compartmental flows of C and N; dashed arrows are flows of C only; dotted arrows designate the N flows.

The biological constraints

A basic constraint that must be applied to any inversion of food web data is that all the solution components are nonnegative (≥ 0). This is because even though the signs (directions) of the flows are determined by the structure of the mass balance equations, the inverse solution to these equations alone (Eq. 2) will generally include negative parameters. Other constraints are listed in Table 3.2. The rationale behind those constraints and the literature sources are given in Vézina and Platt 1988.

Table 3.2. Constraints on the inversion of food web data for sta. E5.

Constraint on	for	Units	Lower bound	Upper bound
Respiration	Autotrophs	% GPP[1]	5%	50%
Respiration	Protozoans	% body C d^{-1}	18%	None
Respiration	Microzoop.	% body C d^{-1}	20%	None
Respiration	Mesozoop.	% body C d^{-1}	6.5%	None
Excretion	Autotrophs	% GPP[1]	5%	50%
Excretion	Protozoans, Metazoans	mgC m^{-2} d^{-1}	1/3 of Resp.	Respiration
Dissolution	Detritus	% d^{-1}	None	1% d^{-1}
Assimilation efficiency	Protozoans, Metazoans		50%	90%
Net production efficiency	Bacteria, Protozoans, Microzoop.		10%	60%
Net production efficiency	Mesozoop.		None	40%
Ingestion	Protozoans	% body C d^{-1}	None	1544%
Ingestion	Microzoop.	% body C d^{-1}	None	206%
Ingestion	Mesozoop.	% body C d^{-1}	None	68%
C:N of respiration (wt/wt)	Bacteria Protozoans, Metazoans		3	13
C:N of egestion (wt/wt)	Protozoans, Metazoans		4	10

[1]GPP: Gross primary production

Finding the least-squares solution

Different levels of k, starting with the maximum possible of 22, were used to compute inverse solutions (Fig. 3.3). The solution complexity falls rapidly as principal components of the mass balance information are removed, i.e. as k is decreased. On the other hand, the residual norm does not begin to increase appreciably until k reaches 19. The contrast between a falling solution norm and a relatively stable residual norm indicates that solutions based on more than 19 principal components track the errors in the data. Therefore, the solution based on $k = 19$ was chosen as the appropriate least-squares solution for this data set.

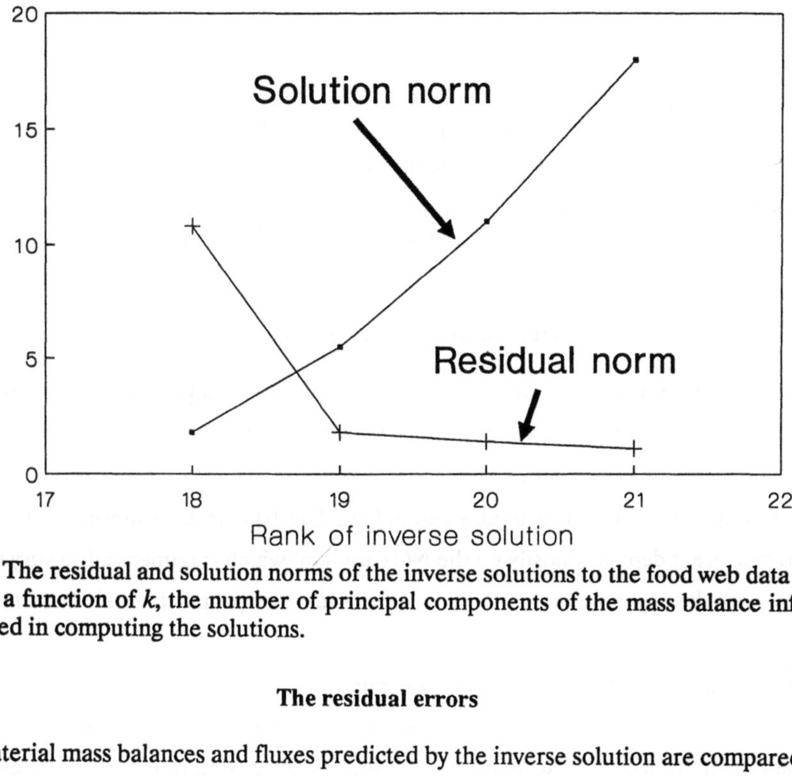

Figure 3.3. The residual and solution norms of the inverse solutions to the food web data of sta. E5 as a function of k, the number of principal components of the mass balance information used in computing the solutions.

The residual errors

When the material mass balances and fluxes predicted by the inverse solution are compared to those expected or observed, an approximately normal frequency distribution of residuals was obtained (Fig. 3.4). Except for NO_3, the predicted absolute rates of standing stock changes vary between .001 and 2% d^{-1}; the inverse solution was approximately consistent with the steady state condition assumed. A disproportionately large negative residual for NO_3 occured because the inverse estimate of the vertical diffusivity was zero (see below). A reasonable value for vertical diffusivity can compensate for the N shortfall in that compartment.

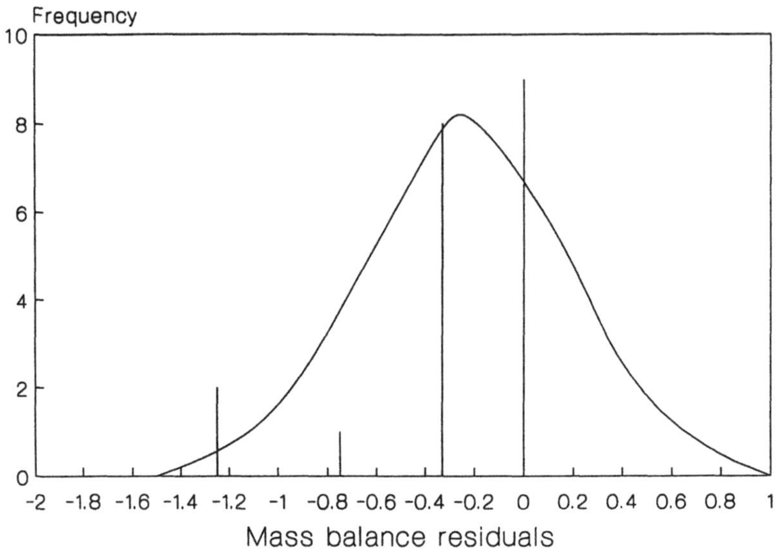

Figure 3.4. Frequency distribution of the mass balance residuals for the inverse solutions for the food web flows at sta. E5 (English Channel) in late august. The solid line is a normal distribution fitted to the empirical distribution.

Resolutions of the mass balance constraints

The resolutions of the mass balance constraints indicate that, as a group, the conditions on N flows have less impact on the solutions than those for C flows (Table 3.3). In particular, the equations for PON and NO_3 balance have virtually no effect on the solutions. N flows through the POM compartment in the same way as C; hence, it is largely redundant. The NO_3 mass balance, on the other hand, is anihilated because it depends mostly on the NO_3 gradient which is numerically a very small variable. This agrees with the observation that these equations leave disproportionately large residuals.

Table 3.3. Resolutions of the mass balance constraints

Mass balance condition on	Resolution	Mass balance condition on	Resolution
Particulate organic C	0.965	Protozoan N	0.884
Autotroph C	0.992	Microzooplankton N	0.886
Bacterial C	0.999	Mesozooplankton N	0.886
Protozoan C	0.999	Detrital N	0.873
Microzooplankton C	0.999	Dissolved organic N	0.879
Mesozooplankton C	0.999	Nitrate	0.030
Detrital C	0.999	Ammonium	0.864
Dissolved organic C	0.999	Gross primary production	1.000
Particulate organic N	0.016	Net primary production	0.992
Autotroph N	0.439	Autotroph N assimilation	0.438
Bacterial N	0.884	C sedimentation	0.981

Resolutions of the flows

The parameter resolutions are as expected: they are highest for the measured or specified fluxes (primary production, sedimentation) and for those that involve large standing stocks (mesozooplankton, detritus, DOM) (Table 3.4). The fluxes through the bacteria, protozoans and microzooplankton compartments are the most sensitive to the *a priori* information. Prior information is not available for a number of these poorly resolved fluxes, and consequently, their inverse estimates should be interpreted with particular caution (Table 3.4). Among them, note the excretion of DON by microheterotrophs, the production of detritus by the autotrophs and the bacteria, and the grazing activity of protozoans. The same applies for the eddy diffusivity and the bacterial NH_4 uptake. On the other hand, estimates of many of the grazing flux estimates seem robust even in the absence of *a priori* information since they are well resolved by the mass balance constraints alone.

Table 3.4. Resolutions of the food web flows.

Food web flow	Resolution	Food web flow	Resolution
Upward diffusion of nitrate	0.000	Detritus dissolution to DOM	0.983
Gross primary production	1.000	Autotroph. detritus product.	0.004
Autotroph. nitrate assimil.	1.000	Bacterial detritus product.	0.012
Autotroph. ammonium assimil.	0.957	Protozoan C egestion	0.005
Bacterial DOM uptake	1.000	Microzoop. C egestion	0.028
Bacterial ammonium assimil.	0.043	Mesozoop. C egestion	0.156
Autotrophic respiration	0.550	Protozoan N egestion	0.028
Bacterial respiration	0.020	Microzoop. N egestion	0.132
Protozoan respiration	0.013	Mesozoop. N egestion	0.156
Microzoop. respiration	0.076	Grazing flows:	
Mesozoop. respiration	0.431	Autotroph. to Protozoan	0.034
Bacterial inorg. N regener.	0.043	Autotroph. to Microzoop.	0.156
Protozoan inorg. N regener.	0.032	Autotroph. to Mesozoop.	0.831
Microzoop. inorg. N regener.	0.149	Bacteria to Protozoan	0.054
Mesozoop. inorg. N regener.	0.724	Bacteria to Microzoop.	0.178
Autotroph. C excretion	0.550	Bacteria to Mesozoop.	0.829
Bacterial C excretion	0.004	Detritus to Protozoan	0.997
Protozoan C excretion	0.005	Detritus to Microzoop.	0.997
Microzoop. C excretion	0.028	Detritus to Mesozoop.	0.999
Mesozoop. C excretion	0.157	Protozoan to Microzoop.	0.168
Bacterial N excretion	0.029	Protozoan to Mesozoop.	0.814
Protozoan N excretion	0.028	Microzoop. to Mesozoop.	0.909
Microzoop. N excretion	0.128	Mesozoop. product. (export)	0.464
Mesozoop. N excretion	0.617	Sedimentation	1.000

The food web flows

The C flows, N flows, and C:N ratios computed from the inverse solution are shown in Figs. 3.5, 3.6 and 3.7 respectively. Note that a number of the fluxes initially assumed to exist were dropped from the inverse solution: they were not required to conserve mass and satisfy basic biological restrictions. One striking feature of this solution is that most of the primary production is grazed by the prorozoans and microzooplankton which in turn constitute an important food source for the mesozooplankton. This agrees with the idea of a "microbial loop" acting as a conduit between small producers (bacteria and picoplankton) and metazoan consumers. However, alternative inverse solutions for this site do not at all support this concept (Vézina *et al.* in prep.). This solution must be viewed as one possible scenario for the maintenance of zooplankton stocks.

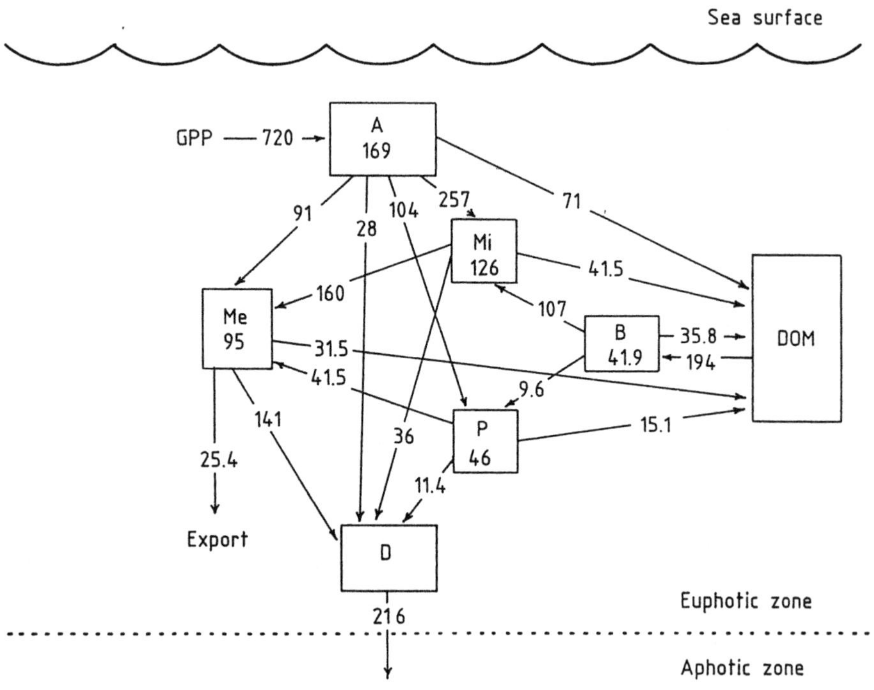

Figure 3.5. C flows in the food web of sta. E5 computed from the inverse solution. The flow units are mgC m^{-2} d^{-1}. The numbers inside the boxes are respiration flows.

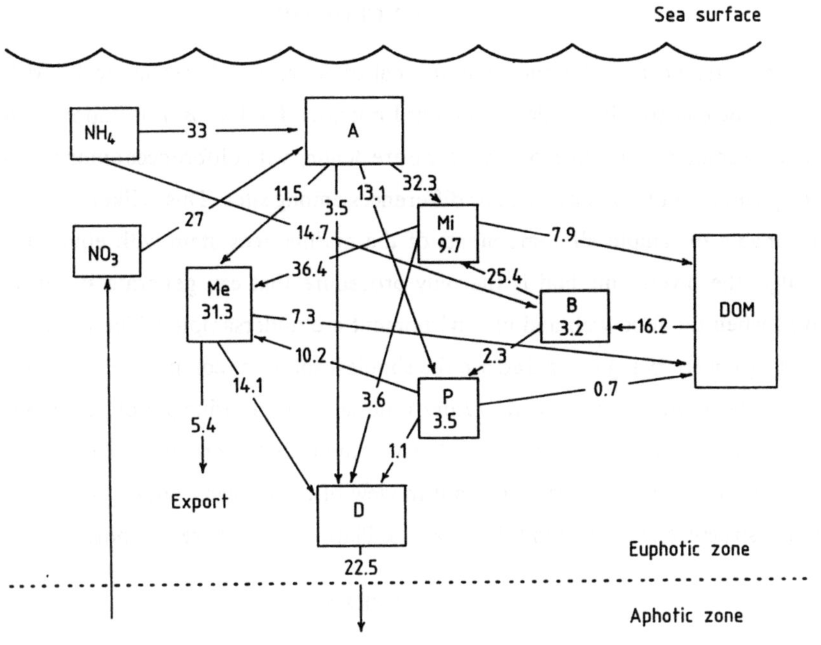

Figure 3.6. N flows (mgN m^{-2} d^{-1}) computed from the inverse solution.

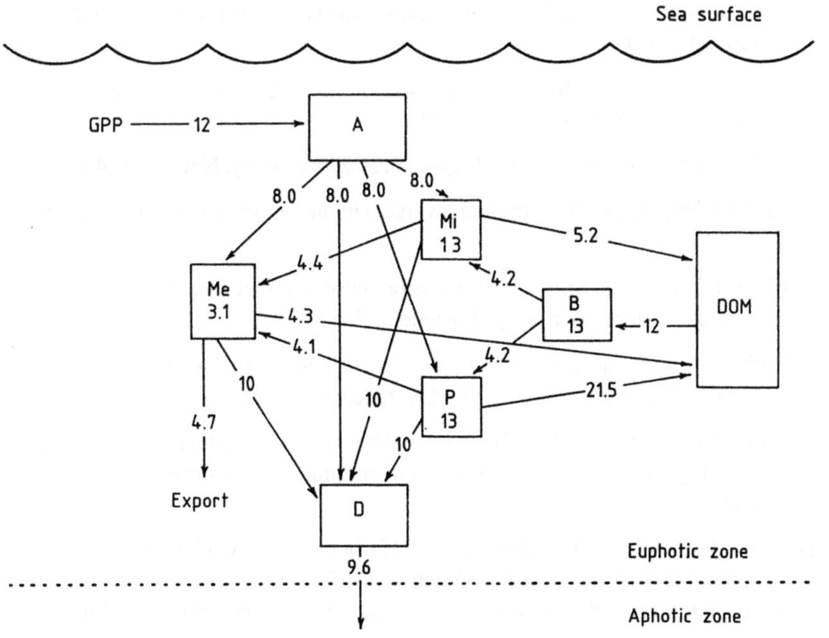

Figure 3.7. C:N ratios (wt/wt) of the food web flows computed from the inverse solution.

CONCLUSION

The inverse method provides the tools to deal effectively and systematically with the limitations inherent to the construction of flow networks for marine food webs. It is important to realize that, in comparative analyses of marine food webs, we are dealing with differences among alternative solutions for the dynamics at one site as much as differences among sites. This is likely to remain so as long as we are unable to sample all components of the marine ecosystem with equal proficiency. To my knowledge, the inverse method is the only procedure that can generate strictly comparable flow networks when the food web and its environment are undersampled. The inverse solutions for the different food webs being compared are the simplest solutions compatible with the site-specific data and generally applicable constraints. Other criteria than the principle of parsimony can be used to derive solutions (Vézina & Platt, in prep.). The point is that alternative solutions can be classified as well as ecosystems. This is very important in view of the growing interest into holistic measures of ecosystem structure and function (Ulanowicz & Platt 1985) that are computed from flow networks.

REFERENCES

Berman, M., E. Shahn & M.F. Weiss. 1962. The routine fitting of kinetic data to models: a mathematical formalism for digital computers. Biophys. J. 2: 275-287.

Betzer, P.R., W.S. Showers, E.A. Laws, C.D. Winn, G.R. DiTullio & M. Kroopnick. 1984. Primary production and particle fluxes on a transect of the equator at 153°W in the Pacific Ocean. Deep Sea Res. 31: 1-11.

Bolin, B., A. Björkström, K. Holmén & B. Moore. 1983. The simultaneous use of tracers for ocean circulation studies. Tellus 35B: 206-236.

Draper, N.R. & H. Smith. 1980. *Applied Regression Analysis.* Wiley, New York. 407 p.

Enting, I.G. 1985. Principles of constrained inversion in the calibration of carbon cycle models. Tellus 37B: 7-27.

Frost, B.W. 1984. Utilization of phytoplankton production in the surface layer. pp. 125-135 In: *Global Ocean Flux Study.* National Academy of Sciences, Wash., D.C.

Fu, L.-L. 1981. General circulation and meridional heat transport of the subtropical south Atlantic determined by inverse methods. J. Phys. Oceanogr. 11: 1171-1193.

Goldman, J.C. & D.A. Caron, 1985. Experimental studies on an omnivorous microflagellate: implications for grazing and nutrient regeneration in the marine microbial food chain. Deep-Sea Res. 32: 899-915.

Holligan, P.M., R.P. Harris, R.C Newell, D.S. Harbour, R.N. Head, E.A.S. Linley, M.I. Lucas, P.G.R. Tranter, & C.M. Weekley, 1984a. Vertical distribution and partitioning of organic carbon in mixed, frontal and stratified waters of the English Channel. Mar. Ecol. Prog. Ser. 14:111-127.

Holligan, P.M., P.J.LeB. Williams , D. Purdie, & R.P. Harris, 1984b. Photosynthesis, respiration and nitrogen supply of plankton populations in stratified, frontal and tidally mixed shelf waters. Mar. Ecol. Prog. Ser. 17: 201-213.

Jackson, D.D. 1972. Interpretation of inaccurate, insufficient and inconsistent data. Geophys. J. Roy. Astr. Soc. 28: 97-109.

Lanczos, C. 1961. *Linear Differential Operators*. Van Nostrand, Reinhold, N.Y. 564 p.

Lawson, C.L. & R.J. Hanson, 1974. *Solving Least Squares Problems*. Prentice-Hall, Englewood Cliffs, N.J. 1054 p.

Leggett, W.C., W.R.Boynton, V.F. Gallucci, L. Johnson, R. Margalef, J. McGlade, D. Mikulecky, R. Rosen, R.O. Ulloa & J. Watson. 1985. I. Hypothesis testing and sampling design in exploited ecosystems. pp. 237-240. In: R.E. Ulanowicz & T. Platt (Eds) *Ecosystem Theory For Biological* Oceanography. Can. Bull. Fish. Aquat. Sci. 213.

Nash, J.C. 1979. *Compact Numerical Methods of Computers: Linear Algebra and Function Minimization*. Halstead Press, Bristol 227 pp..

Parker, R.L. 1977. Understanding inverse theory. Ann. Rev. Earth Planet. Sci. 5: 35-64.

Smith, D.F. 1974. Quantitative analysis of the functional relationships existing between ecosystem components. Oecologia 16: 97-106.

Ulanowicz, R.E. & T. Platt. (Eds) 1985. Ecosystem Theory for Biological Oceanography. Can. Bull. Fish. Aquat. Sci. 213. 260 p.

U.S. GOFS. 1986. Global Ocean Flux Study: Report 3. U.S. GOFS Planning Office, Woods Hole, Mass. 141 p.

Vézina, A.F. & T. Platt. 1987. Small-scale variability of new production and particulate fluxes in the ocean. Can. J. Fish. Aquat. Sci. 44: 198-205.

Vézina, A.F. & T. Platt. 1988. Food web dynamics in the ocean. I. Best-estimates of flow networks using inverse methods. Mar. Ecol. -Prog. Ser. 42: 269-287.

Wiggins, R.A. 1972. The general linear inverse problem: implication of surface waves and free oscillations for earth structure. Rev. Geophys. Space Phys. 10: 251-285.

Wunsch, C. 1978. The north Atlantic general circulation west of 50W determined by inverse methods. Rev. Geophys. Space Phys. 16: 583-620.

Wunsch, C. & J.-F. Minster, 1982. Methods for box models and ocean circulation tracers: mathematical programming and nonlinear inverse theory. J. Geophys. Res. 87: 5647-5662.

Chapter 4

A GENERIC SIMULATION MODEL

FOR TREATING INCOMPLETE SETS OF DATA

Robert E. Ulanowicz
Chesapeake Biological Laboratory
University of Maryland
Solomons, Maryland 20688-0038
USA

INTRODUCTION

As the preceding chapters in this volume have amply demonstrated, much useful knowledge concerning ecosystem structure and function can be obtained by subjecting estimated networks of material and energy flows in marine systems to the analytical techniques described in Kay *et al.*(1989, this volume). Usually, the data required for the network analysis techniques are identical to those necessary to begin the simulation modelling of these same systems-namely, the magnitudes of the stocks and flows. (Ducklow *et al.* 1989, this volume)

To a limited extent the activities of network analysis and simulation modelling can be mutually beneficial exercises. For example, network analysis can provide in-depth analysis of the output from conventional simulation models. Conversely, those simulation techniques with a minimum of *a priori* assumptions can be used to render incomplete or unbalanced network data sets suitable for the analyses described in this book. In addition, there are those practitioners of network analysis who cannot resist the temptation to indulge in "what if" types of questions, e.g.,"What might happen to the fish stocks in my system if the oyster biomasses were doubled?" While the answers to such questions should be sought primarily among the outputs from the network analyses (i.e., among the data themselves), it is an easy enough task to provide the investigator with a very simple simulation package to indulge his/her curiosity. (A strong word of caution is given to those who would use the simulation in this way. The results should never be taken as realistic predictions of how the system would actually respond to the imagined perturbation.)

The primary emphasis here will be upon creating a generic simulation package that will provide the user with some objective means for balancing a data set or for inferring the values of missing data (or for doing both simultaneously) For these purposes it is necessary to choose a highly stable dynamic scheme. Linear donor-controlled kinetics are especially relevant in that they yield models that are inherently stable. Even if the starting network is not balanced, the ensuing simulation will converge to a well-defined, stable steady state, i.e., the program brings the flows smoothly into balance without the user having to make arbitrary decisions to force a balance (like the commonly used tactic of adjusting the respiration). Donor-control kinetics also provide a reasonably acceptable model of "passive" processes occurring in ecosystems, such as respiration, sedimentation, losses incurred by advection, etc. With some discretion one can use this algorithm to estimate missing flows (cf. Vézina 1989, this volume and the example below).

MODEL DESCRIPTION

In the companion package of programs the reader can be provided with a simulation package called AUTOMOD that will accept the network data in exactly the input format used for the other analysis techniques described in Kay *et al.* 1989, this volume. The algorithm will invoke either of two sets of generic constitutive assumptions, prompt the user to describe any perturbations (if necessary) and automatically establish the run parameters.

If one desires only generality and stability, then linear, donor-controlled kinetics will suffice. Knowing the biomass, M_{i0}, present in compartment *i* under nominal conditions (the original network before any perturbations are specified) and the flow from *i* to *j*, F_{ij0}, under the same conditions, then the flow at any later time is modelled by

$$F_{ij} = Kd * M_i, \qquad (1)$$

where:

F_{ij} = the flow form prey *i* to predator *j* at later time *t*,

M_i = the biomass in prey *i* at time *t*, and

Kd = F_{ij0}/M_{i0}, a constant.

Thus, the donor-controlled constants can be readily calculated from the data as presented in the SCOR formatted files.

For those wishing to use the algorithm to accomplish something approaching realistic simulation, the donor-control assumption is usually a poor representation of actual predator-prey dynamics. It sometimes stretches one's belief to think that, as H.T. Odum once opined, "the rabbit forces itself into the fox's jaws." The intensity of predation is often a crucial factor in determining how much passes from prey to predator. Thus, one might try to employ linear, predator-control dynamics in the form

$$F_{ij} = Kp * M_j, \tag{2}$$

where M_j = the biomass of predator j at time t, and

$Kp = F_{ij0}/M_{j0}$, a constant.

Unfortunately, linear predator-control kinetics are inherently unstable. Without any information in (2) about the size of the prey stock in i, a growing predator population is destined to annihilate its prey. One could include information on prey stocks by interjecting their biomass as a factor in (2), resulting in the familiar Lotka-Volterra bilinear relationship. While bilinear kinetics are sometimes stable (Ulanowicz, 1972), they more often are not.

One reasonably successful way of stabilizing predator- control dynamics was suggested by Wiegert (1973), who moderated system behavior by interjecting thresholds into the kinetics. Below a certain threshold of low prey density, predation was assumed to cease, thereby providing the prey a "refuge" level below which it would become immune to capture. Wiegert also assumed that, above a second higher level of prey density, the predator ration reached a saturation value (as do the Michaelis-Menten and Ivlev expressions). Between the two threshold values the predator ration was assumed to rise in a linear fashion. With his two-parameter, piecewise linear expression Wiegert was able to simulate the essential features of predator control in a well-behaved model. The same stabilizing effect achieved by Wiegert can also be accomplished by modifying (2) to include a single-parameter smooth function,

$$F_{ij} = Kp * M_j * e^{(1-X_i)}, \tag{3}$$

where $X_i = M_{i0}/M_i$. The reader will notice that (3) reduces to (2) at the nominal conditions (where all $X_i = 1$.) As prey items become scarce, the argument of the exponent in (3) becomes a large negative number. Thus, the predation by j diminishes in exponential fashion as prey items begin to vanish. For example, by the time the prey biomass has fallen to 0.1 of its nominal level, the predation rate has decreased to $1.23 * 10^{-4}$ of its nominal value. On the other hand, should prey become very abundant, X_i will become insignificant, and the predator's ration will saturate at about 2.72 times its nominal value.

Implementing the modified predator controlled kinetics in algorithmic fashion is straightforward. At the outset, however, one needs to specify whether each compartment is living (i.e., feeding) or dead (passive). If the recipient of a flow is living, the transfer will be modeled by (3), where *KP* and X_i will be determined using the nominal values F_{ij0}, M_{j0} and M_{i0}. If the recipient is not living, then the kinetics will be assumed to be (1), and *Kd* will be calculated using F_{ij0} and M_{i0}. In addition, respirations will be assumed to follow (1). The distinction between living and non-living compartments is achieved in the SCOR format by placing all the non-living elements last in the species list and specifying the total number of living components.

AUTOMOD will model a network using either the linear donor controlled or the modified predator controlled scheme and was written in FORTRAN77 for use on the IBM PC or compatible computers. AUTOMOD first reads the nominal conditions given in the standard format used in this book and immediately sets up one of the kinetic schemes described above. It then queries the user for details on how he/she wishes to perturb the nominal conditions. The response to the perturbations is integrated forward in time using a standard Runge-Kutta fourth order polynomial scheme (Press *et al.* 1986).

For those using AUTOMOD simply to bring their flow data into balance, instructions will appear on the screen as to how to specify the run parameters in order to accomplish this task.

The user can perturb the nominal conditions in several ways. Two of those ways involve changing a nominal biomass or exogenous input level. The response of the predator-controlled dynamics to these changes is usually stable whenever the nominal flows are in balance. One also has the option of changing coefficient values within the model. With the predator controlled dynamics these latter changes often result in the extinction of one or more compartments. (A special extinction subroutine is included to excise any compartment that is about to go extinct before it has the chance to shut down the integration scheme by violating DOS error traps.)

EXERCISING THE MODEL

Three examples serve to demonstrate the use and behavior of the model. The first is exceedingly simple and consists of the hypothetical example portrayed in Figure 4.1a. A producer of 100 units of biomass per unit time has a standing stock of 50 units. It sustains two heterotrophs, the top one feeding at a rate of one unit of producer per unit of predator per unit time and the bottom feeding at triple that rate. If one doubles the primary production and assumes linear, donor control, all the values shown in Figure 4.1a will eventually double in a monotonic, negative exponentially damped fashion.

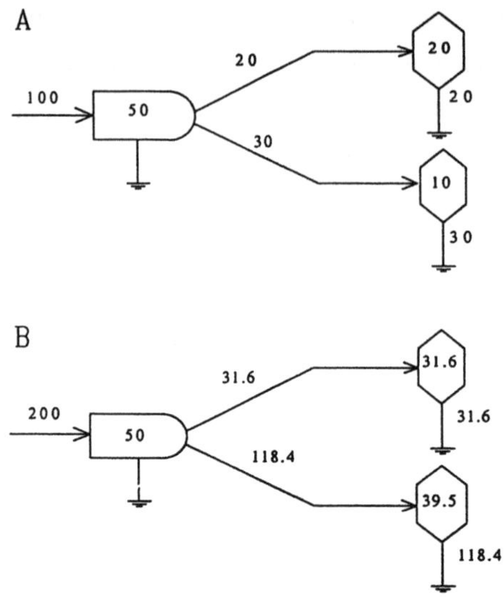

Figure 4.1. (a) Hypothetical flows among a primary producer and two heterotrophs. Units are arbitrary. (b) The result of doubling the primary production when heterotrophs feed according to equation (3).

The response of the modified donor control model (Figure 4.1b) is more interesting. The approach to steady-state involves exponentially damped oscillations. The producer biomass initially increases to about 79.9 units, falls to 48.1 and finally levels out at its nominal value. Fifty percent more production reaches the heterotrophs, but the greater share goes to the more voracious bottom heterotroph. The top heterotroph falls well short of doubling its biomass, as it did in the donor controlled scenario; however, the bottom predator more than compensates by almost quadrupling its biomass. If the two herbivorous transfers were the initial links of two separate loops back to the producer, it becomes obvious that the bottom pathway would be on its way towards displacing the upper one. One perceives a rudimentary mechanism behind the phenomenological observation that enriched systems tend to cycle more, but along more dissipative, trophically lower cyclic pathways (Ulanowicz, 1984).

The second example demonstrates how one might achieve a balanced network starting with data that does not quite add up. In Figure 4.2a are displayed the approximate data for the Cone Spring network shown earlier in Figure 2.1b in Kay *et al.* (1989, this volume). Several flows are depicted as missing, namely all the respirations and the energy transfer from detritus feeders to carnivores. The rates at which these processes occur are known, but not the absolute values of the flows. The approximate values for the respiration coefficients are 7.0, 28.0, 30.0, 12.0 and 0.9 y^{-1} for compartments 1 thru 5, respectively. The detritus feeders were observed to fall prey to the carnivores at the rate of 6.2 y^{-1}.

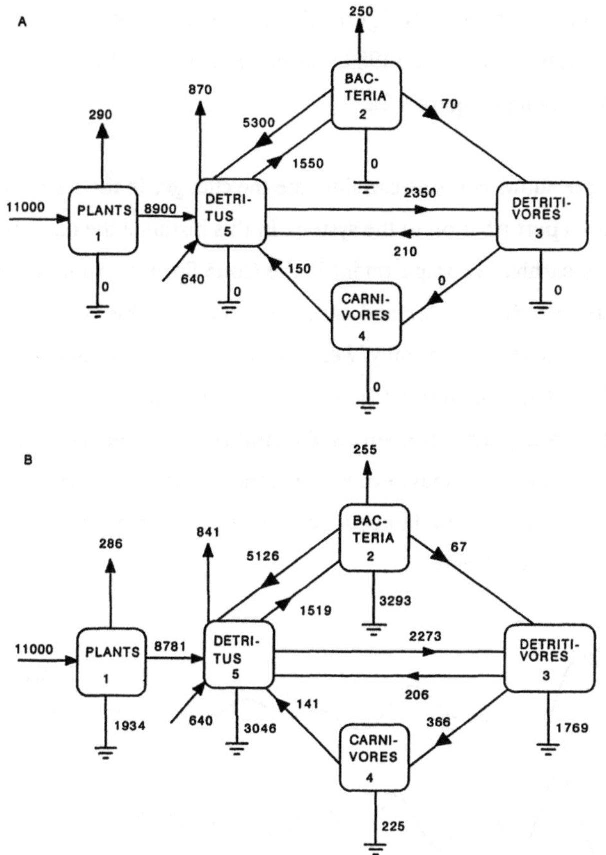

Figure 4.2. (a) Approximate energy flows (kcal $m^{-2}y^{-1}$) through the Cone Spring ecosystem (cf. Figure 2.1b in Kay *et al.* 1989, this volume). Data on respirations and carnivory are available only as rate coefficients. (b) A balanced network produced from 4.2a using the donor-controlled option of AUTOMOD

A balanced estimate of the Cone Spring network is constructed from the approximate data using AUTOMOD as follows: The energy stocks and available flow data are put into a standard input file to be read by AUTOMOD. The program is then run with this data, and the user chooses the linear, donor controlled form for execution. The run parameters are specified so as to allow ample time for the system to come to steady state (about 5 years in this case.) The rate data for the missing flows are entered as if they were perturbations of the rate constants. For example, the respiration coefficient for the plants (compartment 1) is "perturbed" from the default value of 0.0 to the measured value of 7.0 y^{-1}.

The resulting balanced network is shown in figure 4.2b. The correspondence of the balanced flows to the actual data in Figure 2.1b in Kay *et al.* (1989, this volume) is seen to be commensurate with the approximations made in arriving at figure 4.2a.

The third and final example shows how one can simulate the changes in the system's configuration that might initially occur after a perturbation to the system. In this instance the disturbance is a tripling of the energy content of the carnivore compartment in the Cone Spring system. Figure 4.3 displays the responses of the bacteria, detritus feeder and carnivore stocks to this perturbation. As one might expect, the carnivore population is initially well above its carrying capacity and begins to fall precipitously. At first the detritus feeders are being heavily grazed and they, too, rapidly decline. The bacteria, thanks to a relief from grazing pressure and an influx of detrital energy from 3 and 4 start out to rise in numbers. The carnivore biomass eventually undershoots its carrying capacity and starts to recover from its crash. Thereupon follow damped oscillations in all three compartments that eventually converge upon the nominal stock levels.

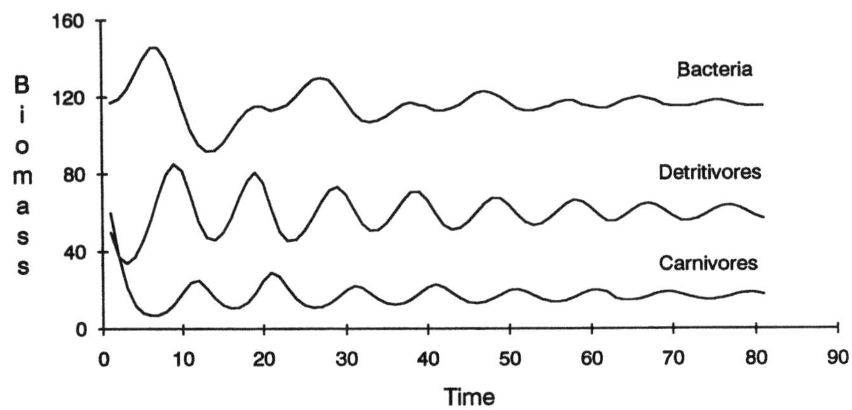

Figure 4.3. Simulated changes in the energy stocks (kcal m^{-2}) of the bacteria, detritus feeders and carnivores resulting from an arbitrary tripling in the stock of carnivores.

DISCUSSION

Despite the extreme simplicity of the linear-donor controlled assumption, models built upon this dynamic scheme are seen to serve as useful aids in helping investigators to estimate the balance of flows in their systems of interest. The available flow data almost never balance around each compartment. However, making arbitrary decisions to force a balance has the effect of changing at least some of the rates away from their values as observed in nature. Using the donor-controlled option in AUTOMOD allows one to retain the correct rates as the flows are brought into balance.

AUTOMOD also allows the user to simulate what might happen in a given network if certain perturbations were forced upon the resident populations. Because there is no easy way to program generic changes in exogenous variables, such as sunlight, temperature, pH, etc., the outputs of these predator-controlled simulations must always be viewed with extreme caution. At best such results represent an extended "back-of-the-envelope" type calculation to conjecture what the initial responses of predator- prey interactions to temporary imbalances might be.

Despite these limitations, AUTOMOD promises to be a useful addition to the arsenal of tools for analyzing networks of marine ecosystem flows.

ACKNOWLEDGMENTS

The author is grateful to Drs. Lee Graham and Richard Wiegert for testing early versions of AUTOMOD and for their suggested changes in the code. The editors of this volume also offered useful suggestions for revision of an earlier draft. The development of the algorithm was supported under a contract with the Department of the Navy (NOOO24-86-C-5188).

REFERENCES

Ducklow, H., M.J.R. Fasham & A.F. Vezina 1989. Derivation and analysis of flow networks for open ocean plankton systems. In: Wulff, F., J.G. Field & K.H. Mann (eds), *Network Analysis in Marine Ecology: Methods and Applications.* Lecture notes on coastal and estuarine studies. Springer-Verlag, New York.

Kay, J.J., L. Graham & R.E. Ulanowicz 1989. A detailed guide to network analysis. In: Wulff, F., J.G. Field & K.H. Mann (eds), *Network Analysis in Marine Ecology: Methods and Applications.* Lecture notes on coastal and estuarine studies. Springer-Verlag, New York.

Press, W.H., B.P. Flannery, S.A. Teukolsky & W.T. Vetterling 1986. *Numerical Recipes: The Art of Scientific Computing.* Cambridge University Press, Cambridge, Mass. 818p.

Ulanowicz, R.E. 1972. Mass and energy flow in closed ecosystems. J. Theor. Biol. 34:239-253.

Ulanowicz, R.E. 1984. Community measures of marine food networks and their possible applications. pp. 23-47 in: Fasham, M.J. R. (ed.) *Flows of Energy and Materials in Marine Ecosystems.* Plenum Publishing Corp., New York.

Vézina, A.F. 1989. Construction of flow networks using inverse methods. In: Wulff, F., J.G. Field & K.H. Mann (eds), *Network Analysis in Marine Ecology: Methods and Applications.* Lecture notes on coastal and estuarine studies. Springer-Verlag, New York.

Wiegert, R.G. 1973. A general ecological model and its use in simulating algal-fly energetics in a thermal spring community. pp. 85-102 in: Geir, P.W., L.R. Clark, D.J. Anderson, and H.A. Nix (eds) *Insects: Studies in Population Management,* Vol. 1. Ecol. Soc. Australia, Canberra.

Chapter 5

DEALING WITH THE "MIXED UNITS" PROBLEM IN ECOSYSTEM NETWORK ANALYSIS

Robert Costanza
Chesapeake Biological Laboratory
University of Maryland
Solomons, Maryland, 20688-0038
USA

Bruce Hannon
Geography Department
and Illinois Natural History Survey
University of Illinois
Urbana, IL 61801, USA

INTRODUCTION

Ecology is often defined as the study of the relationships between organisms and their environment. The quantitative analysis of interconnections between species and their abiotic environment has therefore been a central issue. The mathematical analysis of interconnections is also important in several other fields. Practical quantitative analysis of interconnections in complex systems began with the economist Wassily Leontief (1941) using what has come to be called **Input-Output (I-O) Analysis.** More recently, these concepts, sometimes called **Flow Analysis**, have been applied to the study of interconnections in ecosystems (Hannon 1973, 1976, 1979, 1985a,b,c; Costanza & Neill 1984). Related ideas were developed from a different perspective in ecology, under the heading of **Compartmental Analysis** (Barber *et al.* 1979, Patten & Finn, 1979; Funderlic & Heath, 1971; Hett & O'Neill, 1971). We refer to the total of all variations of the analysis of ecological or economic networks as **Network Analysis** (See Field *et al.* 1989, Chapter 1 of this volume for distinctions among the various forms of network analysis).

In this chapter we: 1) discuss the necessity to deal with multiple commodities (not just carbon, energy, or nitrogen) and joint products (processes that produce more than one type of commodity as output) for a more realistic treatment of ecosystem networks; 2) develop the mathematical form for a general multicommodity system capable of dealing with joint products; and 3) present and briefly discuss a detailed example of these applications.

We find that multicommodity, hierarchial, I-O (flow) analysis can be used to calculate a set of unique net input intensity factors, that can then be used to form more comprehensive and realistic input matrices for the kinds of compartmental analysis used elsewhere in this book. We can think of this as a form of preprocessing that yields a hybrid commodity (embodied energy). This new commodity has the desired characteristic of conservation necessary for compartmental analysis but it also incorporates the production interdependencies (i.e. the ammounts of each commodity required to produce other commodities) between the original commodities.

COMMENSURABILITY

A fundamental issue in any field that deals with the interconnections between dissimilar components is *commensurability* : the ability to measure the components in the same addable (commensurable) units. Both economics and ecology must deal with this "apples and oranges" or mixed units issue, although we feel its importance in ecology has not yet been fully appreciated.

Flow and compartmental analysis differ fundamentally in terms of: (1) their concentration on outputs vs. inputs (respectively) and; (2) what things they require to "add up". The fundamental reasons for these distinctions are summarized in following sections. To obtain a comprehensive understanding of ecosystem interdependence one needs to understand *both* input and output relationships. We therefore seek a way to integrate these approaches.

Flow (output) analysis requires for most of its results only that total output of each commodity in the system equal total input, without ever requiring the addition of different commodities (i.e. apples and oranges, or carbon and nitrogen). Compartmental (input) analysis, on the other hand, requires a "conservative" unit of measure, such as mass or energy or carbon content. A conservative unit is one that obeys the rule that total inputs to any process equal total outputs. One must therefore be able to *add* process inputs and outputs and in real processes these are usually different, noncommensurable commodities. For example,the very different commodities CO_2, water, light energy, and nutrients are all required in the process of photosynthesis to produce the commodity plant biomass. The usual simplification in ecological compartment analysis is to arbitraily choose one commodity (such as carbon) which is conservative and declare that this commodity is a valid tracer for all others. This approach can be misleading, however. Imagine using only the weight (mass) of the interconnecting flows in an economic system to analyze interdependence. Bulk commodity industries like sand and gravel would be given much more "weight" (pun intended) than they deserve under this approach at the expense of service and other less massive industries. Likewise, choosing any one commodity in an ecosystem to analyze interdependence can be misleading.

Before compartmental analysis can yield comprehensive and realistic results for ecosystems one must calculate and apply an appropriate set of intensity factors analogous to prices in economic systems. This chapter demonstrates several related techniques for doing this.

METHODS

Flow Accounting

In the analysis of complex dynamic systems, it is necessary to develop consistent definitions and categorizations of all the identifiable flows. We start with the diagram shown in Figure 5.1.

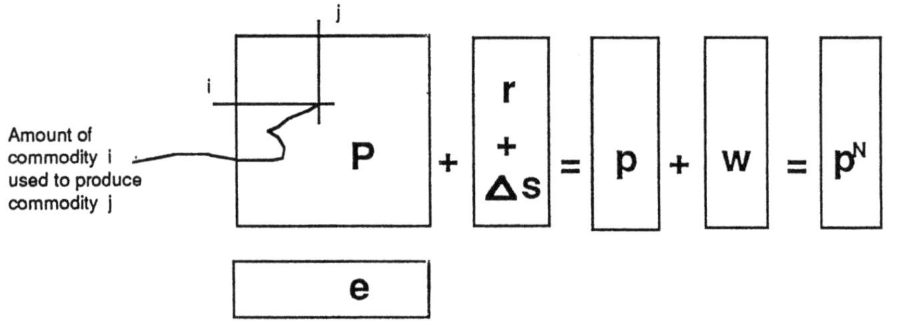

Figure 5.1. Ecosystem flow accounting diagram

A **commodity** is defined as some identifiable unit moving through the system. It can be a simple element (like carbon) or a complex structure (like plant biomass) or even the results of a service (like aerated sediments). Commodities are transformed (produced, consumed, or combined into more complex commodities) in **processes**. Biomass, for example, is produced by the photosynthetic process, which combines (consumes) water, carbon dioxide, sunlight and nutrients into plant material.

In Figure 5.1, **P** is known as the production matrix. (Throughout this Chapter we use bold lowercase letters to indicate vectors and bold uppercase letters to indicate matrices.) It summarizes the direct interdependences between the system's commodities and processes. Its elements are the amount of commodity *i* which is used by process *j* in the given time period. For example, p_{ij} could be the daily amount of nitrogen (*i*) consumed in the photosynthetic process of a particular plant species (*j*). This is a multicommodity system since different, noncommensurable commodities are listed along the rows. The row sums may be calculated (since they are all the same commodity) but the column sums cannot (since that would require adding, for example, nitrogen, sunlight, carbon dioxide, and water).

We begin with a system without joint products, or one in which each process (column) is assumed to produce only one commodity. This ignores, for example, the oxygen produced by photosynthesis as a joint product along with the biomass. Later we relax this requirement.

The vector **r** *is the net output of commodities from the system*. It is composed of two types of commodities: 1) the net export (export minus import) of each produced commodity, during the given period; and 2) the amount of biomass made during the given period used in rebuilding the depreciated stock (we use the basal metabolism as a surrogate for this commodity flows). The importance of this distinction will become clearer later on. Suffice it to say here that it is critical to a general description that allows the desired quantities to be calculated for both open and closed systems. The **Δs** is a vector of net changes in stock during the time period (zero at steady state).

The vector **p** is the total output of commodities by the system. The elements of **p** are the row sums of **P**, plus **r**, and **Δs**. In other words the total amount of each commodity produced equals the sum of the amounts sent to other internal processes, net exports, and change in stock. Excluded from the row summing is the amount of the commodity **i** which is depreciated (given off as heat in non-basal respiration). These "waste heat" or depreciation flows are the elements of the vector **w**. If the vectors **w** and **p** are added, the result is the total input vector $\mathbf{p}^N$. The vector **e** is the net input vector, the amounts of commodities absorbed by the processes of the system but not produced by it, during the given time period.

Flow Analysis

Flow analysis proceeds in the following way given these definitions. Let **s** be a vector of stocks; s_i is the amount of commodity **i** held as a stock in the system. As the time period becomes infinitesimal, the commodity balance along the row becomes:

$$ds_i/dt = p_i - \sum_j p_{ij} - r_i = p^n{}_i - w_i - \sum_j p_{ij} - r_i \qquad (1)$$

Note that the depreciation vector is excluded from the balance in equation 1. This exclusion indicates that **w** has zero *value* to the ecosystem when compared to **p**. In other words **w** and **p** are different commodities. Let us assume that the flow between two components in **P** divided by the receiving component is a constant matrix **G'**. That is: $g'_{ij} = p_{ij}/s_j$. Now equation 1 can be rewritten:

$$ds_i/dt = p_i - \Sigma g'_{ij} s_j - r_i \qquad (2)$$

or in vector-matrix form:

$$\mathbf{ds/dt = p - G's - r} \qquad (3)$$

Equation 1 or 2 could be used in microcosm research to determine the flows (via **G**) from the stock level changes. Assuming an *n* component system and assuming that the outputs **r** can be measured, there are n^2 unknowns. The stock and net output vectors must be measured over at least *n* different ranges to provide sufficient data to determine **G** uniquely.

If we assume further that the ratio of the stock to flow for each component is a constant diagonal matrix **B**, then the elements of this stock/flow matrix are $b_{ii} = s_i/p_i$ and equation 3 can be further rewritten as:

$$\mathbf{B}(d\mathbf{p}/dt) = \mathbf{p} - \mathbf{G'Bp} - \mathbf{r} = (\mathbf{I}-\mathbf{G'B})\mathbf{p} - \mathbf{r}$$

and if we combine the two assumptions, $g_{ij} = p_{ij}/p_j$ and

$$d\mathbf{p}/dt = \mathbf{B}^{-1}(\mathbf{I}-\mathbf{G})\mathbf{p} - \mathbf{B}^{-1}\mathbf{r} \qquad (4)$$

This is the "standard" form for the dynamic equation of the flow analysis approach. It could also be referred to as the recipient-controlled dynamic flow equation. The stability of future **p** when the system is subjected to changes in **r** depends entirely on the matrix which multiplies **p** in equation 4. If the eigenvalues of that matrix are negative, **p** is stable. Since **B** and **G** are always positive, the eigenvalues of the matrix are always positive and future **p** will always respond to changes in **r** in an unstable manner. Equation 4 can be made to respond stably by modifying **r** to include a feedback control in addition to the desired change in **r**. The feedback control can be a function of **p**, and/or the derivative or integral of **p**. The controls can be thought of in two ways: first, they give the ecosystem manager the appropriate coordinated steps to be taken in advance or following the desired changes in **r**; secondly, the ecosystem could be thought of as possessing certain types of these feedback controls. In this case, control theory could be coupled with ecosystem experiment to discern the mathematical form of the natural controls. The details of these control theory approaches are discussed by Hannon (1985a,b,c). If $d\mathbf{p}/dt = \mathbf{0}$, equation 4 becomes:

$$\mathbf{p} = (\mathbf{I}-\mathbf{G})^{-1}\mathbf{r} \qquad (5)$$

which is independent of the stock level. Equation 5 is the steady state relationship between the net and the total output. In an ecosystem with no material flows across the boundary, equation 5 indicates the amount of total output required to make up for the structural depreciation (basal metabolism).

Compartmental Analysis

In compartmental analysis, the inputs to a process are summed and the total output (in this case, $\mathbf{p}^N$ since the non-metabolic heat cannot be considered part of the stock change) is removed, leaving the change in the stock level. The inputs include the net input of non-produced commodities from outside the system. In a form similar to equation 1, the balance equation for the compartmental approach is:

$$ds_i/dt = \Sigma p_{ij} + e_i - p^n_j \qquad (6)$$

We now assume that the intercompartmental flow divided by the stock of the donor compartment is the constant matrix, **G"** whose elements are $g''_{ij} = p_{ij}/s_i$. Equation 6 can now be written

$$ds_i/dt = \Sigma s_i g''_{ij} + e_i - p^n_j$$

or in vector-matrix form:

$$\mathbf{ds/dt = sG" + e - p^n} \qquad (7)$$

Again, we assume that the stock to total output flow ratio is the constant matrix **B'**, whose elements are $b'_{ij} = s_i/p_i^N$. The two assumptions lead, as before, to the "standard" dynamical form of the compartmental analysis approach (donor-controlled):

$$\mathbf{dp^n/dt = - p^n(I\text{-}G^*)B'^{-1} + eB'^{-1}} \qquad (8)$$

where $g_{ij}^* = p_{ij}/p_i^N$.

The matrix which multiplies $\mathbf{P^N}$ determines the stability of the total output in response to changes in **e**. Since **B** and **G** are both positive matrices, the eigenvalues of the critical matrix are always negative, and the predictive solutions to equation 8 are always stable. However, because the method involves the summing of commodities of different qualities, the value of equation 8 is limited to systems measured in commensurable units. As we have pointed out commensurable units can be achieved by ignoring everything in the system except one commodity, or through the use of a weighting function (see below).

The steady state version of compartmental analysis is:

$$\mathbf{p^n = e(I\text{-}G^*)^{-1}} \qquad (9)$$

which is similar to equation 5 in the flow analysis approach, except that here the equation is based on the net input **e** (rather than net output **r**) and a "donor controlled coefficient matrix **G*** (rather than the "recipient controlled" coefficient matrix **G**).

To fully understand the physical meaning of the compartmental approach, we draw a picture which represents the balance given by the steady state version (equation 9):

Figure 5.2 clearly reveals both forms of the mixed commodity problem in compartmental analysis. The production matrix **P** is summed down the columns, adding together different commodities of different qualities. Even though these commodities may be measured in the same physical units, they cannot be added together because of their differences in quality. For the same reasons, the quantity $\mathbf{p}^N$ cannot exist. It is obtained by adding physical units representing both valuable structures and waste heat.

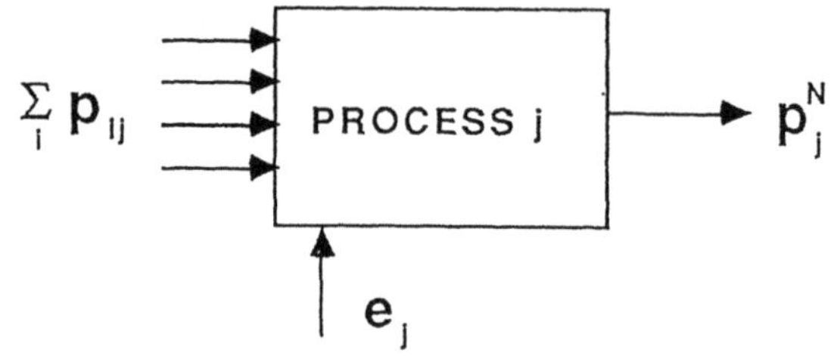

Figure 5.2. Balance diagram for the compartmental analysis approach, where Σp_{ij} represents the amount of commodity i used in process j, $\mathbf{e}_j$ = net input of the jth process, and $\mathbf{p}_j^N$ is a hypothetical quantity produced by process j.

To allow the addition required in the compartmental analysis approach, a weighting function must introduced. Let this function be the vector ϵ. We assume that the ecosystem acts as though it is perfectly competitive in the economic sense: no producer or consumer acting alone is significant enough to affect the ϵ values; information about the stocks and flows in the system is "communicated" directly or indirectly to every process in the system. Such conditions produce a unique set of weighting factors. Then, the diagram in Figure 5.2 becomes:

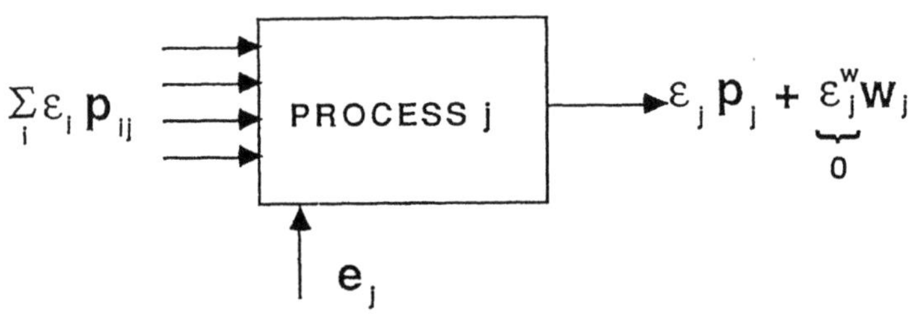

Figure 5.3. Weighted balance for the compartmental analysis approach. Symbols as in Fig 5.2. and where $\Sigma\epsilon_i$ is a weighting factor for commodity i, and ϵ_j, ϵ_j^W are weighting factors for process j and its associated metabolic loss $\mathbf{W}_j$, respectively.

In Figure 5.3, $\epsilon^W = 0$ means that the non-basal metabolic heat has no value to the system or, in other words, no value relative to the commodity flows represented in **P** or **r**. The equation which represents the balance shown in Figure 5.3 is:

$$\Sigma\epsilon_i p_{ij} + e_j = \epsilon_j p_j$$

or in vector-matrix form:

$$\epsilon \mathbf{p} + \mathbf{e} = \epsilon \mathbf{D} \mathbf{p} \tag{10}$$

where $\mathbf{D_p}$ is the diagonalized $\mathbf{p}$ vector

$$\epsilon = \mathbf{e}(\mathbf{Dp} - \mathbf{p})^{-1}$$

and since

$$\mathbf{G} = \mathbf{p}\mathbf{D_p}^{-1}$$

then

$$\epsilon = \mathbf{e}\mathbf{D_p}^{-1}(\mathbf{I}-\mathbf{G})^{-1} \quad (11)$$

Equation 10 is the solution for the needed weighting factors which allows a recognition of commodity quality in the compartmental analysis approach. It is derivable directly from the flow analysis approach: the $\mathbf{p}$ and $\mathbf{G}$ in equation 10 are the total output vector and the constant flow coefficients matrix used in the flow analysis approach. The analogy is obviously extendable to the dynamic case.

In compartmental analysis, the weighting factors, ϵ are usually implicitly assumed to be one. This is not likely to ever be a valid assumption, since, at minimum, from the second law of thermodynamics one would expect some waste heat production. If the waste heat w_j is included in the p_j, the resulting ϵ are all ones (Hannon, 1985a). The ϵ therefore, may be thought of as weighting factors among a diverse set of flows, each of whose thermodynamic quality is the same along its row in $\mathbf{Pi} + \mathbf{r}$. The quantity $\epsilon_i p_i$ is directly comparable to the quantity $\epsilon_j p_j$: the ϵ are like economic "prices" but they arise from thermodynamic and production considerations.

When the compartmental analysis approach is weighted using net input intensity factors to more closely represent reality, it turns into the flow analysis approach. Consequently, we have combined the two approaches into one: if the desire is to calculate the stock and total output change due to a net output change, the flow analysis approach is the more directly useful one. When the ϵ vector is introduced into the standard compartmental approach, the resulting output vector is identical to the one obtained from the flow analysis approach.

The flow analysis approach requires that a given commodity have the same value everywhere in the system. Otherwise, summing along the rows in a matrix of physical units would not be appropriate. This assumed uniformity of commodity weighting is a result of a broader assumption of system competition. For a system to be competitive in this sense, it must have many, small consumers and producers each possessing all the current information about the variables which affect them. Under these conditions, no single consumer can directly affect the unit weight of a commodity.

Under certain conditions, these values can be calculated on the basis of the net inputs alone, as occurs in equation 10. If each producer has only one type of output, if all producers of the same commodity use the same inputs per unit output (common production recipe), and if the system has only one net input, then the ϵ in equation 10 are the commodity unit values (Samuelson 1966). The system produces all the desired quantities of each commodity at that unit value. We can always reduce the system to one net input (see the global ecosystem discussion and Table 5.7, below*). We can also deal with systems in which each producer (process) has multiple commodity outputs, or in other words where there are joint products (see Figure 5.4, below).

Joint Products in Flow Analysis

In the discussion so far, we have avoided reference to ecological processes which produce more than one useful output. Our analysis recognized that each process produced only two types of output: **p** and **w**. Only one of these, **p**, is considered usable by the system; **w** is therefore a "product" which is not absorbed by any other process in the system. It is dropped from consideration or equivalently, it is assigned an ϵ which is zero. Now we complicate the picture a little further (attempt to better approximate reality) by introducing the multiple (useful) output format. The result is parallel to the

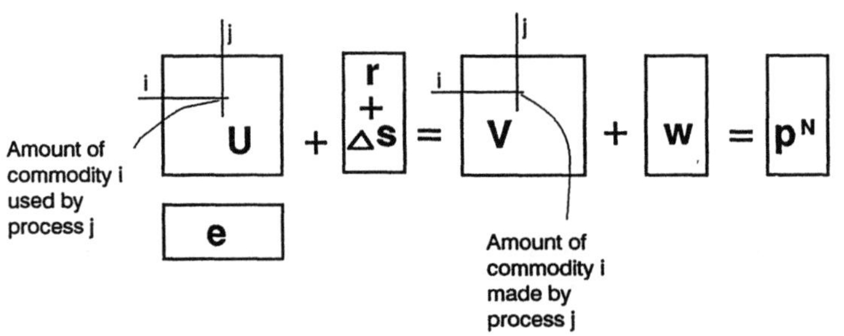

Figure 5.4. Multiproduct ecosystem flow accounting diagram where **U** is an input matrix, **e** is a vector of inputs required by the processes, **r** is a vector of output commodities, and **Δs** the changes in biomasses of commodities; **V** is an output matrix, **w** a product not absorbed by any other process, and $\mathbf{p}^N$ a vector of usable commodities produced.

previous development: the column-summing and row-summing approaches become equivalent under the appropriate assumptions. In Figure 5.4, we show the expanded accounting system.

Interpret the diagram in Figure 5.4 as follows: the **i**th row sum of **U** plus the **i**th element of **r** plus Δs_i equals the **i**th row sum of **V** (which is equal to the **i**th element of **p**, the total output without the non-basal respiration), all for the given time period. The additional step taken in the multiple product formulation is the elaboration of the **p** vector into the "Make" or output matrix, **V**. Two assumptions are possible for the linear model. One of them requires summing down the columns of **U** which we have argued

*All tables are shown at the end of the chapter.

above is not appropriate. Under conditions of perfect competition, each commodity has a similar unit value or weight to each of its consumers, and summing commodity flows along the rows is appropriate.

A row from the accounting diagram in Figure 5.4 can be stated in continuous time as:

$$\Sigma V_{ij} = \Sigma U_{ij} + r_i + ds_i/dt \qquad (12)$$

or in vector-matrix form,

$$\mathbf{Vi} = \mathbf{Ui} + \mathbf{r} + \mathbf{ds/dt} \qquad (13)$$

where i is a vector of ones. Rewriting equation 13 and solving for ds/dt gives;

$$\mathbf{ds/dt} = (\mathbf{V} - \mathbf{U})\mathbf{i} - \mathbf{r} \qquad (14)$$

or

$$\mathbf{ds/dt} = (\mathbf{I}-\mathbf{UV}^{-1})\mathbf{Vi} - \mathbf{r}$$

With the stock-flow assumption used above, Equation 14 becomes;

$$\mathbf{dp/dt} = \mathbf{B}^{-1}(\mathbf{I}-\mathbf{UV}^{-1})\mathbf{p} - \mathbf{B}^{-1}\mathbf{r} \qquad (15)$$

By a procedure identical to equation 10, the ϵ or net input intensities are (with dp/dt = 0)

$$\epsilon = \mathbf{e}(\mathbf{V}-\mathbf{U})^{-1} \qquad (16)$$

$$\epsilon = \mathbf{e}((\mathbf{I}-\mathbf{UV}^{-1})\mathbf{V})^{-1} \qquad (17)$$

The intensities are evaluated below for the global system. Equations 15 and 16 (or 17) are the multiple input-output equivalent of the output flow analysis approach.

It is important to note that in the joint product flow analysis formulation it is possible to have negative entries in the $(V-U)^{-1}$ matrix (the commodities required directly and indirectly as input to processes). These negative values are a consequence of joint production. A negative value in the $(V-U)^{-1}$ matrix implies that the column process directly or indirectly *produces* (rather than requires) the row commodity. The less significant joint production is in the system, the fewer and smaller will be the negative elements in the $(V-U)^{-1}$ matrix.

The existence of negatives in the $(V-U)^{-1}$ matrix may also give rise to negative elements in the net input intensity vector (ϵ). Negative energy intensities would imply that production of an additional unit of the commodity in question would require less energy (directly and indirectly) and not more energy. We argued elsewhere (Hannon *et al.* 1985) that these negative intensities are most likely the result of poor data or poor system specification, and can be eliminated by better data or better specification of the system.

If all positive intensities are obtained for a system using equation 16 or 17, it is possible to combine them with an additional assumption to derive a commodity by commodity flow matrix (**P**) and coefficient matrix (**G**) whose elements are guaranteed to be non-negative and that can then be analysed using the methods described earlier. The first step is to calculate "evaluated" **U** and **V** matrices. These matrices are defined as:

$$\mathbf{U^*} = \mathbf{D\epsilon U} \tag{18}$$

$$\mathbf{V^*} = \mathbf{D\epsilon V} \tag{19}$$

where $\mathbf{D\epsilon}$ means the ϵ vector diagonalized.

A characteristic of the $\mathbf{U^*}$ and $\mathbf{V^*}$ matrices is that they are in commensurable units and both their row and column totals are meaningful. They are analogous to dollar flow matrices in economic applications. From the definition of ϵ the total evaluated commodity inputs to each process should equal the total evaluated commodity outputs, or:

$$\mathbf{iU^*} + \mathbf{e} = \mathbf{iV^*}$$

The column-summing assumption is a way of allocating inputs to joint products in order to derive an equivalent single-product system (Hannon *et al.* 1985). In the context of the multiple input and output view, this assumption is related to compartmental analysis, except that the production matrix **U** and the use matrix **V** have been weighted first by ϵ. It assumes that input commodities are allocated to output commodities in proportion to the proportion which the output commodity represents of the *total* output of the process. This situation of course requires that we can calculate the total output of processes which in turn requires that we measure these outputs in commensurable units. Given that we know $\mathbf{V^*}$ from the previous analysis, we can calculate a matrix to perform this allocation:

$$\mathbf{Q} = \mathbf{V^*}(\mathbf{D_{iV^*}})^{-1} \tag{20}$$

The elements of **Q** are simply the percentages which each commodity flow represents of the total input to each process. A commodity by commodity flow matrix (**P**) can then be calculated as:

$$\mathbf{P} = \mathbf{UQ}^{T} \tag{21}$$

Standard flow analysis techniques as outlined above can then be applied to the system, and will yield the same ϵ as that obtained using equation 16 or 17 (Hannon *et al.* 1986).

Similarly, one can calculate a process by process flow matrix (call it $\mathbf{P^*}$) as:

$$\mathbf{P^*} = \mathbf{Q^{*T}U} \tag{22}$$

where $\mathbf{Q^*}$ is a matrix of the percentages which each commodity flow represents of the total outputs of that commodity (from the evaluated $\mathbf{V^*}$ matrix).

HIERARCHICAL NET INPUT INTENSITIES USING A GLOBAL MULTICOMMODITY INPUT-OUTPUT MODEL

Multicommodity flow analysis is employed to calculate the net input intensity factors that can then be used to weight different commodities to make them commensurable. If there is more than one net input to the system, however, this method does not yield complete weightings. For example, consider an ecosystem with boundaries defined such that there are net inputs of both sunlight and water. In such a case we could calculate sunlight intensities *and* water intensities but would have no way of comparing or adding the two. One approach to this problem is to define an hierarchial series of models with ever more expansive boundaries. At the level of the whole globe, there is effectively only one net input - sunlight. A global multicommodity flow analysis model can thus be used to evaluate the relative solar energy intensities of commodities at this scale. These estimates can then be used at lower levels to put different net inputs into the same units. If we can calculate the solar energy cost of fresh water via the hydrologic cycle at the global level we can use that information to equilibrate net inputs of water and sunlight at the ecosystem level. We can speak of the net input of sunlight and the sunlight "embodied" in the water, and produce a single set of more detailed net input intensities at the ecosystem level.

Of course, the level of detail of the description of the global system will limit the number of different commodities that can be evaluated. A description with only aggregate nitrogen cannot be used to evaluate the different species of nitrogen. The nine process by nine commodity global model in the following section is intended only as a simplified example. Practical applications would require a more elaborate global model with probably several hundred commodities and processes. This more detailed level of description would allow distinctions between, say, ammonia and nitrate, different levels of water purity, and perhaps regional differences (i.e. fresh water in high rainfall vs. desert areas).

Tables 5.1 and 5.2 are examples of a global multicommodity flow table set up in the multiple output, commodity-process format discussed earlier. Table 5.1 shows the inputs of the the commodities listed along the left, to the processes listed along the top. This is the **U** matrix. Note that the units for the commodities are different (they are not commensurable) so one cannot add down the columns of U. The outputs of each commodity from each process are listed in Table 5.2. This is the **V** matrix. Note that conservation holds for each commodity but not for the processes, which are each measured in different units. The total output of each commodity from all the processes which produce it (the row sum of V) is equal to the total input of that commodity to all the processes that use it (the row sum of U), including the amount of commodity that is "depreciated" or exported (**r**). Processes (columns in V) that contain more than one entry represent "joint products." At the global level joint products are significant and unavoidable.

Using the methods described earlier we can calculate the direct and indirect use matrix, $(\mathbf{V}\text{-}\mathbf{U})^{-1}$, and the net input (in this case sunlight) intensity factors. This matrix and vector are shown in Table 5.3. Note that Table 5.3 contains many negatives. This reflects the fact that joint products are significant. There is no guarantee that using this method will produce an all-positive ϵ vector Appropriate data for **U** and **V** have to be provided to yield an all-positive ϵ vector.

Tables 5.4 and 5.5 are the evaluated **U** and **V** matrices (**U*** and **V*** using equations 15 and 16) in embodied solar energy units. Note that the total inputs to processes listed in Table 5.4 equal the total outputs from processes listed in Table 5.5. Table 5.6 is the commodity input distribution matrix (**Q**) calculated from **V*** using equation 21.

Table 5.7 is the commodity by commodity flow matrix (**P**) calculated from **U** and **Q** using equation 22. Table 5.7 shows the amount of each row commodity required to produce a unit of column commodity in the global process taken as a whole. In Table 5.7 each commodity is measured in its own units. It is a simple matter to convert this table into a commensurable table by multiplying through each commodity by its energy intensity. The result is Table 5.8, which is a commodity by commodity flow matrix with all flows in embodied solar energy units. Since column sums equal row sums for this matrix, compartmental analysis is directly applicable (the **G*** and $(\mathbf{I}\text{-}\mathbf{G}^*)^{-1}$ matrices and all of the network indices outlined elsewhere in this book can be calculated from Table 5.8)

There is an additional way of looking at this system to make it a closer analog to the usual starting point for compartmental analysis. Table 5.8 shows the interdependence among *commodities* within the global *process* taken as a whole. We might also want to investigate the interdependence among the different global *processes* with the *commodities* aggregated and treated as a single hybrid commodity. To do this we apply equation 22 to the global **U** and **Q*** matrices and multiply through by the energy intensities. The results are shown in Table 5.9. This matrix indicates the flow of a single hybrid commodity (again measured in embodied solar energy units) between the global processes. This hybrid commodity incorporates the interdependencies in the disaggregated commodities through the calculation of the energy intensities.

Tables 5.8 and 5.9 are interesting for what the say about the relative importance of the various commodities and processes in the global system. For example, Table 5.8 indicates the importance of fresh water as an input to the production of manufactured goods. Fresh water represents 252.24 of a total of 757.65 x 10^{18} solar calories/yr (or 33%) of direct inputs to manufactured goods. This is slightly more than fossil fuels (29%) and turns out to be the most significant single input. Fresh water is an even more significant input to agricultural products and natural products.

Looking at the total input row in Table 5.8 indicates the relative importance of the hydrologic and carbon cycles in the global system. Water is three and carbon is two orders of magnitude larger than manufactured goods in terms of total inputs.

Table 5.9 yields a related but unique picture. Here the surface ocean and atmosphere (the two main processes in the hydrologic cycle) dominate the total input and output figures. The urban economy is shown to be very dependent on inputs from soil, atmosphere, and deep geology (for fossil fuels).

Whether one uses the commodity by commodity (table 5.8) or process by process (table 5.9) matrices as the starting point for further analysis depends on what questions are of interest. If the questions center on the interdependence between commodities (i.e. the effect of changes in the carbon cycle on the nitrogen cycle) then the commodity by commodity matrix is most appropriate. If the questions center on the interdependence between processes (i.e. the degree of dependence of the shallow ocean on land-based processes) then the process by process matrix is most appropriate. For calculating total system interdependence measures (i.e. Finn's (1976) cycling index or Ulanowicz's (1980) ascendency it is not clear which starting point is preferable. One can speak of a "commodity ascendency" and also of a "process ascendency." Each of these may reveal different aspects of system organization.

Ascendency of the Global System

By way of example, we have calculated both full and internal ascendency (A) capacity (C) and relative ascendancy (A/C) for the commodity by commodity (table 5.8) and process by process (table 5.9) descriptions of the global system (see Kay *et al.* 1989, Chapter 2 of this volume for a description of the methods for these calculations). The results are given in Table 5.10. They indicate a relative ascendency of .477 to .495. This means about half of the capacity of the global system for flow organization is being utilized. The differences between relative ascendency (A/C) for the commodity by commodity and process by process descriptions and between full and internal variants of the measures seem minor. Both the total ascendency and capacity are significantly lower for the process by process than for the commodity by commodity descriptions, however (on the order of 25%). Total system throughput is slightly less for the process by process than the commodity by commodity description and this may account for the discrepency, since ascendency and capacity both include T as a factor. At any rate, relative ascendency (A/C) seems to be a more stable and reliable quantity and would thus be preferable for comparisons between different systems and for the same system over time.

SUMMARY AND CONCLUSIONS

In this paper we have compared compartmental and flow analysis and have shown that flow analysis is the more appropriate approach if the system is not in commensurable units. Comprehensive analysis of ecosystems must include several different commodities measured in physical (noncommensurable) units. Flow analysis can be used to derive a set of unique intensity factors that perform the same function as prices in economic systems. They allow physical flows to be put in commensurable units and the "mixed units" problem to be solved.

We have also elaborated methods to deal with hierarchial intensity factors and joint products in ecosystems, and to convert commodity by process joint product systems to equivalent commodity by commodity single process systems and process by process single commodity systems. We have given a detailed description of how this works out for the global system, including what are probably the first attempted calculations of the ascendancy and relative ascendancy of the biosphere as a whole. While no conclusions can be drawn from these calculations alone, the methods discussed allow a more realistic and comprehensive treatment of ecosystem interdependence and can form the basis for comparative analysis of whole system performance once comparable calculations have been performed for enough systems.

REFERENCES

Barber, M., B. Patten & J. Finn 1979. Review and evaluation of I-O flow analysis for ecological applications, in: Matis, J., B. Patten & G.White, (Eds)*Compartmental Analysis of Ecosystem Models,* Vol 10 of Statistical Ecology, Internatl Coop. Publish. House, Bertonsville, Md.

Costanza, R. & C. Neill 1981. The energy embodied in the products of the biosphere. pp. 745-755 in: Mitsch, W. J., R. W. Bosserman, and J. M. Klopatek, (Eds) *Energy and Ecological Modeling.* Elevier, Amsterdam.

Costanza, R. & C. Neill 1984. The energy embodied in the products of ecological systems: a linear programming approach. J. Theor. Biol. 106: 41-57

Finn, J. 1976. The cycling index. J. Theor. Biol. 56: 363-73

Funderlic, R & M. Heath 1971. Linear Compartmental Analysis of Ecosystems. Oak Ridge Natl Lab, ORNL-IBP-71-4.

Hannon, B. 1973. The structure of ecosystems, J. Theor. Biol. 41:535-46.

Hannon, B. 1976. Marginal product pricing in the ecosystem. J. Theor. Biol. 56:256-267.

Hannon, B. 1979. Total energy costs in ecosystems. J. Theor. Biol. 80:271-293.

Hannon, B. 1985a. Ecosystem flow analysis. In: Ulanowicz, R.E. & T.Platt (Eds), *Ecosystem theory for biological oceanography,* Can. Bull. Fish. Aquat. Sci. 213: 97-118.

Hannon, B. 1985b. Conditioning the ecosystem. Mathematical Biology, 75: 23-42.

Hannon, B. 1985c. Linear dynamic ecosystems. J. Theor. Biol. 116: 89-98.

Hannon, B. 1986. Ecosystem control theory, submitted.

Hannon, B., R. Costanza, & R. Herendeen 1986. Energy cost and value in ecosystems. J. Environ. Ecol. Manage. 13: 391-401.

Hett, J & R. O'Neill 1971. Systems analysis of the aleut ecosystem. US-IBP, deciduous forest biome Memo Report, 71-16, September.

Hippe, P. 1983. Environ analysis of linear compartmental systems: The dynamic time invariant case. Ecol. Mod. 19:1-26.

Kay, J J., L. Graham & R. E. Ulanowicz 1989. A detailed guide to network analysis. In: Wulff, F., J.G. Field & K. H. Mann (Eds), *Network Analysis in Marine Ecology: Methods and Applications*. Lecture notes on coastal and estuarine studies. Springer-Verlag, New York.

Kercher, J. 1983. Closed-form solutions to sensitivity equations in the frequency and time domains for linear models of ecosystems. Ecol. Model.18: 209-221. .

Leontieff, W. 1941. *The Structure of American Economy. 1919, 1929: An empirical application of equilibrium analysis.* Harvard University Press. Cambridge.

Samuelson, P. 1966. A new theorem on non-substitution, in: Stiglitz, J., (ed.), *The Collected Papers of P. A. S.*, 1, MIT Press, Cambridge, Mass., Chapter 37.

Stone, R. 1963. Input-output relationships 1954-1966; A programme for growth, MIT Press, Cambridge, Mass. 1-41.

Ulanowicz, R. 1980. An hypothesis on the development of natural communities. J. Theor. Biol.. 85: 223-245.

Vernberg, J. F. 1978. *Physiological Ecology of Organisms*. Columbia University Press, N.Y.

Table 5.1. Global multicommodity input ("use") matrix (U), along with the vectors for net export + stock replacement (**r**), total output excluding waste heat (**p**), and the net input vector (sunlight, **e**). Units are: 1. Manufactured goods - 10^{12} \$/yr; 2. Agricultural products - 10^{15} g dry wt/yr; 3. Natural products - 10^{15} g dry wt/yr; 4. Nitrogen - 10^{12} gN/yr; 5. Carbon - 10^{15} gC/yr; 6. Phosphorous - 10^{12} gP/yr; 7. Water vapor - km^3/yr; 8. Fresh water - km^3/yr; 9. Fossil fuel - 10^{15} gC/yr; e. Sunlight - 10^{18} kcal/yr. Complete references are given in Costanza & Neill (1981).

	Processes										
Commodities	Urban Econ. (1)	Agri- culture (2)	Natural Plants (3)	Ani- mals (4)	Soil (5)	Deep Ocean (6)	Surface Ocean (7)	Atmos- phere (8)	Deep Geology (9)	Net Output r	Total Output p
Manufact. Goods (1)	2.71	0.08	0	0	0	0	0	0	0	1.19	3.98
Ag. Products (2)	1.28	4.55	0	3.27	0	0	0	0	0	0	9.1
Natural Products (3)	1.18	0	0	27.9	103.4	34.6	0	0	0.16	0.06	167.3
Nitrogen (4)	55	62.4	208	0	493.6	0	168	389.5	0	0	1376.5
Carbon (5)	0	8.2	147	0	0	15.6	37.2	110.3	0	0	318.3
Phosphorous (6)	12.6	28.5	1345.7	0	8.4	0	21	9.5	13	0	1438.7
Water Vapor (7)	0	0	0	0	0	0	0	496100	0	0	496100
Fresh Water (8)	1008	15490	51226	0	111419	0	424700	0	0	2000	605843
Fossil Fuels (9)	5	0	0	0	0	0	0	0	0	-4.93	0.07
Net Input (Sunlight) (e)	0	23	227	0	0	0	606	0	0		

Table 5.2. Global multicommodity output ("make") matrix (V), along with the vector for total input. Units are the same as Table 5.1.

	Processes									
Commodities	Urban Econ. (1)	Agri-culture (2)	Natural Plants (3)	Ani-mals (4)	Soil (5)	Deep Ocean (6)	Surface Ocean (7)	Atmos-phere (8)	Deep Geology (9)	Total Input
Manufact. Goods (1)	3.98	0	0	0	0	0	0	0	0	3.98
Ag. Products (2)	0	9.1	0	0	0	0	0	0	0	9.1
Natural Products (3)	0	0	163.4	3.9	0	0	0	0	0	167.3
Nitrogen (4)	80	31	0	295	340.5	0	182	448	0	1376.5
Carbon (5)	5	6.1	73.6	14	46.5	15.6	49.5	108	0	318.3
Phosphorous (6)	14.2	0	0	0	241.3	1161.1	0	9.5	12.6	1438.7
Water Vapor (7)	79	5931	50740	0	14650	0	424700	0	0	496100
Fresh Water (8)	929	9829	0	0	98985	0	0	496100	0	605843
Fossil Fuels (9)	0	0	0	0	0	0	0	0	0.07	0.07

Table 5.3. Matrix $(V\text{-}U)^{-1}$ for the U and V matrices shown in Tables 5.1 and 5.2, respectively. Entries in this matrix indicate the *direct and indirect* inputs (or outputs if negative) of the row commodities to (or from) the column processes. Net input (solar energy) intensties for the commodities based on this matrix and the direct sunlight input vector listed in Table 5.1 (using equation 14) are shown as the last column on the right.

	Processes									
	Urban Econ.	Agri-culture	Natural Plants	Ani-mals	Soil	Deep Ocean	Surface Ocean	Atmos-phere	Deep Geology	Net Input (Sunlight) Intensities
Commodities	(1)	(2)	(3)	(4)	(5)	(6)	(7)	(8)	(9)	(ε)
Manufact. Goods (1)	0.79869	0.01329	-0.00029	0.00018	-0.00112	-0.00001	0.00002	0.00002	-0.00070	**190.365**
Ag. Products (2)	0.17925	0.21105	-0.00453	0.00281	-0.01771	-0.00013	0.00037	0.00037	-0.01112	**13.997**
Natural Products (3)	0.07260	-0.02471	0.00916	0.00063	-0.00336	0.00027	0.00051	0.00051	0.02249	**39.158**
Nitrogen (4)	-0.06323	-0.01735	-0.00619	0.00384	-0.02421	-0.00018	0.00050	0.00050	-0.01520	**0.630**
Carbon Dioxide (5)	-0.00592	-0.03113	0.00256	-0.00020	0.00142	0.00008	0.00052	0.00052	0.00629	**57.123**
Phosphorous (6)	0.11335	-0.01646	0.01109	0.00086	-0.00289	0.00119	0.00051	0.00051	0.03216	**1.167**
Water Vapor (7)	0.28014	0.02434	0.06136	0.00070	0.09619	0.00183	0.00070	0.00070	0.15070	**0.550**
Fresh Water (8)	0.24934	0.01992	0.05349	0.00069	0.08183	0.00159	0.00067	0.00067	0.13136	**0.550**
Fossil Fuels (9)	57.04948	0.94962	-0.02038	0.01265	-0.07970	-0.00061	0.00165	0.00165	14.23566	**96.171**

Table 5.4.Evaluated U matrix in embodied solar energy units. Each element in this matrix is calculated by multiplying the physical flow listed in Table 5.1 by the commodity's energy entensity (ε) listed in Table 5.3

	Processes								
Commodities	Urban Econ. (1)	Agri-culture (2)	Natural Plants (3)	Ani-mals (4)	Soil (5)	Deep Ocean (6)	Surface Ocean (7)	Atmos-phere (8)	Deep Geology (9)
Manufact. Goods (1)	515.89	15.23	0.00	0.00	0.00	0.00	0.00	0.00	0.00
Ag. Products (2)	17.92	63.69	0.00	45.77	0.00	0.00	0.00	0.00	0.00
Natural Products (3)	46.21	0.00	0.00	1092.50	4048.91	1354.86	0.00	0.00	6.27
Nitrogen (4)	34.65	39.31	131.03	0.00	310.95	0.00	105.83	245.37	0.00
Carbon Dioxide (5)	0.00	468.41	8397.05	0.00	0.00	891.12	2124.97	6300.64	0.00
Phosphorous (6)	14.70	33.26	1570.26	0.00	9.80	0.00	24.50	11.09	15.17
Water Vapor (7)	0.00	0.00	0.00	0.00	0.00	0.00	0.00	273068.23	0.00
Fresh Water (8)	555.03	8529.11	28206.08	0.00	61349.57	0.00	233848.46	0.00	0.00
Fossil Fuels (9)	480.86	0.00	0.00	0.00	0.00	0.00	0.00	0.00	0.00
Net Input (Sun) (e)	0.00	23.00	227.00	0.00	0.00	0.00	606.00	0.00	0.00
Total Inputs	1665.24	9172.00	38531.42	1138.27	65719.23	2245.97	236709.77	279625.32	21.43

Table 5.5 Evaluated **V** matrix in embodied solar energy units. Each element in this matrix is calculated by multiplying the physical flow listed in Table 5.2 by the commodity's energy entensity (ε) listed in Table 5.3.

	Processes								
Commodities	Urban Econ. (1)	Agri-culture (2)	Natural Plants (3)	Ani-mals (4)	Soil (5)	Deep Ocean (6)	Surface Ocean (7)	Atmos-phere (8)	Deep Geology (9)
Manufact. Goods (1)	757.65	0.00	0.00	0.00	0.00	0.00	0.00	0.00	0.00
Ag. Products (2)	0.00	127.37	0.00	0.00	0.00	0.00	0.00	0.00	0.00
Natural Products (3)	0.00	0.00	6398.38	152.72	0.00	0.00	0.00	0.00	0.00
Nitrogen (4)	50.40	19.53	0.00	185.84	214.50	0.00	114.65	282.22	0.00
Carbon Dioxide (5)	285.61	348.45	4204.24	799.72	2656.21	891.12	2827.58	6169.26	0.00
Phosphorous (6)	16.57	0.00	0.00	0.00	281.57	1354.86	0.00	11.09	14.70
Water Vapor (7)	43.48	3264.60	27928.81	0.00	8063.80	0.00	233767.54	0.00	0.00
Fresh Water (8)	511.53	5412.05	0.00	0.00	54503.16	0.00	0.00	273162.76	0.00
Fossil Fuels (9)	0.00	0.00	0.00	0.00	0.00	0.00	0.00	0.00	6.73
Total Outputs	1665.24	9172.00	38531.42	1138.27	65719.23	2245.97	236709.77	279625.32	21.43

Table 5.6. Commodity input distribution matrix (Q). Each element in this matrix indicates the percentage of the total inputs to the column process to be allocated to the production of the row commodity.

	Processes								
Commodities	Urban Econ. (1)	Agri-culture (2)	Natural Plants (3)	Ani-mals (4)	Soil (5)	Deep Ocean (6)	Surface Ocean (7)	Atmos-phere (8)	Deep Geology (9)
Manufact. Goods (1)	.454981	.000000	.000000	.000000	.000000	.000000	.000000	.000000	.000000
Ag. Products (2)	.000000	.013887	.000000	.000000	.000000	.000000	.000000	.000000	.000000
Natural Products (3)	.000000	.000000	.166056	.134164	.000000	.000000	.000000	.000000	.000000
Nitrogen (4)	.030264	.002129	.000000	.163263	.003264	.000000	.000484	.001009	.000000
Carbon Dioxide (5)	.171515	.037991	.109112	.702573	.040418	.396761	.011945	.022063	.000000
Phosphorous (6)	.009950	.000000	.000000	.000000	.004284	.603239	.000000	.000040	.685929
Water Vapor (7)	.026113	.355931	.724832	.000000	.122701	.000000	.987570	.000000	.000000
Fresh Water (8)	.307178	.590062	.000000	.000000	.829333	.000000	.000000	.976888	.000000
Fossil Fuels (9)	.000000	.000000	.000000	.000000	.000000	.000000	.000000	.000000	.314071

Table 5.7 Commodity by commodity flow matrix (**P**) calculated from the **U** matrix in Table 5.1 and the **Q** matrix in Table 5.6. $\mathbf{P} = \mathbf{U}\mathbf{Q}^T$. The elements in this matrix indicate the amounts of the row commodities required *directly* to produce the column commodities. The units are the same as Table 5.7, as are the net output vector (**r**) and the total output vector (**p**).

	Commodities								
Commodities	Manuf. Goods (1)	Ag.-Products (2)	Natural Products (3)	Nitrogen (4)	Carbon Dioxide (5)	Phos-phorous (6)	Water Vapor (7)	Fresh Water (8)	Fossil Fuels (9)
Manufact. Goods (1)	1.23	0.00	0.00	0.08	0.47	0.03	0.10	0.88	0.00
Ag. Products (2)	0.58	0.06	0.44	0.58	2.69	0.01	1.65	3.08	0.00
Natural Products (3)	0.54	0.00	3.74	4.93	37.71	21.44	12.72	86.12	0.05
Nitrogen (4)	25.02	0.87	34.54	3.88	65.05	2.68	400.89	843.57	0.00
Carbon Dioxide (5)	0.00	0.11	24.41	0.15	25.42	9.41	146.21	112.59	0.00
Phosphorous (6)	5.73	0.40	223.46	0.49	150.88	9.08	1007.65	36.93	4.08
Water Vapor (7)	0.00	0.00	0.00	500.70	10945.25	19.67	0.00	484634.38	0.00
Fresh Water (8)	458.62	215.11	8506.39	632.85	15927.19	487.39	475762.24	101853.20	0.00
Fossil Fuels (9)	2.27	0.00	0.00	0.15	0.86	0.05	0.13	1.54	0.00

Table 5.8. Global commodity by commodity flow matrix in embodied solar energy units (10^{18} Embodied solar calories/yr)

	Commodities										
Commodities	Manuf. Goods (1)	Ag.-Products (2)	Natural Products (3)	Nitrogen (4)	Carbon Dioxide (5)	Phos-phorous (6)	Water Vapor (7)	Fresh Water (8)	Fossil Fuels (9)	Net Output	Total Output
Manufact. Goods (1)	234.15	0.00	0.00	15.23	89.47	5.71	19.04	167.52	0.00	226.53	757.65
Ag. Products (2)	8.12	0.84	6.16	8.12	37.65	0.14	23.10	43.11	0.00	0.00	127.23
Natural Products (3)	21.15	0.00	146.45	193.05	1476.65	839.55	498.09	3372.29	1.96	2.35	6551.52
Nitrogen (4)	15.76	0.55	21.76	2.44	40.98	1.69	252.56	531.45	0.00	0.00	867.20
Carbon Dioxide (5)	0.00	6.28	1394.37	8.57	1452.07	537.53	8351.95	6431.48	0.00	0.00	18182.25
Phosphorous (6)	6.69	0.47	260.78	0.57	176.08	10.60	1175.93	43.10	4.76	0.00	1678.96
Water Vapor (7)	0.00	0.00	0.00	275.39	6019.89	10.82	0.00	266548.91	0.00	0.00	272855.00
Fresh Water (8)	252.24	118.31	4678.51	348.07	8759.95	268.06	261669.23	56019.26	0.00	1191.57	333305.21
Fossil Fuels (9)	218.31	0.00	0.00	14.43	82.71	4.81	12.50	148.10	0.00	0.00	480.86
Net Input (Sun) (e)	1.24	0.78	43.49	1.34	46.81	0.06	852.60	0.00	474.14		
Total Input	757.65	127.23	6551.52	867.20	18182.25	1678.96	272855.00	333305.21	480.86		

Table 5.9. Global process by process flow matrix in embodied solar energy units (10^{18} Embodied solar calories/yr)

Processes	Urban Econ. (1)	Agri-culture (2)	Natural Plants (3)	Ani-mals (4)	Soil (5)	Deep Ocean (6)	Surface Ocean (7)	Atmos-phere (8)	Deep Geology (9)	Net Output	Total Output
Urban Economy (1)	518.88	37.99	197.30	0.00	110.19	13.99	390.53	167.90	0.15	229.89	1,666.82
Agriculture (2)	29.71	244.43	734.68	45.77	1,233.21	17.82	4,721.59	2,861.60	0.00	0.00	9,888.81
Natural Plants (3)	45.15	108.20	1,939.72	1,067.37	3,955.79	1,529.55	490.87	29,308.41	6.13	153.21	38,604.38
Animals (4)	8.20	26.99	363.40	21.85	146.28	62.74	107.22	303.55	0.13	94.63	1,134.99
Soil (5)	101.50	1,473.93	6,119.73	0.00	10,078.43	130.10	38,457.80	9,174.41	2.55	380.92	65,919.37
Deep Ocean (6)	11.86	49.79	1,678.66	0.00	7.91	43.66	123.90	317.68	12.24	0.00	2,245.70
Surface Ocean (7)	4.57	78.26	1,327.24	0.00	41.05	139.01	345.46	234,761.69	0.00	823.14	237,520.43
Atmosphere (8)	465.93	7,157.14	26,000.96	0.00	50,346.43	302.09	192,276.82	2,215.74	0.11	345.89	279,111.10
Deep Geology (9)	481.01	0.33	15.70	0.00	0.10	0.00	0.25	0.11	0.15	0.00	497.65
Net Input (e)	0.00	711.75	227.00	0.00	0.00	6.73	606.00	0.00	476.20		
Total Inputs	1,666.82	9,888.81	38,604.38	1,134.99	65,919.37	2,245.70	237,520.43	279,111.10	497.65		

Table 5.10. Summary of information indices for the commodity (Table 5.8) and process by process (Table 5.9) descriptions of the global system

	Type of Description	
Information Index	Commodity by Commodity	Process by Process
Full Ascendency	553,190	409,000
Full Capacity	1,150,800	826,710
Full A/C	.481	.495
Internal Ascendency	543,930	393,360
Internal Capacity	1,139,900	821,790
Internal A/C	.477	.479

PART III

APPLICATIONS

Chapter 6

NETWORK INFORMATION INDICES WITH AN ESTUARINE MODEL

James N. Kremer
Marine Biology Section,
Department of Biological Sciences
University of Southern California
University Park
Los Angeles, California 90089-0371
USA

INTRODUCTION

Our insight into complex ecological systems benefits from diverse perspectives. Historically various methods of observation and analysis have been used to reveal patterns and principles of structure and function. The question is not whether there is an all-powerful single best approach; such an expectation is obviously unrealistic for complex real systems. Rather, we must develop and test several tools and evaluate each in terms of specific uses for which it is best suited. The goal of the Santa Cruz SCOR Workshop on Ecological Flow Analysis was to explore the utility of a number of tools for the analysis of networks in the development of ecosystem theory (see Field *et al.* 1989, this volume).

This chapter is one of a group focusing on time series. Ecological data frequently take this form, and much effort is typically devoted to characterizing changes or similarities in a system through time. Various powerful techniques for the analysis of networks were discussed at the workshop and are presented throughout this volume (Wulff *et al.* 1989). But some of the methods are likely to be less useful than others in certain cases. Some of the techniques are most appropriate near steady state, others are especially informative for numerous compartments with diverse connections. In this chapter, I have focussed on the information indices of *development capacity*, *ascendency*, and their component parts. I believe them to be the subset of network tools most appropriate to the time series of a functionally complex but structurally simplified network.

Simulation modeling is a tool that has been used in ecosystem analysis for some time. Simulation can test and help to refine the conceptual models that we must use in studies of complex dynamic systems. In this paper, I have explored the behavior of the information indices with results from an estuarine simulation model. The advantage of using of a model system is that it is completely defined. Thus we can explicitly depict the state of the system, *and* the causal mechanisms at work beneath the summary flow diagram that serves as input for the flow analyses.

While single indices can never convey the rich spectrum of relevant information about a system, they are of use presumably because they may capture some essence of the system that is especially relevant. My goal was to ask what features of the ecosystem these information indices respond to, and whether they can be related to the dominant processes taking place during different conditions of the model's annual cycle.

Theoretical Basis of Information Indices for Networks

The merit of the Shannon-Wiener index developed in information theory is that it responds to both richness and evenness, two facets that were also recognized as relevant to ecological diversity. In this case, we are using the same theory to characterize the diversity of flows in the network (rather than species abundances). As a standard against which to index the state of a system, Ulanowicz (1980; Hirata and Ulanowicz 1984) has proposed an idealized condition where the flow has the highest possible diversity. The most "fully developed" system is one with maximum evenness, richness (absolute number of flows), and no redundancy (no parallel paths) (see also Kay *et al.* 1989, this volume).

In any data set for a real system the specific number of flows and their magnitude fix the richness and throughput. While greater total throughput and richness would allow a more fully developed system, this cannot be specified *a priori*. Thus, Ulanowicz's theoretical standard *development capacity* is when the flows connecting all compartments are unique and equal in size. *Ascendency* measures how close the real system is to this standard configuration. The residual or shortfall is composed of *overhead* and *redundancy*. Overhead, in turn, has the components: *inputs, respiration* and other *exports* (Kay *et al.* 1989, this volume). Note that these indices do not respond directly to the stock sizes of compartments.

THE MODEL

The simulation model of Narragansett Bay (Rhode Island, USA) has been discussed extensively elsewhere (Kremer & Nixon 1978). It was designed to simulate the spatial patterns of plankton and inorganic nutrients throughout the annual cycle. The model includes formulations of functional relationships important in controlling rates of productivity and material flux in a plankton-based coastal ecosystem (Fig. 6.1). While the specific case study was Narragansett Bay, the processes included in the model are quite typical of many marine systems, and provide a useful analog on which to test the network indices.

The complete Bay Model includes anthropogenic and river inputs, tidal mixing among regions of the bay, and offshore exchanges. For this study, this physical context was simplified to a single region with an average flushing rate of -2%. All of the biotic interactions were retained, including grazing and remineralization by the benthos and carnivores, for which biomasses were not modeled but were forced using typical seasonal patterns for lower Narragansett Bay. Of course any interpretation of a model depends totally upon its makeup, yet a detailed description is inappropriate here (see Kremer & Nixon 1978, or the synopsis in Kremer 1979). For these purposes, it will suffice to indicate the kinds of factors that are included in the formulations for the major deterministic compartments. The following word equations express the main processes of the model, with the determining factors for each process in parentheses. For the phytoplankton (P), zooplankton adults (ZA) and juveniles (ZJ) and dissolved inorganic nutrients (N, representing NH_4, $NO_2 + NO_3$, PO_4, Si):

dP/dt = Production(Temp, Light, N, P, Si) - Grazing(Zooplankton, Benthos, Temp) - Flushing

dZA/dt = Ingestion(Temp,Food) + maturing ZJ - Resp(Temp) - Predation(Temp, Carnivore stocks) - Flushing

dZJ/dt = Ingestion(Temp,Food) - maturing ZJ - Resp(Temp, Size) - Predation(ZA,Temp, Carnivore stocks) - Flushing

dN/dt = Inputs - Uptake(Phyto) + Remineralization(Zoopl., Carnivores, Benthos) - Flushing

Benthic grazing = f(Temp, forced bivalve stocks)

Benthic remineralization = f(Temp, according to empirical data)

Temperature ranges from 3 - 20°C. Initial conditions represent the winter state of Narragansett Bay, prior to the onset of the dramatic winter-spring phytoplankton bloom.

Additionally, it is important to note that zooplankton reproduction is modeled as egg production followed by juvenile development of daily cohorts. The goal was to represent accurately the time lags associated with predator-prey cycles within the plankton. Adult zooplankton are assumed to prey upon their own eggs and juveniles (Fig. 6.1). This elaborate population structure of the zooplankton allowed a test of the effect of aggregation on the patterns of the information indices.

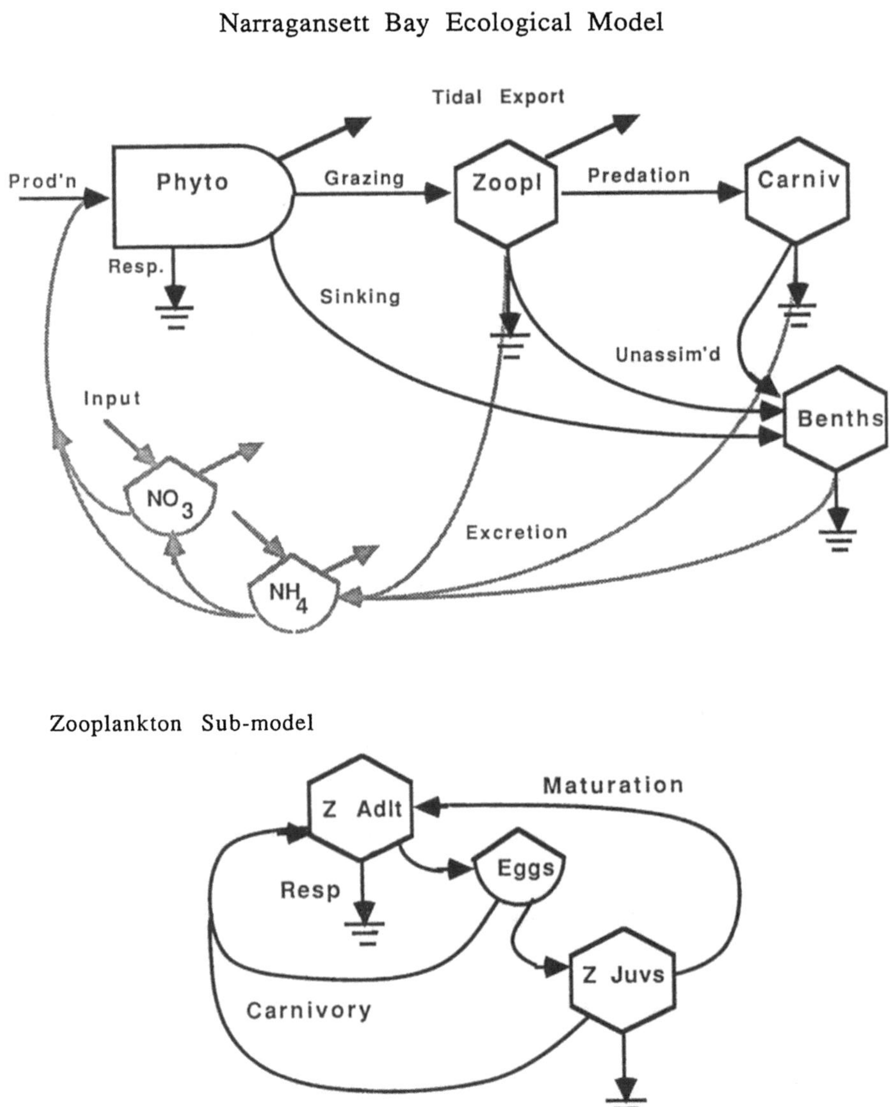

Figure 6.1 Network flow diagram for the Naragansett Bay simulation model. Arrows connecting living compartments were the carbon flows considered in the analyses; stippled compartments and flows were added in the nitrogen cases. Insert below represents the detailed resolution within the zooplankton compartment.

TYPICAL RESULTS OF THE MODEL

Overall description

The model results used in these analyses are similar to those of the "standard run" presented elsewhere (Kremer & Nixon 1978; Kremer 1979), with differences due to the changes in the physics and outside exchanges mentioned earlier. The run begins with a phytoplankton bloom which peaks and then declines due to the onset of nutrient limitation, then declines due to flushing plus increasing zooplankton grazing pressure (Fig. 6.2). Increasing benthic regeneration supports additional primary productivity in the late spring even though nutrient stocks remain low. The summer is characterized by tightly coupled phyto-zooplankton cycles, which are suppressed during the fall as temperature and light decrease.

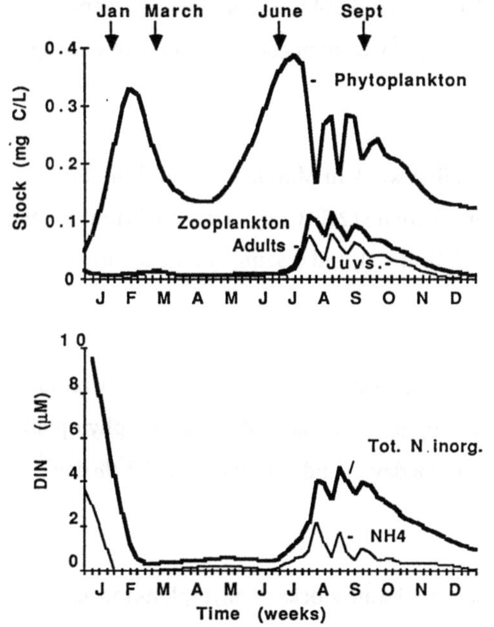

Figure 6.2. Results of a one-year run of the model for the major compartments of phytoplankton, zooplankton, ammonia, and total inorganic nitrogen. Snapshots of the flow network were saved for the four times indicated by the arrows.

The annual cycle may be seen as a sequence of different states of the system, each with a set of stocks and fluxes of different relative and absolute magnitude. In these different states, the ecosystem may appear to be near steady state, with strongly damped internal cycles, or it may exhibit stable oscillations, or other behaviors (Kremer 1979). Snapshots of the state during each season were used in the network analysis with information indices. For carbon and separately for nitrogen, average standing stocks and flows among the living and non-living compartments were saved in a computer file for specified time intervals. These files were then used for input to the network analysis programs.

Identification of 4 times of interest

Four intervals were chosen for subsequent analysis (Fig. 6.2). One week averages for stocks and flows were used except for the September case, when one month was averaged to span the period of the predator-prey cycles.

1. The first data set represents conditions early in the annual run, during the initial phytoplankton bloom. In the last week of January, standing stocks of diatoms are growing rapidly (net productivity over 7 % day^{-1}), nutrients are still abundant (4 μM total dissolved inorganic nitrogen), and zooplankton grazing is not a serious loss ($<$ 1 % day^{-1}), even in combination with grazing by the dormant benthos (1% day^{-1}).

2. About a month later, during the first week in March, nutrient limitation has terminated the bloom (0.3 μM N). Phytoplankton net productivity (2.5 % day^{-1}) is insufficient to meet losses due to benthic grazing(1 % day^{-1}), tidal flushing (2 % day^{-1}), and grazing by a small but growing population of juvenile zooplankton (1 % day^{-1}).

3. In the last week of June, warm temperatures (18°C), increasing irradiance, and added nutrients recycled from the benthos and zooplankton (0.6 μM N) allow phytoplankton to increase again. Flushing plus grazing by benthos (4 % day^{-1}) and zooplankton (2.5 % day^{-1}) cannot restrain the high net productivity (10 % day^{-1}).

4. During the month of September, moderate stocks of phytoplankton and zooplankton are oscillating out of phase, in classical predator-prey cycles. Nutrients are available (3 μM N), phytoplankton productivity (20-30 % day^{-1}), and grazing by zooplankton (15-20 % day^{-1}) and benthos (4 % day^{-1}) are all high.

For input to the network analysis software, flows were characterized as *exogenous inputs*, *exports*, *dissipations* (respiration), and *internal exchanges* (see Kay *et al.* 1989, this volume). The comparison of indices using flows of carbon vs. nitrogen to portray identical states of the system were based on important differences. Since the inorganic CO_2 pool was not modeled, respiration was seen as a dissipation and primary production as an input in the carbon case. For nitrogen, however, excretion was not viewed as dissipation (i.e. not lost from the system) so these flows became internal exchanges. Primary production was also an internal exchange, and real inputs by sewage and rivers took place.

My analysis was used to ask how important these arbitrary conventions were for the conclusions drawn from the information indices.

An important caveat in my analyses results from the fact that the benthos was not fully simulated in this version of the model. Thus the benthic stock remained constant, with biomasses based on estimates of macrofauna. Grazing and remineralization rates varied only with temperature.

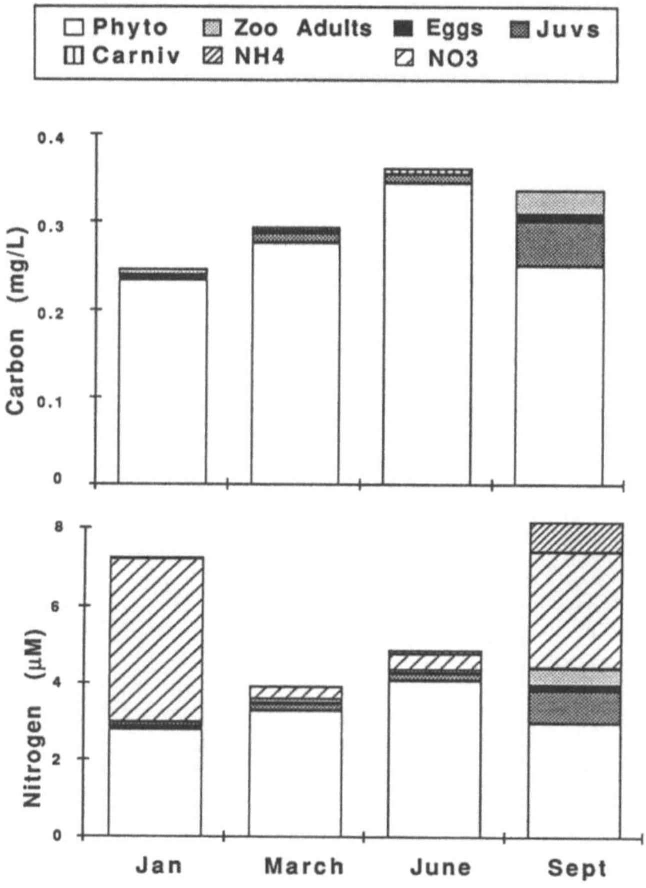

Figure 6.3. Distribution of standing stock during the four seasonal snapshots. *Top:* Carbon Biomass in the five dynamic biotic compartments. *Bottom:* Nitrogen stock in the seven dynamic living and nonliving compartments. Benthic stock was constant and is not shown.

Carbon Cases

Phytoplankton biomass was the dominant form of living carbon in the water column (Fig. 6.3) during all seasons, with a total biomass maximum in the June sample. Note that benthic stocks represented

1.4 mg C l^{-1} but were not modeled dynamically. These stocks and their related flows were included in the analysis, however they are not shown on the histograms as they are constant and dominate the figures. Zooplankton stocks were generally less than 5 % of the total, but reached nearly 30 % during the Fall cycles when adults and juveniles assumed almost equal importance. Because of the restricted area in lower Narragansett Bay represented in this version of the model, carnivores were unimportant, with small stocks of ctenophores and fish larvae present only in June.

In terms of carbon flows, algal production was the single input (Fig. 6.4 top). The 8-12 internal exchanges were relatively small except in September. Exports were never large, and respiration by the six living compartments increased in importance throughout the year.

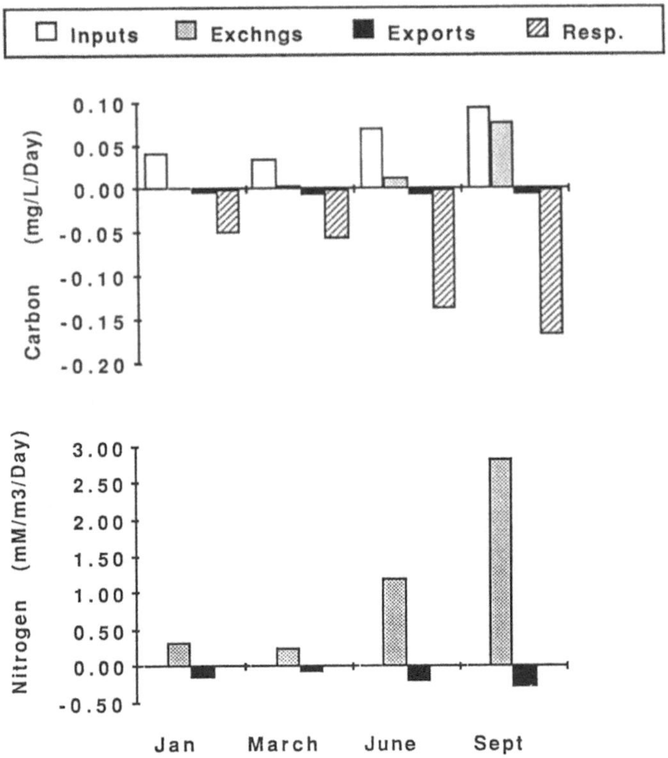

Figure 6.4. Distribution of flows in the network for the four seasons. *Top:* Carbon flows grouped into four categories. *Bottom:* Nitrogen flows.

Nitrogen Cases

Snapshots of the same four time intervals using nitrogen were different (Fig. 6.3). The underlying pattern of biomass seen in the carbon case was still visible, but inorganic nitrogen forms (NH_4 and

$NO_2 + NO_3$) exceeded the living stocks in the water column in January and September. Overall totals are greatest in September. Benthic nitrogen biomass (not shown) converts to 23.4 $\mu M\ N \cdot l^{-1}$.

For flows in the nitrogen case (Fig. 6.4 bottom), exogenous inputs of inorganic forms were small. There were no dissipations, as excretion is recycled. The bulk of the activity was in the 14-20 internal exchanges, which were low in January and March, higher in June, and reach a peak in September.

THE BEHAVIOR OF THE NETWORK INDICES

I find that the indices are easier to relate to when they are normalized to development capacity (Kay *et al.* 1989, this volume). Throughout this paper I have used the relative form of the indices, so that values of *ascendency* refer to A/C, *overhead* to O/C, *redundancy* to R/C, etc.

Carbon Case - 6 Compartments

Development capacity (C) depends on total system throughput and the diversity of flows. *C* for the model increased through the four seasons (Fig. 6.5 top). *Total system throughput (T)* accounted for much of the pattern, but the divergence of the 2 lines reveals that increased flow diversity was especially important in September.

Despite the fourfold change in *C* associated with significant changes in the community structure and function by conventional measures (Figs. 6.2-4) the *relative ascendency* index *(A/C)* was constant (Fig. 6.5 bottom). The relative flow of carbon to *respiration (Resp./C)* remained constant, and the other overhead costs *(exports(E/C) and redundancy(R/C))* were small.

Nitrogen Case - 10 Compartments

When the snapshot was taken in terms of nitrogen, the differences were not dramatic. The change in units is reflected in the absolute values of *C* and *T* (Fig. 6.6 top). C looked similar to the carbon case, but declined subtly in March, and increased a total of eight fold by September, twice the span of the carbon case. The relative contribution of *T* to *C* is similar, suggesting seasonal changes in the flow diversity of both carbon and nitrogen.

The overhead allocation is qualitatively different (Fig. 6.6 bottom). Average *ascendency* is unchanged, but A/C now showed some variation, with a peak in June. The max of A/C was attributable to decreased *redundancy*. Overhead associated with *exports* declined, and recycled *excretion* replaced *respiration*.

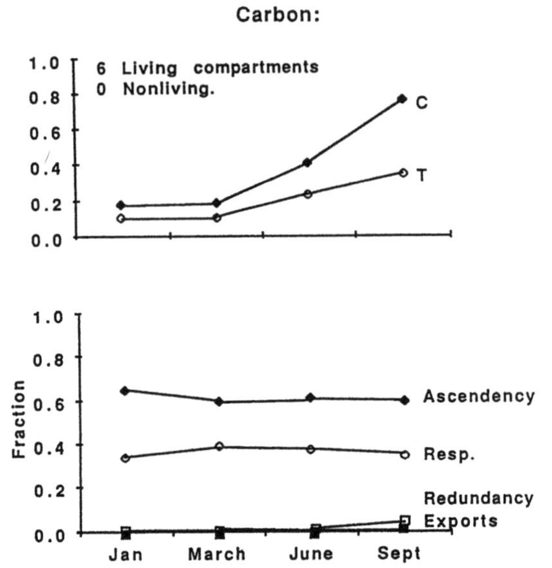

Figure 6.5. Throughput and information indices in four seasons for simulated carbon flows as fractions of Development Capacity (C). *Top:* The divergence of total system throughput (T) from development capacity (C) reflected changes in the diversity of carbon flows. *Bottom:* Yet, there were only subtle seasonal changes in the four factors contributing to C.

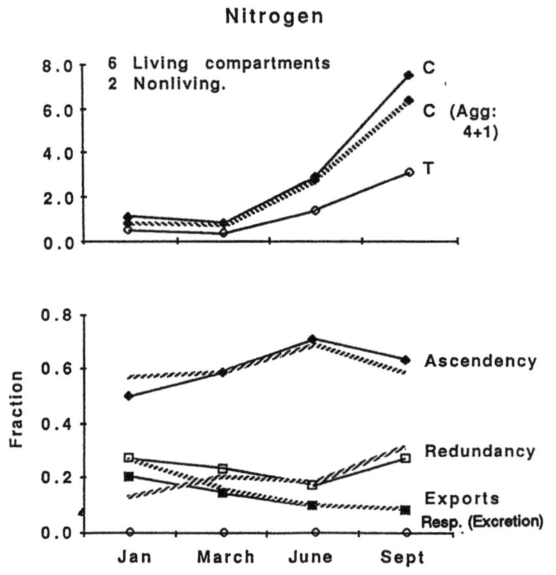

Figure 6.6. Throughput and information indices in four seasons for simulated nitrogen flows as fractions of Development Capacity (C). *Top:* The divergence of total system throughput (T) from development capacity (C) was qualitatively similar to the carbon case. *Bottom:* Though average ascendency was similar to the carbon case, its seasonality and the contribution of the other components of C were very different. Each line is paired with a dashed one for the network generated when nitrogen was aggregated from eight compartments into five.

Effect of Aggregation

Aggregation was most meaningful for nitrogen. Three stages of zooplankton and two of inorganic nitrogen were combined, for a total of four living and one non-living compartment, leaving *total system throughput* unchanged. The result was a small decrease in *development capacity*, most apparent in September, although the January decrease was by a greater proportion (Fig. 6.6 top).

Relative Ascendency in January was increased by a large decrease in *relative redundancy* (e.g. parallel paths from excretion to DIN to phytoplankton) despite higher export costs (Fig. 6.6 bottom). But *relative redundancy* increased in September when zooplankton populations were largest, even though parallel paths within zooplankton were hidden.

Interpretation

Ascendency was quite constant throughout all cases. A value of about 0.6 seems representative of the Narragansett Bay model. Unexpectedly, this value persisted throughout four seasons, considering flows of carbon or nitrogen, and despite aggregation. If this index varies markedly in other systems, my result suggest that the index may capture some fairly conservative property of the flow network. The question of the real dynamic range of *A*/*C* for a variety of systems needs to be pursued.

The minor changes in *A*/*C* were not readily interpretable. The peak in June for the nitrogen case (Fig. 6.6 bottom) does not coincide with extremes of stocks or total flow (Fig. 6.3, 6.4). The minimum redundancy in June, and the fact that *A*/*C* was unchanged by aggregation, point away from zooplankton and DIN. June is the time of greatest carnivore impact, and a significant number of flows are associated with this compartment. The carnivore flows were analogous in the carbon network, however, where *A*/*C* peaked in January, not in June.

The divergence of *development capacity* from *throughput* (Fig. 6.5 & 6.6 top) exposes the increased diversity (richness and evenness) of the network as the season progresses. This would not be readily apparent from the flow diagram, or even from the table of data without careful study. This effect is greatest for nitrogen, as expected from the richness of flows associated with recycling and the absence of respiration. Comparing *C* and *T* for carbon vs. nitrogen demonstrates that any insight about flow diversity depends on the absolute numerical value of flows, in this case on the units of measure. Aggregation also alters these patterns. I conclude that comparison of *C* for different systems using flows reported by different workers should be undertaken with caution. (An analogous problem has plagued interpretation of species diversity, where the statistic has been applied to different subsets of the organisms present in a community.)

Impressions drawn from patterns in the overhead factors also varied with carbon vs. nitrogen and with aggregation. The omission of respiration from the nitrogen network is the main culprit here, but the general effect is real, and could complicate comparisons attempted across different systems.

DISCUSSION

The tools of network analysis are admittedly most powerful with large and complex systems where the sheer size of the input/output flow matrix makes it difficult to comprehend. This is especially true with the cycle and trophic analyses. This model is fairly simple with so few compartments as to make the trophic cycle analyses marginally interesting. Thus I emphasized the information indices in this study. Application of these indices based on diversity theory to this idealized model system clearly revealed some of their strengths and weaknesses.

How would these indices be the most useful? They could summarize massive quantities of information. They could help reveal anomalous or interesting parts of the data. They could provide a basis for comparing systems, a basis that might be related to ecologically important factors, like productivity or stability. How well are these goals met in this test case?

Most of these potential benefits seem viable. Even in this modest model network with limited structural diversity, variations in the flow networks were hard to present concisely - to provide the complete information would have required four tables of 40 values, not much by "real network" standards, yet still resistant to easy inspection. The comparisons by seasons for the two elements did indeed summarize relevant features of the flow. Even as a novice in their interpretation, they led me to insights that I had not identified independently.

Comparisons of different systems, with measurements by different workers and with differing degrees of aggregation, may be more tenuous. Comparing the same system through time, or replicate experimental systems is probably the safest application. The apparent consistency of the ascendancy may be a notable exception worthy of further attention.

Whether these indices relate to grand ecological concepts remains to be seen. The theoretical foundation for ascendency suggests it is a measure of the competitive vigor of a network (Ulanowicz 1980; Hirata and Ulanowicz 1984). This is predicated on the postulate that the most fully developed and efficient configuration is one with maximum total diversity and no parallel paths. This assumption is not intuitively clearcut, and is a difficult idea to test. The general implications of species diversity have proved elusive (Odum 1983, p. 416 and refs. therein). We may find that the qualities of diverse flow networks or high ascendency are similarly hard to interpret.

REFERENCES

Field, J., F. Wulff & K. H. Mann 1989. The need to analyze ecological networks. In: Wulff, F., J.G. Field & K. H. Mann (eds) *Network Analysis in Marine Ecology.* Lecture Notes in Coastal and Estuarine Studies, Springer Verlag, New York.

Kay, J.J., L. Graham & R. E. Ulanowicz 1989. A detailed guide to network analysis. In: Wulff, F., J.G. Field & K. H. Mann (eds) *Network Analysis in Marine Ecology.* Lecture Notes in Coastal and Estuarine Studies, Springer Verlag, New York.

Hirata, H. & R .E. Ulanowicz 1984. Information theoretical analysis of ecological networks. Int. J. Systems Sci. **15**(3): 261- 270.

Kremer, J. N. & S.W. Nixon 1978. *A Coastal Marine Ecosystem -- Simulation and Analysis.* Springer-Verlag, New York. 217 p.

Kremer, J.N. 1979. An analysis of the stability characteristics of an estuarine ecosystem model, pp 189-206. In: Dame, R. (Ed) *Marsh-Estuarine Systems Simulation.* Univ. S. Carolina Press, Columbia.

Odum, E. P. 1983. *Basic Ecology.* Saunders College Publ., New York. 613 p.

Ulanowicz, R. E. 1980. *Growth and Development: Ecosystems Phenomenology.* Springer-Verlag, New York. 203 p

Wulff, F., J.G. Field & K. H. Mann (eds) 1989. *Network Analysis in Marine Ecology.* Lecture Notes in Coastal and Estuarine Studies, Springer Verlag, New York.

Chapter 7

NETWORK ANALYSIS OF SIMULATED SUCCESSION AFTER AN UPWELLING EVENT

J.G. Field, C.L. Moloney, C.G. Attwood
Marine Biology Research Institute
Zoology Department
University of Cape Town
Rondebosch 7700
South Africa

INTRODUCTION

Traditional approaches to ecological succession have viewed succession as a gradual change in community composition characterized by a progressive change in species structure, an increase in living and dead biomass and a tendency towards a balance between community production and respiration (Smith 1980). Odum (1969), in his classic paper on ecosystem development, identified 24 attributes of the total system which may be used to determine whether a system is in the early stages of development or approaching the mature, climax stage. In upwelling marine systems there is commonly a period of rapid upwelling of nutrients followed by a period of vertical stability, during which a plankton community may be seen to change from one showing characteristics of early development to one showing more maturity. We were interested in applying network analysis to the upwelling ecosystem at various stages of its development, to see how the indices changed.

In practice it is impossible to measure all the flows among all components of an ecosystem at time intervals short enough to follow succession in a plankton community. An alternative and more feasible approach is to develop a simulation model which mimics salient features of a plankton community, and to use the biomasses and flows generated by the model as input to a network analysis of the model plankton system. This is the approach adopted in the present chapter.

The Benguela upwelling system has been studied for a number of years by Shannon, Hutchings and co-workers (e.g. Andrews & Hutchings 1980, Shannon & Field 1985, Chapman & Shannon 1985, Shannon & Pillar 1986, Crawford *et al.* 1987). The southern and northern sub-systems are separated from one another by a large cold plume of almost perennial upwelling at about 26°S off southern Namibia. It has been established that the southern Benguela upwelling system has varying conditions in different geographical sub-regions:

a) a southern Agulhas region characterised by stable, stratified waters (spawning grounds of sardine and anchovy populations),

b) a central region of pulsed upwelling off the Cape Peninsula past which fish larvae are rapidly transported in the Benguela jet current over the edge of the narrow shelf, and

c) a northern region from 32-28°S characterized by steadier belt upwelling for much of the year, the recruitment grounds of the anchovy and sardine populations.

The maturing anchovy and sardine populations migrate southwards inshore in the southern Benguela region on their return to the spawning grounds at the southern tip of Africa. Field observations have shown varying biomasses and populations of phytoplankton, zooplankton and bacterioplankton after upwelling (Barlow 1982, Brown & Hutchings 1985, Verheye-Dua & Lucas 1988).

The fieldwork has been supplemented by experimental studies of phytoplankton, bacteria and bactivory in the laboratory (Lucas 1986, Lucas *et al.* 1986, 1987; Painting *et al.* in press). Experimental microcosm studies have demonstrated that there is a succession of microbial organisms after a phytoplankton bloom following upwelling of nutrient rich water. The microcosm experiments have been limited to small organisms and short time scales because of the small size of the containers (60 l). An even more serious limitation for studying the dynamics, is that it is impossible to measure all the flows of any nutrient or energy among all the compartments to analyse the time series of flows during successional change.

The simulation model used here was designed to investigate the succession observed in laboratory microcosm experiments and to explain the variability observed in the plankton community in the field. It is based on particle-size and simulates the biomasses and flows of carbon and nitrogen in a plankton community. The model has been validated using data obtained in fieldwork by Hutchings and his co-workers in a series of upwelling-monitoring cruises in the southern Benguela region (e.g. Brown & Hutchings 1985, Shannon & Pillar 1986, Armstrong *et al.* 1987) and experimental data of Lucas, Painting and their co-workers (e.g. Lucas 1986, Lucas *et al.* 1987, Painting *et al.* in press). The model has been kept as simple as possible in order to explore the relationships between changing flows, changing biomasses and the new and less familiar measures of network analysis. The simulation model is therefore not intended to be a fully realistic imitation of all the factors affecting the growth and decline of a plankton bloom, it is based simply on varying concentrations of nitrogen in the euphotic zone assuming temperature to remain constant at 10°C after an upwelling event and light to be non-limiting in the euphotic zone.

In this chapter, a simulation model of a developing plankton bloom after an upwelling event (Moloney 1988, Moloney& Field in press b) is used to provide a time series of nitrogen flows at daily intervals which are then analyzed using several network measures of inter-component relationships and of whole system structure. This chapter is therefore devoted to explore the usefulness of that measures such as Ascendency (A), Diversity (D) and Resource Niche Breadth (S), Dependency coefficients and Lindeman Spines (Ulanowicz (1980, 1984,1986); Kay 1984) in providing insight in the growth and development of a plankton community during a period of vertical stability.

THE SIMULATION MODEL

The model depicts a plankton community in the size range 0.2 μm - 2000 μm. The model has three main categories, autotrophs (phytoplankton), heterotrophs, and a non-living nitrogen pool. The living components are subdivided into size categories on the basis that the main trophic interactions in the pelagic community are size-based (Azam *et al.* 1983) and so are nutrient-uptake rates, respiration rates, and sinking rates. The derivation of the size-based rates for planktonic organisms is given in Moloney & Field (in press a). The size categories have been arranged on a scale of logarithms to the base ten (Table 7.1), giving two phytoplankton categories and four heterotroph categories from bacteria to meso-zooplankton. Table 7.1 also shows the assumed values of wet mass per particle (calculated from the geometric mean equivalent spherical diameter, e.s.d.), C:N ratios (Strickland 1960, Fenchel & Blackburn 1979, Stuart 1986), and the proportion of biomass that sinks out of the euphotic zone per day after the phytoplankton bloom peak.

Table 7.1. Living compartments in the model, showing geometric mean equivalent spherical diameter (esd), size range, wet mass, C:N ratio and sinking rate (S). (Note: 1 μm^3 = 1 pg)

Compartment	Mean esd (μm)	Range (μm)	WetMass (pg)	C:N ratio	S $\%d^{-1}$
Nano Phyto	6.32	2.0-20	132 x103	6.0	00.1
Net Phyto	63.2	20-200	132 x106	6.0	15.1
Bacteria	.632	.2-2.0	132	4.0	-
Zoo-Flagellates	6.32	2.0-20	132 x103	4.5	-
MicroZoopl.	63.2	20-200	132 x106	4.5	-
MesoZoopl.	632	200-2000	132 x109	4.5	-

The model is described in detail by Moloney (1988) and Moloney & Field (in press b) and is summarised here. Figure 7.1 gives a diagrammatic representation of the compartments and the nitrogen flows among the compartments, including sinking of phytoplankton and of faeces which are assumed to sink out of the euphotic zone rapidly and to be utilized below the euphotic zone, at or below the pycnocline. For simplicity, regenerated and new nitrogen are put in the same compartment, although a separate record is kept of each.

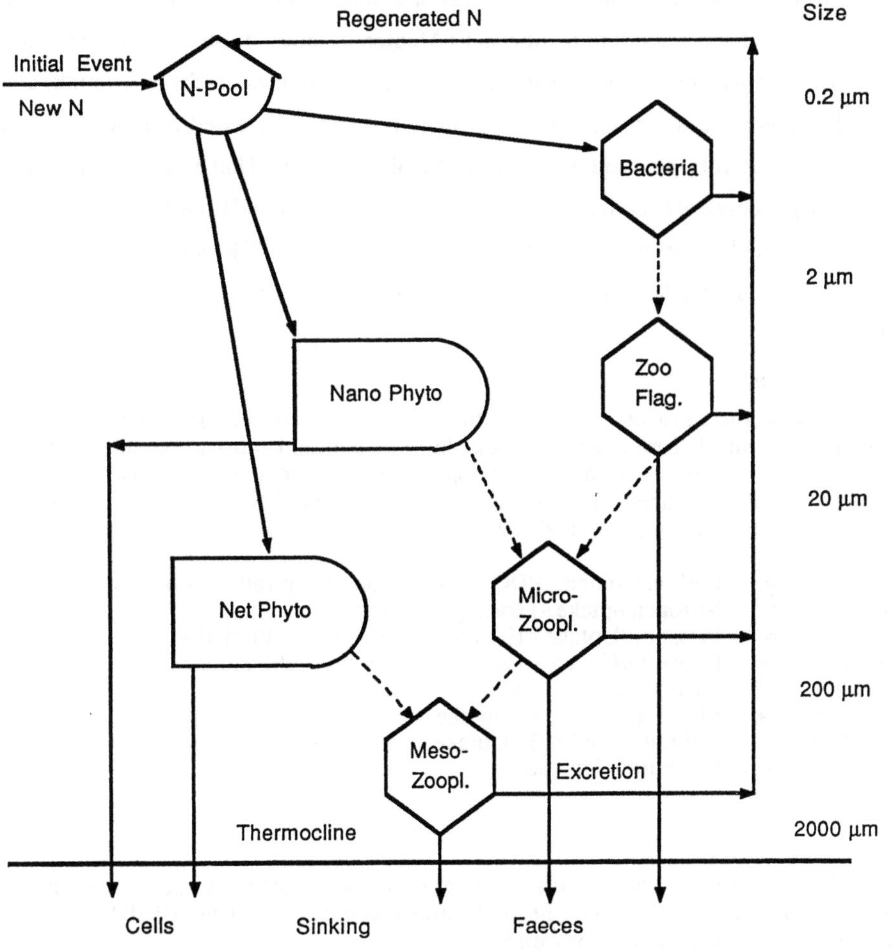

Figure 7.1. Flow diagram of the microplankton community used for the simulation and network analysis of changes after an upwelling event. The initial upwelling event adds new nitrogen to the nitrogen pool and flushes most of the plankton community out of the system. Organisms are arranged by size-class from top (0.2 μm) to bottom (2000 μm). There are two size classes of phytoplankton (bullets) and four of heterotrophs (hexagons). Arrows depict flows of nitrogen, carbon flows are not illustrated but are modeled. Broken arrows indicate predator-prey interactions. Vertically downward arrows show sinking out of the euphotic zone.

The model uses dual currencies of nitrogen and carbon. The main processes in the model are given by the equations in Table 7.2. which are solved numerically using a second-order Runge-Kutta method. Light is assumed to be non-limiting in the euphotic zone, and phytoplankton photosynthesis is limited by the supply of nitrogen only and modeled according to Michaelis-Menten kinetics (MacIsaac & Dugdale 1969). Carbon uptake is calculated to maintain a constant C:N ratio in phytoplankton. Bacteria are limited either by the supply of photosynthetically produced dissolved organic carbon (PDOC) or by the supply of dissolved nitrogen available from the nitrogen pool, whichever is the slower; both modeled according to Michaelis-Menten kinetics. Table 7.3 gives the values of half saturation constants for dissolved nutrient uptake and saturation values for both uptake and ingestion rates. Also shown are the proportions of primary production released as PDOC, the assimilation efficiencies of heterotrophs, and mass-specific respiration rates. Hall *et al.* (1976) were not able to show size-dependence of assimilation efficiencies, which are assumed to be 90% for all particle-feeding aquatic heterotrophs (Barthel 1983, Miller & Landry 1984) and 100% for bacteria. This does not imply that bacteria do not recycle nitrogen, since nitrogen excretion occurs after it has been assimilated into tissue.

Table 7.2. Equations describing changes in the main variables of the simulation model of a plankton community. PC and PN represent phytoplankton carbon and nitrogen respectively, HC and HN represent heterotroph carbon and nitrogen, DET represents detritus, PDOC represents photosynthetically produced dissolved organic carbon, NEWN and REGN represent new and regenerated nitrogen respectively and *j* denotes the size class of living components.

dPC_j/dt	=	Carbon fixation - PDOC Production - Respiration - Grazing - Sinking
dPN_j/dt	=	Nitrogen uptake - Grazing - Sinking
dHC_j/dt	=	Ingestion/Uptake - Egestion - Respiration - Predation
dHN_j/dt	=	Ingestion/Uptake - Egestion - Excretion - Predation
dDET/dt	=	Faeces egested - Sinking
dPDOC/dt	=	PDOC Production - Uptake
dNEWN/dt	=	(Initital upwelling) - Uptake
dREGN/dt	=	Excretion - Uptake

Table 7.3. Table of living components and rates of nitrogen uptake (V_{max} and K_s), ingestion (I_{max}), percent nitrogen assimilation (AE), percent extracellular carbon release (PER) and respiration (R), based on size-specific metabolic processes at 10°C.

Compartment	V_{max} d^{-1}	K_s $\mu gN\,l^{-1}$	I_{max} d^{-1}	PER %	AE %	R d^{-1}
Nano Phyto	1.49	.713	-	30	-	.487
Net Phyto	.265	66.28	-	30	-	.087
Bacteria	8.38	.00072	-	-	100	2.74
Zoo-Flagellates	-	-	22.3	-	90	4.30
Microzooplankton	-	-	3.97	-	90	.764
Mesozooplankton	-	-	.707	-	90	.136

Large heterotrophs obtain both carbon and nitrogen by ingesting, in this simple version of the model, only from the size class immediately smaller (Table 7.4), irrespective of whether it is autotrophic or heterotrophic. Table 7.4 also gives the half-saturation constants of the predator-prey interaction curves; these constants depend upon prey size (Moloney 1988). Predator size, on the other hand, determines the saturation levels of the model predator-prey interactions (see Table 7.3).

Table 7.4. Matrix of predator- prey interactions, giving Ks values (half saturation concentration, g C l^{-1}) for prey of a particular mean size.

	PREDATOR		
PREY	Flagellates	Microzoopl.	Mesozoopl.
Nano Phyto	-	64.5	-
Net Phyto	-	-	112
Bacteria	37.1	-	-
Flagellates	-	64.5	-
MicroZooplankton	-	-	112

Note: Ingestion rate is given by: I (μgC l^{-1} d^{-1}) = (I_{max})(Predator Biomass)(Prey Biomass)/ (Ks + Prey Biomass).

Changes in the model state variables (the biomasses and nitrogen pool concentrations) are given in Fig. 7.2. The model starts after a simulated upwelling event in which high concentrations of "new nitrogen" (25 μg at. nitrate-N l^{-1}) are brought to the euphotic zone, with very low levels of biomass in all living components except the net phytoplankton which consistently has standing stocks of some 50 mgC. m^{-3} (G. Pitcher, Sea Fisheries Research Institute, pers. comm). Nitrogen concentrations decline rapidly as nano- and net-phytoplankton build up to peaks on days 6-7. The phytoplankton standing stocks drop due to reduced production rates in response to a decline in nitrogen concentration, coupled with metabolic losses and increased grazing and sinking (Fig. 7.2. a, b). The nitrogen pool is slightly topped up by regenerated nitrogen on days 8-12. Bacteria peak on day 7. The bactivorous flagellates remain at low biomasses, but increase after the bacteria and keep the bacterial populations in check, resulting in bacterial decline from day 8 (Fig. 7.2b). Fig. 7.2c shows a succession of major predators peaking after their prey: nano-phytoplankton peak on day 8, followed by a peak of micro-zooplankton on day 9 which declines gradually as the meso-zooplankton build up to a peak on days 12-25. Note that the micro-zooplankton also prevent the bactivorous flagellates from becoming abundant (Fig. 7.2. b,c).

The model system gradually runs out of nitrogen as sinking of net- phytoplankton and faeces removes nitrogen from the euphotic zone. The model was run for 40 days, although it is unlikely that there would be neither upwelling to the surface (induced by local winds) nor mixing (caused by distant or local storms) for longer than 20 days. During the spring-summer upwelling season off the Cape Peninsula,

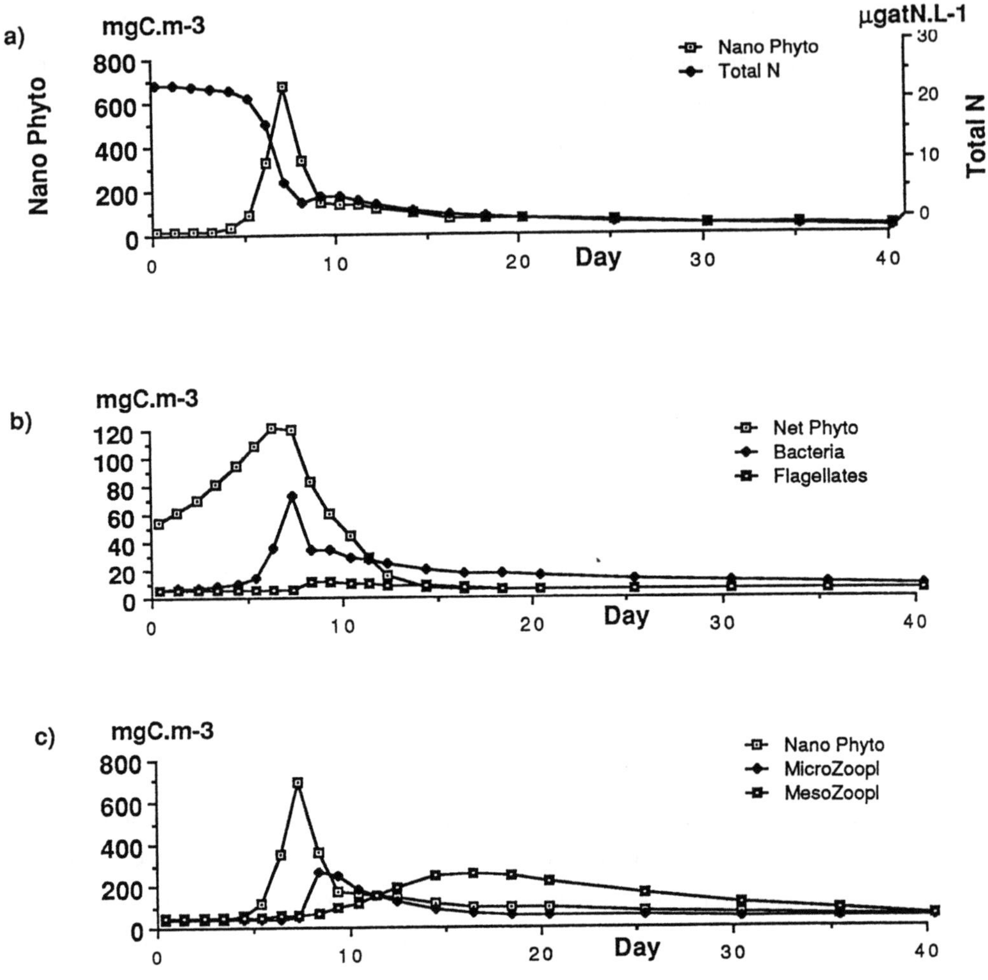

Figure 7.2. Simulation model output showing standing stocks of the components of the model from the starting values (day 0) to day 40 of the simulation.

a) Total nitrogen concentration in the nitrogen pool (μgat-N.l^{-1}), nano-phytoplankton (mgC.m^{-3})

b) Net-phytoplankton, bacteria, zoo-flagellates

c) Micro-zooplankton, meso-zooplankton, nano-phytoplankton.

the pattern is likely to be repeated after upwelling events at a frequency of 5-20 days (Wulff & Field 1983, Brown & Hutchings 1985), so that meso-zooplankton would not often get an opportunity to develop large populations in the nearshore areas. In spite of excluding larger predators such as euphausids and other macro-zooplankton, anchovy, sardines and other planktivorous pelagic school fish, the model gives results which are consistent with both field observations and microcosm experiments in both time-scale and standing stocks of the components. In short, nitrogen concentrations decline rapidly 5 days after upwelling, with nano- and net-phytoplankton biomass peaking on days 6-7, bacteria on day 7, zoo-flagellates and micro-zooplankton on days 8-10 and meso-zooplankton on days 12-25.

Table 7.5. Exchange matrices for days 1 and 8 showing flows of nitrogen from rows to columns (mg N $m^{-3}day^{-1}$), as output from the simulation model. Corresponding biomasses of the components of the model are shown in Fig. 7.2.

DAY 1

FROM: \ TO:	Nano Phyto	Net Phyto	Bacteria	Flagellates	MicroZpl.	MesoZpl.	N-Pool	Export
NanoPhyto	-	-	-	-	.000519	-	-	-
NetPhyto	-	-	-	-	-	.210	-	-
Bacteria	-	-	-	-	-	-	.000383	-
Flagellates	-	-	-	-	.000301	-	.0249	.000153
MicroZpl.	-	-	-	-	-	.00300	.0983	.000447
MesoZpl.	-	-	-	-	-	-	.174	.127
N-Pool	.0249	1.87	.556	-	-	-	-	-

DAY 8

FROM: \ TO:	Nano Phyto	Net Phyto	Bacteria	Flagellates	MicroZpl.	MesoZpl.	N-Pool	Export
NanoPhyto	-	-	-	-	38.1	-	-	.341
NetPhyto	-	-	-	-	-	1.03	-	14.3
Bacteria	-	-	-	8.30	-	-	31.9	-
Flagellates	-	-	-	-	6.71	-	5.22	3.32
MicroZpl.	-	-	-	-	-	1.09	13.9	23.2
MesoZpl.	-	-	-	-	-	-	.754	1.10
N-Pool	48.1	.827	33.7	-	-	-	-	-

As biomasses increased and decreased, the flows of nitrogen among components of the simulation model were integrated each day and saved for network analysis. Table 7.5 gives two examples of flows among the components of the model and exported by sinking through the thermocline. Day 1, immediately after the upwelling event, has very small flows, except to net-phytoplankton, because all the biomasses are so small in the newly upwelled water. On day 8, when nano-phytoplankton are at their peak, there are large flows amongst several components of the model reflecting the rapid dynamics and changing biomasses seen in Fig 7.2. Matrices such as those of Table 7.5 formed the input to the network analyses described below.

NETWORK ANALYSIS OF THE MODEL OUTPUT

Flows among the model compartments on different days of the simulation were saved as matrices and analysed following Kay *et al.* (1989, this volume Chap. 2) using Ulanowicz' program NETWRK and the program of Kay (1984) ECOSYSII. The main measures of whole system network properties calculated by Ulanowicz are:

The sum of all the individual compartment throughputs, the **Total System Throughput**,

$$T = \Sigma T_j \qquad (1)$$

The proportion of T that is due to cycling T_c is the **Finn Cycling Index**,

$$ci = T_c / T \qquad (2)$$

The diversity of various throughputs can be considered a measure of network complexity, formulated as $(-\Sigma Q_i \log Q_i)$, where Q_i is the fraction of T contributed by each compartment, i. The product of size and complexity is the **Development Capacity**,

$$C = -(T)\, \Sigma\, Q_i \log Q_i \qquad (3)$$

The average mutual sustenance of the system, **Ascendency**, is given by:

$$A = (T)\, \Sigma\Sigma\, f_{ij}\, Q_i \log (f_{ij} / Q_j') \qquad (4)$$

where f_{ij} is that proportion of flow leaving compartment i that directly contributes to compartment j. C serves as an upper bound to A.

In addition Kay (1984) calculates the flow resource niche breadth and flow diversity, restricting the system to living components only, and in the present analysis, omitting the nitrogen pool from calculations. **Flow resource niche breadth** is calculated as:

$$S = \Sigma - (T_j / (T)\, \Sigma\, (T_{ij} / T_j) \log (T_{ij} / T_j) \qquad (5)$$

and **flow diversity** as:

$$D = \Sigma - [T_j/(T)] \log [T_j/(T)] \qquad (6)$$

where T is total system throughput, T_j the throughput of compartment j, and T_{ij} the flow from i to j

Relationships among components

Outflows

Summed nitrogen flows out of each of the model components given by the program ECOSYSII are displayed in Figure 7.3. These are related to the standing stocks shown in Fig. 7.2 above, but give an indication of the dynamics of the system as opposed to a "snapshot" of the various stores in each component. Fig. 7.3a shows that the N-pool of freshly upwelled new nitrogen has little outflow until the populations of phyto-plankton and bacteria build up four days after the upwelling event (see Fig. 7.2). The N-pool outflows decline rapidly after day 8 but the decline is arrested as regenerated nitrogen is added to the pool by heterotroph excretion. The nano-phytoplankton outflows lag behind the nitrogen pool by 1-2 days and closely follow them.

Fig. 7.3b shows the net-phytoplankton outflows reaching a peak on day 8 as its standing stock declines. Likewise the bacterial outflows peak on day 8, since bacteria depend upon PDOC exuded by the nano- and net-phytoplankton. Bacterial flows peak to the same level as those of nano-phytoplankton although bacterial standing stocks are an order of magnitude smaller than nano-phytoplankton (see Fig. 7.2). Outflows from zoo-flagellates, which prey exclusively upon bacteria, lag behind the bacterial flows.

Fig. 7.3c displays the large micro-zooplankton outflows in relation to nano-phytoplankton and meso-zooplankton. The micro-zooplankton depend on inflows from both nano-phytoplankton and zoo-flagellates (both in the 2-20 μm size range). Note that although the micro-zooplankton standing stocks are lower than those of the meso-zooplankton from day 12 (see Fig. 7.2), the summed nitrogen outflows from micro-zooplankton to all other components remain higher, owing to their faster metabolic rates. Insights into dynamics such as these are not given by conventional measurements or model outputs of biomasses.

Indirect diets

Ulanowicz's program NETWRK gives a matrix of **total dependency coefficients** which describe the extent to which each component depends upon every other component for its supply of the nutrient being modeled, either by direct, indirect or recycled flows. Some examples are presented graphically in Figs 7.4 and 7.5.

Figure 7.4a shows that bacteria depend almost 100% upon the nitrogen pool for their supply of nitrogen, since all nitrogen is recycled via the N-pool. On day 8 almost 100% of the nitrogen used by bacteria is also recycled via micro-zooplankton, but this tapers off later in the simulation as more nitrogen is cycled via meso-zooplankton. Fig. 7.3b depicts the small contribution of net-phytoplankton to bacterial nitrogen nutrition, which declines from an initial value of some 6%, to zero by day 12. The contribution by both classes of phytoplankton is indirect, occurring via grazers which excrete nitrogen into the dissolved pool. Bacterial dependence on nitrogen cycled via zoo-flagellates increases from day 7 to 9, then gradually declines, and dependence on nano-phytoplankton peaks to 55% on day 9, before it too, declines.

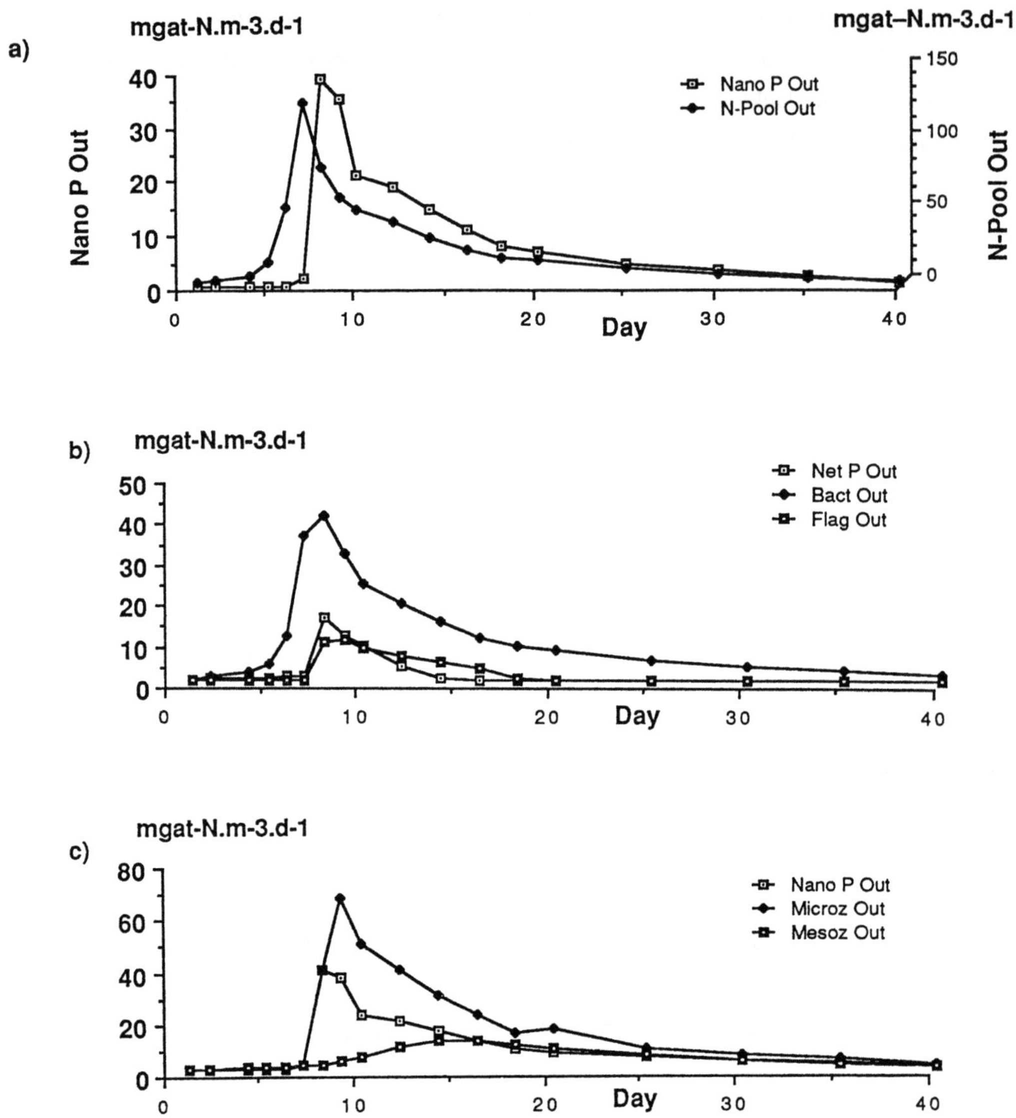

Figure 7.3. Network analysis results: summed outflows of nitrogen (mgat-N.m^{-3}.day^{-1}) from each compartment a) from the nitrogen pool, nano-phytoplankton b) from bacteria, zoo-flagellates, net-phytoplankton c) from nano-phytopllankton, micro-zooplankton, meso-zooplankton.

Fig. 7.4 c,d illustrates zoo-flagellate dependence on other components. These can be seen to change considerably during the succession, with all flows coming via the nitrogen pool (and bacteria) on days 6-7, micro-zooplankton contributing over 80% on day 9 and meso-zooplankton contributing considerably after about day 15. The dependence of flagellates upon bacteria is the same as their dependence upon the N-pool since they both reflect the same pathway of nitrogen - bacteria-flagellates. Similarly, the flagellate dependence upon nano-phytoplankton follows the same trends as that upon micro-zooplankton (Fig. 7.4 c,d).

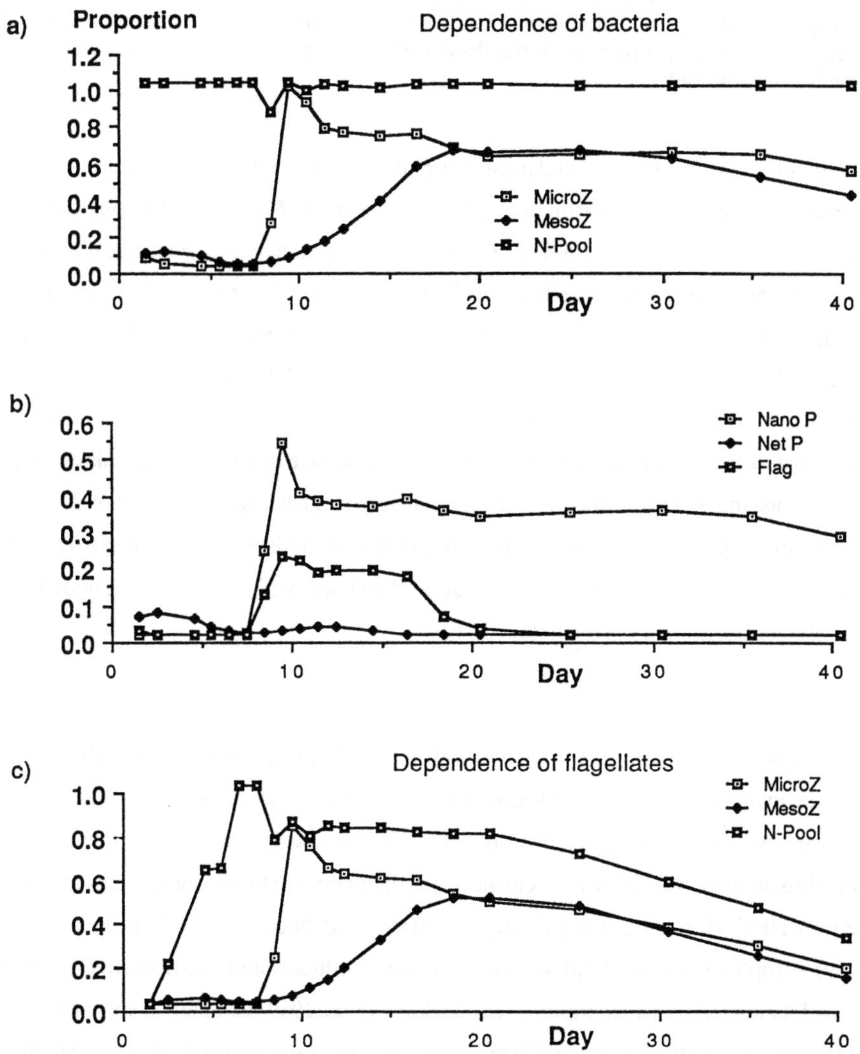

d)

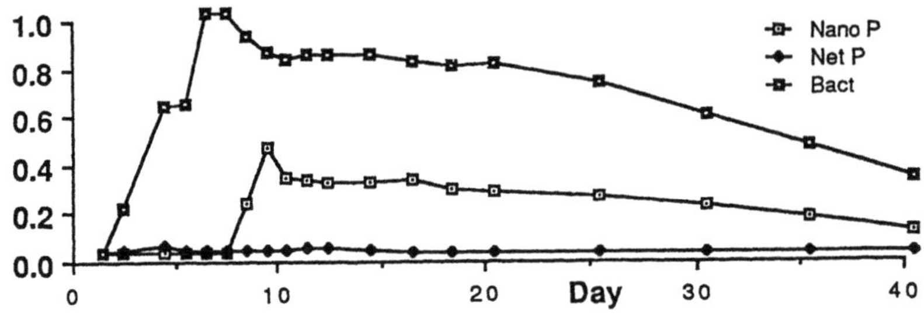

Figure 7.4. Network analysis results: graphical presentation of the dependency of bacteria (a, b) and zoo-flagellates (c, d) upon the flows from other components of the model system. The vertical axis gives the proportion of the flows through bacteria (or flagellates) which come via each compartment.

Fig. 7.5 depicts the indirect diets of micro- and meso-zooplanktonankton in the model succession. The nitrogen requirements of micro-zooplankton depend heavily upon the N-pool to nano-phytoplankton pathway, peaking at about 100% on days 6-8 (Fig. 7.5 a,b). Dependence upon bacteria is at a lower level, peaking at some 60% on day 8. Flagellates peak at 20% on day 8, then are of less importance, and net-phytoplankton are an unimportant indirect source of nitrogen for micro-zooplankton throughout. Fig. 7.5b,c illustrates that meso-zooplankton depend largely upon the N-pool to net-phytoplankton nitrogen route peaking at day 7, and then switch to a pathway via nano-phytoplankton and micro-zooplankton. The indirect dependence on micro-zooplankton reaches a peak on day 14 as meso-zooplankton standing stock is growing fastest (see Fig. 7.2). Meso-zooplankton are surprisingly dependent upon bacteria as an indirect source of nitrogen, over 20% dependent from days 2-20 and less dependent upon flagellates which contribute about 10% of the meso-zooplankton nitrogen supply from days 8-16.

Effective Trophic Position

Ulanowicz (1986) describes the mapping of a food web into a simplified "**Lindeman Spine**" or food chain. Each of the living components is allocated an "effective trophic position" relative to the non-living components with an arbitrary position of 1. In Ulanowicz's examples, the non-living component is detritus, in this case it is the nitrogen pool. Perhaps because this is a nearly closed nitrogen cycle, the program NETWRK gives the primary producers and bacteria (which utilize dissolved nitrogen directly) a trophic position of 2.0, whereas primary producers and detritus are all allocated to position 1.0 in Ulanowicz's own examples (see Ducklow *et al.* 1989, Chapter 8, this volume, for explanation of trophic "jump" when f-ratio of "new" production to total production is very small.)

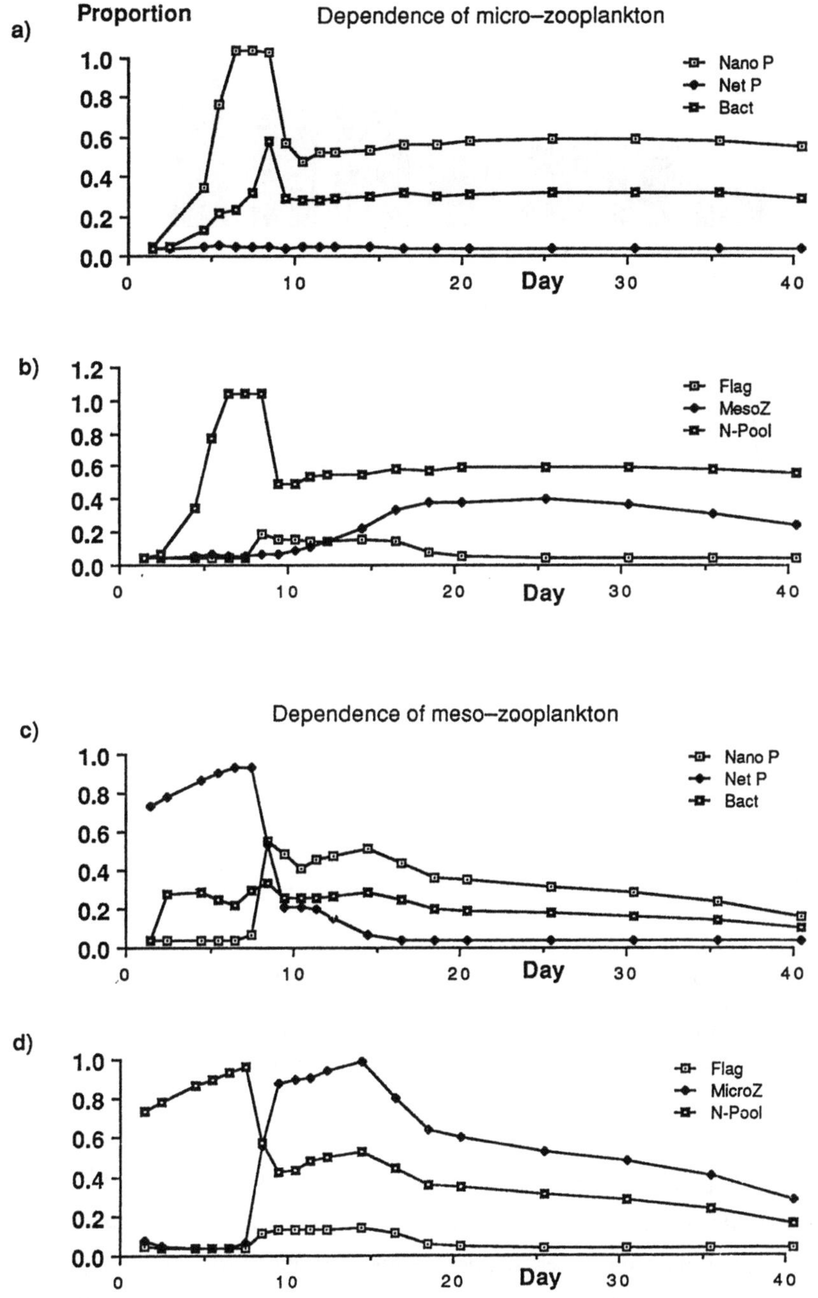

Figure 7.5. Network analysis results: graphical presentation of the dependency of micro-zooplankton (a, b) and meso-zooplankton (c, d) upon the flows from other components of the model system. The vertical axis gives the proportion of the flows through micro-zooplankton (or meso-zooplankton) which come via each compartment.

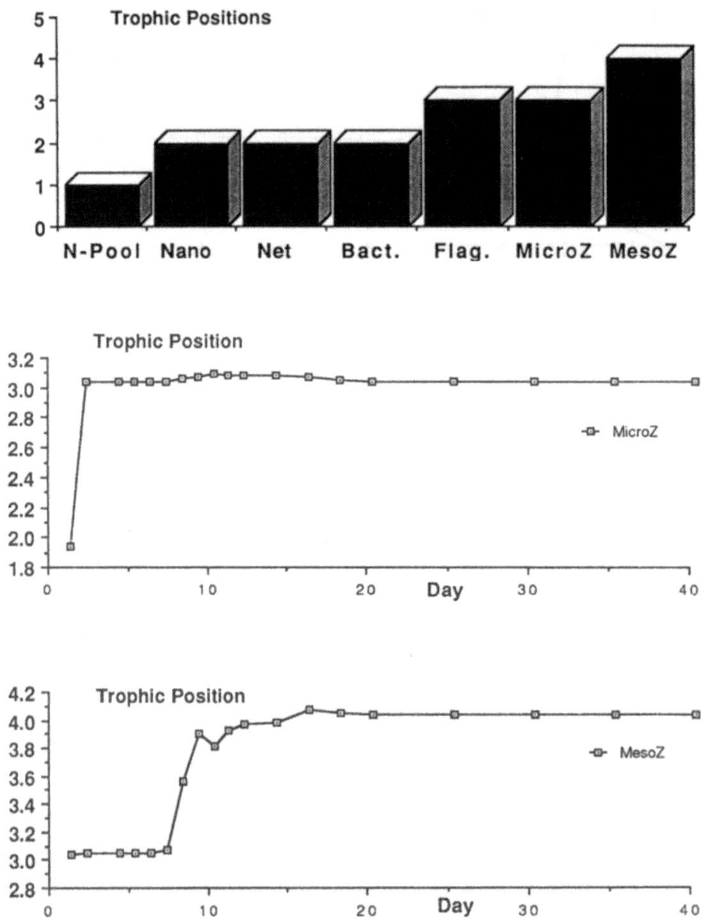

Figure 7.6. Effective trophic positions of the model components, relative to the nitrogen pool which is allocated an arbitrary position of 1.0. The top panel shows all components at day 40 at the end of the simulation. Middle and bottom panels depict the changes in effective trophic position of micro-zooplankton and meso-zooplankton, respectively, during the simulated succession.

The output is presented graphically in Fig. 7.6. The top panel shows trophic positions of all seven model components on day 40 at the end of the simulation. Most of the components do not change position during the simulation of plankton succession, but the lower panels show that the trophic position of micro-zooplankton does change dramatically from day 1 to day 2 as bacteria and zoo-flagellates enter the food web leading to micro-zooplankton. Similarly, the meso-zooplankton change trophic position as the net-phytoplankton population declines after day 7, and meso-zooplankton rely increasingly on micro-zooplankton for food.

Figures 7.7 and 7.8 show a selection of Lindeman Spines at different stages of the succession. On day 1 the Lindeman Spine consists of only four trophic categories with small flows of nitrogen between the categories and very little recycling. Category I consists only of the N-pool component, with 0.236 mgat-N.m^{-3}day^{-1} recycling into the pool. Category II consists of nano- and net-phytoplankton and bacteria, which are found only in that category (100%). The inflow to category III divided by the inflow to category II gives the trophic efficiency of category II (8.5%). The low efficiency is explained by the fact that the model system is not in steady state, and nitrogen is accumulating in the expanding standing stocks of phytoplankton and bacteria at this early stage of the succession. Nitrogen is being immobilized into particulate organic matter. The lack of an upward export arrow from category II is due to the absence of sinking phytoplankton early in the bloom. Category III contains micro- and

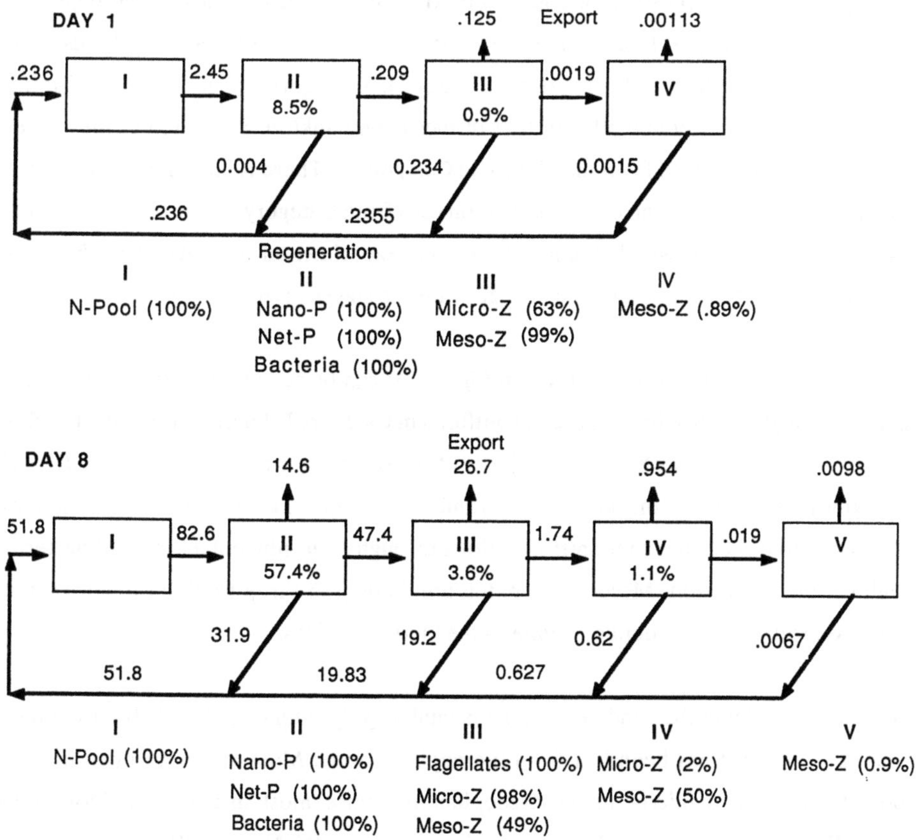

Figure 7.7. Lindeman Spines depicting hypothetical trophic categories (boxes numbered in roman numerals). Arrows indicate flows of nitrogen (mgat-N.m^{-3}.day^{-1}), with regenerated nitrogen recycling below the boxes. The trophic efficiency of grazing transfer through each trophic category along the spine is given as a percentage inside the boxes. Vertical arrows indicate export. Model components are listed below the categories with the percentage of their

meso-zooplankton; it shows small exports as faeces and a low efficiency because the stocks are growing. Category IV in the food chain consists only of meso-zooplankton, with only 0.9% of the meso-zooplankton flows falling into this category. The fourth category is minute because so little of the meso-zooplankton nutrition is dependent upon micro-zooplankton at this stage. Flagellates do not appear in the Lindeman Spine on day 1 because nitrogen flows resulting from upwelling have not yet reached them.

On day 8 of the succession the Lindeman Spine looks quite different. Flows from category to category are larger by orders of magnitude. Category I is shrinking in standing stock with outflows exceeding inflows. Category II contributes largely to recycling nitrogen (via bacteria), and to the through-flows of nitrogen (via phytoplankton). The through-flow contribution is indicated by the 57% efficiency, as the phytoplankton standing stocks are grazed down. Category III includes all the flagellate activity, most of the micro-zooplankton and half of the meso-zooplankton activity. It also contributes considerably to recycling (19.2 mgat-N m^{-3} day^{-1}). Category IV is small, including 2% of micro-zooplankton activity and 50% of that of meso-zooplankton. Some 1% of meso-zooplankton activity is five steps removed from the N-pool in Category V. There is considerable export by sinking of faeces from Category III, and of phytoplankton cells from Category II, both depicted by the upward arrows. This analysis suggests that, contrary to most experimental measurements, bacteria may be important as regenerators of inorganic nitrogen at some stages in succession.

The Lindeman Spine on day 10 is shown in Fig. 7.8. It can be seen that recycling is building up the nitrogen pool slightly, since inflows exceed outflows in Category I. There is less export of sinking cells from category II than on day 8 and the majority of the recycling is now from category III, as is export from sinking faeces. More of the micro- (4.6%) and meso-zooplankton (79%) activity is now in category IV, compared to day 8, since more nitrogen flows through nano-phytoplankton and bacteria and less through net-phytoplankton. Meso-zooplankton are feeding in three trophic categories, and overall there is less export from the euphotic zone as the bloom subsides.

On day 20 of the bloom the Lindeman Spine is qualitatively similar to day 10, but the flows have all decreased as nitrogen has been lost from the euphotic zone. Most of the nitrogen flow is through nano-phytoplankton to micro- to meso-zooplankton, since most microzooplankton activity is in category III, and most meso-zooplankton activity in category IV. Nearly half the nitrogen recycling is by bacteria (category II), a fifth by micro-zooplankton and flagellates (category III) and nearly a third by meso-zooplankton (category IV). Category III is more efficient in the food chain, but since the three outflows considerably exceed the inflows, the standing stocks in the category can be seen to be decreasing.

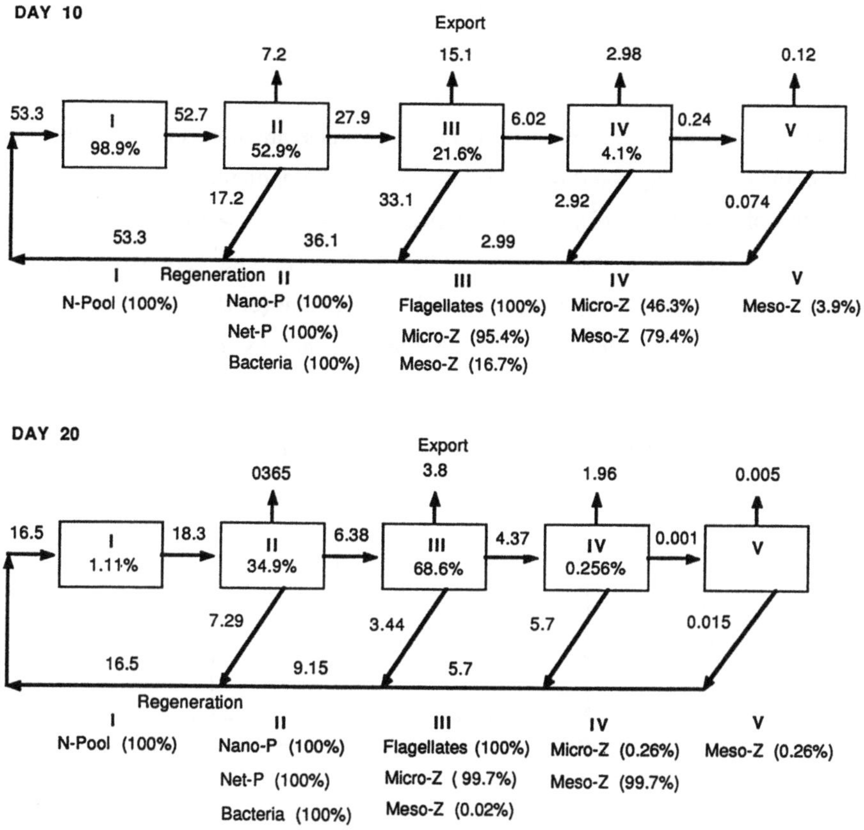

Figure 7.8. Lindeman Spines depicting hypothetical trophic categories (boxes numbered in roman numerals). Arrows indicate flows of nitrogen (mgat-N.m^{-3}.day^{-1}), with regenerated nitrogen recycling below the boxes. The trophic efficiency of grazing transfer through each trophic category along the spine is given as a percentage inside the boxes. Vertical arrows indicate export. Model components are listed below the categories with the percentage of their activity in each hypothetical trophic category. The Spine for day 10 of the simulated succession is shown in the upper panel, and that for day 20 in the lower panel.

Global measures of ecosystem organization

According to the hypotheses of Ulanowicz (1980, 1986) and Kay (1984), as an ecosystem matures in succession it will tend towards a configuration of flows in which its components are least ambiguously linked, i.e. there is little uncertainty as to which component is feeding on which. If stressed, the system will be forced back to a configuration of greater connectance in which the fate of a unit of flow is less certain. This is consistent with traditional concepts of resource niches - the system most likely to evolve is that which shows greatest specialization and hence its components have the narrowest niches (Margalef 1968; E.P. Odum 1969). The process of community succession manifests itself in two ways:

firstly, species diversity tends to increase resulting in a more even and diverse distribution of resources, and secondly, each species will feed upon fewer other components as its diet becomes more specialized.

Kay (1984) measures these tendencies using D, the evenness and diversity over which resources are used, and S, the uncertainty about which resource a living component is using as food. An ecosystem shows positive organization if $dD/dt > 0$ and $dS/dt < 0$, where $D > S > 0$. Ulanowicz suggests that the state of an ecosystem is characterized by the product (termed Ascendency, A) of its size (as measured by Total System Throughput, T) and development (or the average mutual sustenance of its components). As an ecosystem grows and develops Ascendency should increase.

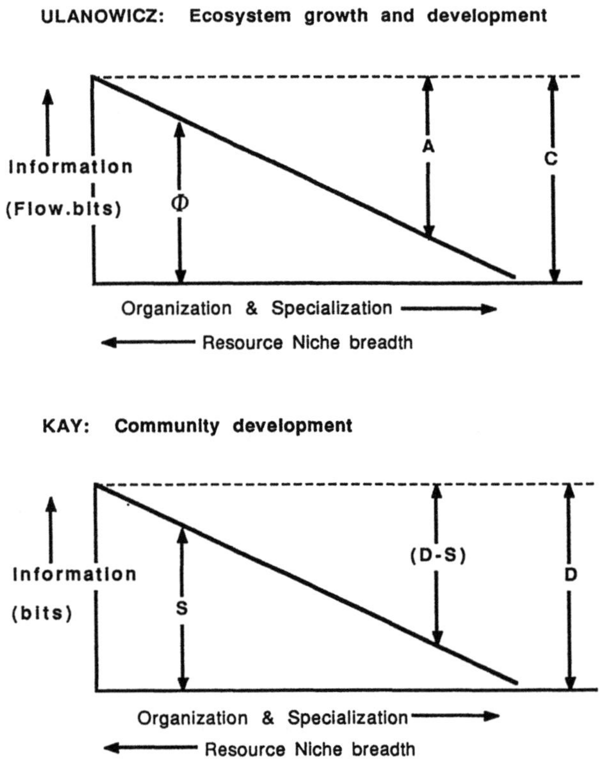

Figure 7.9. Diagrammatic comparison of the information theory measures of ecosystem growth and development proposed by Ulanowicz (1980, 1986) and Kay (1984). The upper panel shows the measures proposed by Ulanowicz: Development Capacity (C), Ascendency (A) and Overheads (Φ) with units of flow.bits. The lower panel depicts Kay's corresponding measures of flow diversity and evenness (D) and resource niche breadth (S) based on the living components only. The difference (D-S) is also shown for comparison with A above. The x-axes show increasing organization and specialization from left to right. In both hypothetical cases the structural uncertainty of the flow network (C and D) is assumed to remain constant, i.e. the network topology does not change.

Figure 7.9 provides a graphical comparison between the measures of whole system functioning proposed by Ulanowicz (1980, 1986) and Kay (1984). It should immediately be noted that Ulanowicz' measures include all components in the system, whereas Kay deliberately confines his measures to flows among the living components of an ecosystem, omitting the nitrogen pool in the present case. A second fundamental difference is that Ulanowicz is concerned with growth and development, and therefore scales his measures by the total system throughput, T, whereas Kay is only concerned with development and his measures are independent of T.

The structure (or topology) of a particular system determines its potential to develop and to attain self-organization. The maximum potential evenness and diversity of flows is termed Development Capacity (C) by Ulanowicz, and corresponds to the evenness and diversity with which resources are used by the living components, which Kay terms Flow Diversity (D). These are shown by broken lines in Fig. 7.9. The lines are parallel to the x-axis because the structure of the system is assumed to be fixed, i.e. flows are spread evenly among all components.

Ulanowicz (1986) shows that Ascendency (A) is a measure of the total uncertainty of a particular ecosystem structure (C) that is resolved by knowledge of the flow structure. The difference between C and A is the residual uncertainty or ambiguity of connectance, termed the Overhead (Φ). The overhead corresponds to Kay's average resource niche breadth (S). Ascendency corresponds to (D-S). It can be seen from Fig. 7.9 that if the structure of the system is fixed, then Ascendency should increase as a system becomes more organized in succession, as should (D-S). Conversely, both the uncertainty measured by Overheads (Φ) and resource niche breadth (S) should decrease with increasing resource specialization.

The simulation of succession in a plankton community provides an opportunity to observe the changes in various measures of whole system behaviour. This is explored below.

Fig. 7.10a shows the changes in Finn Cycling Index (CI) as succession proceeds, juxtaposed against the microzooplankton standing stock as a reference. The CI is an index of the amount of flow that is recycling in relation to the total throughflow. It increases initially, then decreases as the total throughflow speeds up during the bloom, and increases again from day 6-9 as heterotroph populations become active in recycling nitrogen. The CI remains at a high level of about 65-70% for the remainder of the succession.

Ulanowicz's measures are shown together in Fig. 7.10b. The dependence of Capacity and Ascendency upon T is clear, and all three measures peak on day 8 of the succession, and thereafter decline, as the system runs out of nitrogen. Fig. 7.10c depicts resource niche breadth (S) and flow diversity (D). Both increase from days 7-12, thereafter flow diversity (D) decreases gently, suggesting fairly even flow distribution among the compartments. In contrast, resource niche breadth (S) decreases rapidly from days 12-20 indicating much greater specialization of diets in the plankton community. This apparent paradox occurs because the relative proportions of compartmental throughflows remain similar (D nearly constant), but the exchanges become more concentrated in fewer linkages (S decreases).

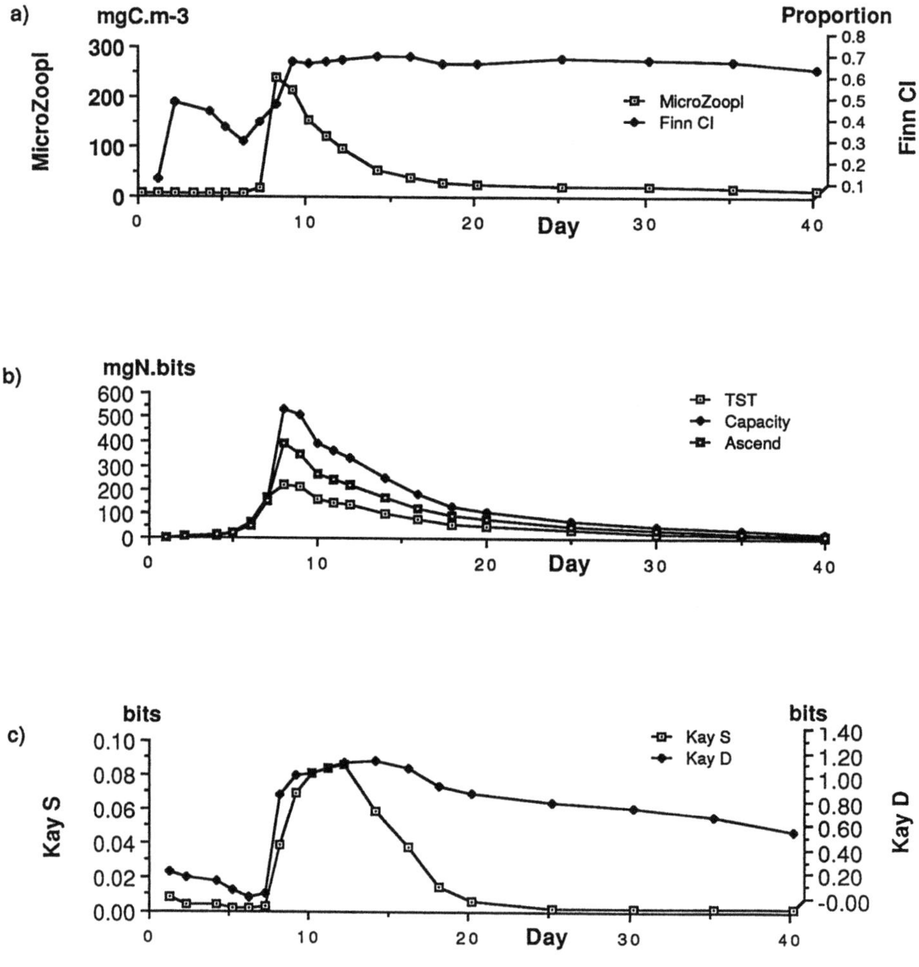

Figure 7.10. Graphical presentation of changes in global measures of ecosystem organization during simulated succession after upwelling. (a) Finn's cycling index (CI) with micro-zooplankton standing stock ($mgC.m^{-3}$) as a reference for succession. (b) Total System Throughput ($mgat\text{-}N.m^{-3}.day^{-1}$), Ascendency and Development Capacity (both in mgat-N.bits). (c) Kay's measures of flow diversity (D) and resource niche breadth (S) (bits).

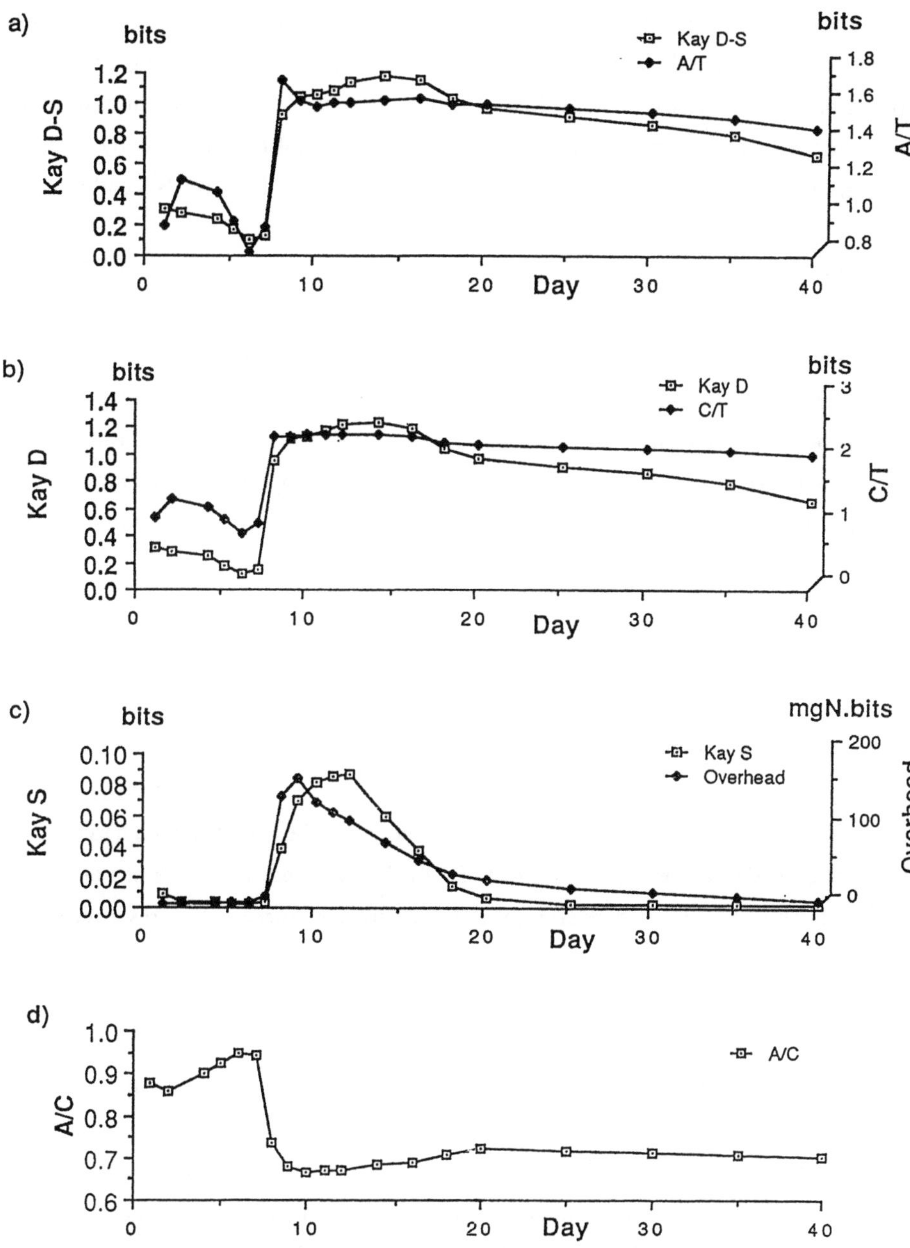

Figure 7.11. Comparison of global measures of ecosystem organization during simulated succession after upwelling. (a) Kay's measures (D-S) compared with Ulanowicz's (A/T) both in bits. (b) Kay's diversity (D) compared with Development Capacity (C) divided by Total System Throughput (T) (bits). (c) Ulanowicz's Overheads (mgat-N.bits) compared with Kay's resource niche breadth (S) (bits). (d) Relative Ascendency (A/C).

The Ulanowicz and Kay measures of whole system function are compared in Fig. 7.11. In the top panel, the effect of flow magnitude is removed from Ascendency by dividing by throughput (T) and this is compared with the difference between Kay's D and S (see Fig. 7.9). Both functions decrease to days 6-7 and then increase steeply at day 8, before leveling off and declining gently. Presumably the differences between the graphs are due to exclusion of the N-Pool from Kay's S and D.

In similar fashion, the structural components are compared between Development Capacity divided by throughput (C/T), and Kay's flow diversity (D). Both show a similar pattern to the panel above. Overheads are compared with Kay's S (Fig. 7.11c) with fairly similar trends. Again, differences may be due to omission of the N-Pool from Kay's measures. In Fig. 7.11d, Relative Ascendency (A/C) is shown over the progression of a succession. The trend here is the reverse of other measures of ecosystem organization and Relative Ascendency would appear to be inappropriate for dynamic studies of a single system, but it may be useful for comparing organization relative to the potential organization of different systems.

CONCLUSION

Global ecosystem indices

Contrary to expectations, the global measures did not simply increase or decrease regularly during the successional development of the plankton community. The Finn Cycling Index fluctuated initially before increasing at the end of the phytoplankton bloom, but then remained high as did Kay's (1984) measure of diversity and evenness of flows, D. Kay's other index, S increases at the end of the phytoplankton bloom and then decreases towards the end of the succession as resource niches become very narrow and specialized. These three measures appear to be suitable as indices of succession in a single ecosystem followed with time.

Ascendency (A) is designed as a single measure of both growth and development. In the present study in which flows varied so greatly during the growth and decline of the plankton bloom, it was found to be dominated by the growth component , T (see Fig. 7.10). When growth is removed by dividing A by T, the resulting function has much the same shape as the Finn cycling index and Kay's comparable (D - S) (see Fig. 7.11). These show a dramatic increase on days 7-8 of the succession, as phytoplankton biomass declines and the heterotrophs come into action. The fact that both the "structure" indices C and D show major changes at this time, indicates that the flow structure of the ecosystem changes then, and that this change is detected whether or not the size of the system (Total System Throughput) is included in the index.

Relative Ascendency (A/C) is dimensionless, with both the structure and the size of the system cancelling out between numerator and denominator. Since the structure or topology of the flows and their magnitude both depend upon the modeler's concept of the system and how it is aggregated, it is suggested that Relative Ascendency is most likely to be appropriate for comparisons between one system and another. In the present example, Relative Ascendency decreases when the indices of structural organization (D and C) increase (see Fig. 7.11).

The Upwelling Ecosystem Network

Network analysis of a very simple model of the southern Benguela ecosystem has helped us to look closely at the dynamics of flows in the system, not only at the biomasses of species of interest. Studies of the outflows of model compartments (Fig. 7.3) throw new light on the relative dynamics of various components which are not obvious from the biomass changes alone. Similarly, the dependencies of components on each other are not obvious without the graphs of Figs. 7.4 and 7.5. Mapping the size-classes of organism in a food web into simplified effective trophic positions has been useful in showing the changing structure of the food web, even in the very simple model studied here. For example, the changes in position of micro-zooplankton and meso-zooplankton (depicted in Fig. 7.6) prompted us to focus on the changes in the food web from days 2-3 and 7-9 respectively.

The Lindeman Spine diagrams (Figs. 7.7- 7.8) provide much insight into the changes in system structure and relationships between components after an upwelling event. The magnitude and distribution of flows among the compartments of the model are illustrated simply and clearly. Initially flows are small because the standing stocks are too small to take up the abundant new nitrogen. By day 8 there are massive flows to the primary producers and bacteria, and most recycling is by bacterial excretion whilst most nitrogen export is by the sinking faeces of microzooplankton and slightly less by sinking net-phytoplankton cells. On day 10 more flows go through to larger heterotrophs and most recycling of nitrogen is from flagellates and micro-zooplankton. By day 20 there is much less nitrogen flowing through the system, and the major part of recycling is by bacteria and meso-zooplankton. Thus at different stages of the succession, different sizes of organism are responsible for grazing flows, nitrogen cycling, and for loss of nitrogen through sinking out of the euphotic zone. The flow analyses are based upon an idealized system, with a massive upwelling event followed by a 40-day period in which there is no further mixing. The early stages of succession in the model are consistent with both laboratory and field studies in the southern Benguela ecosystem, the later stages probably occur very seldom and are more difficult to observe in nature.

We have been convinced by this study that ecosystems need to be studied at different levels and from different aspects. Simulation models have an important role to play in complementing experimental and field work to test hypotheses of processes that form part of ecosystem functioning. Similarly, network analyses give a view of one aspect of ecosystem structure, that of the flows among ecosystem components, and therefore a glimpse into the dynamics of the system integrated over the period of measurement. Together, the output of simulation models used as input into network analyses can provide insight into a series of snapshots of ecosystem structure and function which are nearly impossible to obtain by direct measurement.

ACKNOWLEDGEMENTS

We are grateful to Drs R. E. Ulanowicz (University of Maryland), J.J. Kay (University of Waterloo), and M. I. Lucas (University of Cape Town) for helpful discussions. We thank S. Tolosana and C. A. Villacastin-Herrero for assistance in preparing the manuscript. The work was supported by S.C.O.R. and the South African Foundation for Research Development through its Main Research Support and Benguela Ecology Programmes.

REFERENCES

Andrews, W . R .H. & L. Hutchings 1980. Upwelling in the southern Benguela current. Prog. Oceanogr. 9: 1-81

Armstrong, D.A., B.A. Mitchell-Innes., F. Verheye-Dua, H. Waldron & L. Hutchings 1987. Physical and biological features across an upwelling front in the southern Benguela. In: Payne, A. I. L., J. A. Gulland & K. H. Brink (eds) The Benguela and Comparable Ecosystems. S. Afr. J. mar. Sci. 5:171-190

Azam, F., T. Fenchel, J. G. Field, J. S. Gray, L-A. Meyer-Reil & F. Thingstad 1983. The ecological role of water column microbes in the sea. Mar. Ecol. Prog. Ser. 10: 257-263

Barlow, R. G. 1982.Phytoplankton ecology in the southern Benguela current. III. Dynamics of a bloom. J. Exp. Mar. Biol. Ecol. 63: 239-248

Barthel, K.-G. 1983. Food uptake and growth efficiency of *Eurytemora affinis* (Copepoda:Calanoida). Mar. Biol. 74: 269-274

Brown, P. C. & L. Hutchings 1985. Phytoplankton distribution and dynamics in the Southern Benguela Current. International Symposium on upwelling off the west coast of Africa. Instituto de Investigaciones Pequeras. Barcelona. 1: 319-344

Chapman, P. & L.V. Shannon 1985. The Benguela ecosystem. 2. Chemistry and related processes. Oceanogr. Mar. Biol. Ann. Rev. 23: 183-251

Crawford, R. J. M., L.V. Shannon & D.E. Pollock 1987. The Benguela ecosystem. 4. The major fish and invertebrate resources. Oceanogr. Mar. Biol. Ann. Rev. 25: 353-505

Ducklow, H. W., M. J. R Fasham & A. F. Vézina 1989. Derivation of flow networks for open ocean plankton systems. In: Wulff, F., J. G. Field & K. H. Mann(Eds), *Network Analysis in Marine Ecology: Methods and Applications*. Lecture Notes on Coastal and Estuarine Studies. Springer-Verlag, New York.

Fenchel, T & T.H. Blackburn 1979. *Bacteria and mineral cycling*. Academic Press, London, 225 pp

Hall, D.J., S.T. Threlkeld, C.W. Burns & P.H. Cowley 1976. The size-efficiency hypothesis and the size structure of zooplankton communities. Ann. Rev. Ecol. Syst. 7: 177-208

Kay, J.J. 1984. Self-organization in living systems. PhD thesis, University of Waterloo, Ontario, Canada.

Kay, J.J., L.Graham & R.E.Ulanowicz 1989. A detailed guide to network analysis. Wulff, F., J.G. Field & K.H. Mann (eds) *Network Analysis in Marine Ecology: Methods and Applications*. Lecture notes on coastal and estuarine studies. Springer-Verlag, New York.

Lucas, M.I. 1986. Decomposition in the pelagic zone. J. Limnol. Soc. sth. Afr. 12: 99-122

Lucas, M.I., S.J.Painting & D.G. Muir 1986. Estimates of carbon flow through bacterioplankton in the S. Benguela upwelling region based on 3H-Thymidine incorporation and predator free incubations. Proceedings of the Second International Colloquium of Marine Bacteriology, Actes de Colloques 3. Brest. Centre National de la Recherche Scientifique/IFREMER, 357-383

Lucas, M. I., T. A. Probyn & S. J. Painting 1987. An experimental study of microflagellate bacterivory: Further evidence for the importance and complexity of microplanktonic interactions. In: Payne, A. I., J. A. Gulland & K. H. Brink (eds) *The Benguela and Comparable Ecosystems*. S. Afr. J. mar. Sci. 5: 791-808

MacIsaac, J.J. & R.C. Dugdale 1969. The kinetics of nitrate and ammonia uptake by natural populations of marine phytoplankton. Deep Sea Res. 16: 45-57

Margalef, R. 1968. *Perspectives in ecological theory*, Univesity of Chicago Press, 112 p.

Miller, C. A. & M. R. Landry 1984. Ingestion-dependent rates of ammonium excretion by the copepod *Calanus pacificus*. Mar. Biol. 78: 265-270

Moloney, C. L. 1988. A size-based model of carbon and nitrogen flows in plankton communities. PhD Thesis, University of Cape Town

Moloney, C. L. & J. G. Field, in press a. Modelling carbon and nitrogen flows in a microbial plankton community. In: Reid, P. C., P. H. Burkhill & C. M. Turley (eds) *Protozoa and their role in Marine Processes*. NATO ASI series, Springer-Verlag, New York.

Moloney, C. L. & J. G. Field, in press b. General allometric equations for nutrient uptake, ingestion and respiration in planktonic organisms. Limnol. Oceanogr. (submitted).

Odum, E. P. 1969. The strategy of ecosystem development. Science. 164: 262-270

Painting, S. J., M. I. Lucas & D. G. Muir, in press. Fluctuations in heterotrophic bacterial production, activity and community structure in response to development and decay of phytoplankton in a microcosm. Mar. Ecol. Prog. Ser.

Shannon, L.V. & J. G. Field 1985. Are fish stocks food limited in the southern Benguela pelagic ecosystem? Mar. Ecol. Prog. Ser. 22: 7-19

Shannon, L. V. & S. C. Pillar 1986. The Benguela ecosystem. 3. Plankton. Oceanogr. Mar. Biol. Ann. Rev. 24: 65-170

Smith, R. L. 1980. *Ecology and Field Biology*, 3rd edition. Harper & Row, New York

Strickland, J. D. H.1960. Measuring the production of marine phytoplankton. Bull. Fish. Res. Bd Can. 122: 1-172

Stuart, V. 1986. Feeding and metabolism of *Euphausia lucens* (Euphausiacea) in the southern Benguela current. Mar. Ecol. Prog. Ser. 30: 117-125

Ulanowicz, R. E. 1980. An hypothesis on the development of natural communities. J. theor. Biol. 85: 223-245

Ulanowicz, R. E. 1984. Community measures of marine food networks and their possible applications. In: Fasham, M.J.R. (ed.) *Flows of Energy and Materials in Marine Ecosystems*. Plenum Publishing Corp, New York, p. 23-47

Ulanowicz, R. E. 1986. *Growth and Development: Ecosystems Phenomenology*. Springer-Verlag, New York

Verheye-Dua, F. & M. I. Lucas 1988. Southern Benguela frontal region. I. Hydrology, Phytoplankton and Bacterioplankton. Mar. Ecol. Prog. Ser. In press.

Wulff, F. & J. G. Field 1983. The importance of different trophic pathways in a nearshore benthic community under upwelling and downwelling conditions. Mar. Ecol. Prog. Ser. 12: 217-228

Chapter 8

DERIVATION AND ANALYSIS OF FLOW NETWORKS FOR OPEN OCEAN PLANKTON SYSTEMS

Hugh W. Ducklow
Horn Point Environmental Laboratories
Box 775
Cambridge, MD 21601
USA

M.J.R. Fasham
Institute of Oceanographic Sciences
Brook Road
Wormley, Godalming, Surrey
GU8 5UB, UNITED KINGDOM

Alain F. Vézina
Département d'Océanographie
Université du Québec à Rimouski.
310, Allée des Ursulines, Rimouski
Québec, Canada G5L 3A1

INTRODUCTION

Within the past decade a revolution has occurred in our understanding of the structure and functioning of marine plankton systems. Among the major developments contributing to a new paradigm for planktonic communities were the elaboration by Dugdale & Goering (1967) and Eppley & Peterson (1979) of the concept of new and regenerated production and the role of these processes in setting the balance between inputs to and outputs from the upper ocean ecosystem. Another important factor contributing to new ideas about the organization of the marine plankton was the discovery of the large and dynamic stocks of autotrophic and heterotrophic microbes and their dominant role in cycling carbon and nitrogen (Pomeroy 1974; Azam & Hodson 1977; Hobbie *et al.* 1977; Johnson & Sieburth 1979; Waterbury *et al.* 1979; Williams 1981; Ducklow 1983).

Early models of the marine euphotic zone included just a few compartments, because the roles of microbial plankton groups were still unclear (Steele 1974). These models usually consisted of phytoplankton, macrozooplankton herbivore and predator compartments, with direct export to the benthos or to economic harvest. Some current models have retained this simple flow structure, but may have addressed greater complexity in animal behavior or in physical-biological interactions (e.g. Franks *et al.* 1986). Landry (1977) and Steele & Frost (1977) suggested the importance of diverse size classes, trophic groups, and alternative pathways for structuring plankton communities. Williams (1981) and Joint & Morris (1981) were among the first to emphasize flows of dissolved organic matter through bacterioplankton, and of bacterial biomass through protozoans. Pace *et al.* (1984) presented a complex simulation model which included a large number of plankton size classes and functional groups. In addition they explored the effect of variations in the input of new nitrogen on community structure and showed the importance of the microbial groups in different scenarios. This process of evolutionary development in the modeling of plankton systems has narrowed the gap between observation and conceptual understanding, but has not yet resulted in a widespread dialog between modelers and observationalists in biological oceanography.

One of the reasons for the essentially one-way communication from observation to model may be the emphasis in simulation modeling on the variations in compartment size (biomass) rather than on flow structure (rate processes) (Platt *et al.* 1981; Fasham 1984). Another reason is that neither complex simulations of simple systems with few compartments, nor more complex, many-compartment models adequately serve the needs of the experimentalists who now measure bacterial production and herbivory, but still lack the capability to make routine observations of the many other flows in the complex models. Thus flow analysis of aggregated networks appears to offer a useful tool both for modelers and observationalists. Input-output flow analysis of plankton networks was introduced by Fasham (1985) who presented generic, six compartment models of carbon and nitrogen flows in an open ocean system. A similar approach was adopted by Frost (1985), Vézina & Platt (1987) and Michaels & Silver (1988), in models which addressed the influence of foodweb structure on variations in export (sinking) fluxes.

Recognizing that plankton systems are grossly undersampled with respect to information on flows, our principal objective in this chapter is to reconstruct flow networks from the few known flows, using two different modelling approaches. Then we use flow analysis routines to explore some of the properties of the flow networks. In particular we will investigate the varying importance of microbial foodwebs in response to variations in cycling implicit in the input data. Our results suggest two mechanisms for the seasonal evolution of foodwebs toward lower f-ratios (as indexed by NO_3-N uptake divided by tot-N uptake of phytoplankton) and a greater proportion of recycling: a uniform increase in the cycling efficiency of all components, or irregular increases in cycling efficiency accompanied by changes in the actual pathways linking different compartments.

METHODS

Our aim in this exercise was to produce a model of an open ocean flow network which was simple enough to be useful for interpretation of extant data sets and for the design of future measurement programs; yet sufficiently complex to include a realistic depiction of a foodweb, and to produce new insight into variations in foodweb structure. Furthermore the model had to be sufficiently generic or versatile to allow comparisons among different systems. We followed the theme presented in Vézina & Platt (1988), which stated that plankton networks are greatly undersampled, and that direct estimates of most of the flows are not available. Models must be used to derive objective estimates of unknown flows, the range of possibilities for which is constrained by measured flows and by suitable choices of other model parameters (assimilation efficiencies, respiration coefficients, etc.). Our first task was to agree on a generalized flow structure.

The generic foodweb model elaborated by Fasham (1985) was modified following extensive discussions at a SCOR workshop (see introduction to this volume, Field *et al.* 1989).

This model differs from the earlier version presented by Fasham primarily in that it has three kinds of dissolved nitrogen fluxes: an input of new nitrogen as nitrate (NO_3^-), and compartments and associated intercompartmental flows for dissolved organic nitrogen (DON) and regenerated dissolved inorganic nitrogen (NH_4^+ and Urea). Our assumptions about the model structure (choice of compartments and intercompartmental flows) are discussed in Appendix A (see also Vézina & Platt 1988; and Vézina 1989, this volume). Our model foodwebs are depicted in Figure 8.1.

We have followed two complementary approaches to deriving models of flow networks. In the first approach called *a priori*, steady-state, input-output flow analysis was used to derive a system of linear equations relating intercompartmental flows to total compartmental throughputs as described in Appendix A, *using specific choices for parameter values.* Parameter evaluation was based on sensitivity analysis as described in the Appendix. The second, called *inverse* approach, described in detail in Vézina & Platt (1988), Vézina (1989, this volume) and in Appendix 8B, differs in that it *used numerical inverse methodology to find within specified ranges of parameter values* a solution best fitting certain *a priori* criteria. The main difference between these approaches is that the *a priori* approach uses only flow information, whereas the *inverse* approach also relies on biomass data to constrain the solutions. In the *a priori* approach, the parameter choices, and thus to an extent the network solutions, are chosen explicitly by the modelers. In the *inverse* methodology, parameter values are estimated directly from the available data on stocks and flows.

Our next requirement was for a suitable data set to use as a foundation for deriving realistic flow networks under a range of environmental conditions. A survey of the literature revealed that there are few, if any data sets with extensive, redundant measurements of more than a few of the flows shown in Figure 8.1. In most cases, measurements other than primary production have been made on a piecemeal basis resulting in a general lack of comparability and synopticity amongst the flow estimates. We chose the data sets associated with the Warm Core Rings Program (Joyce & Wiebe 1983; Joyce 1985; Wiebe & McDougall 1986) as being representative of an open sea location and having a large core of synoptic, state-of-the art measurements of different flows. The model-building and -testing effort described in the Appendices was based on a distillation of data from various investigators in the Warm Core Rings Program. We have used data from three different months in order to examine a range of variability in upper ocean properties and in ring evolution (Table 8.1). Our overall approach was to start with the five flow estimates for which both synoptic data and reasonably unequivocal interpretations were available (Table 8.2), and to use these in our models to derive estimates of the other flows as represented in Figure 8.1 and discussed in detail below. Biomass data were also used to derive networks with the inverse method; these are given in Table 8.3.

The data from which the average flow estimates were derived, and details on sampling and measurements are discussed in the following references: Primary production: Hitchcock *et al.* (1985); Nitrogen fluxes: McCarthy & Nevins (1986); Bacterial production and bacterivory: Ducklow (1986); Herbivory by macrozooplankton: Roman *et al.* (1986). We obtained the data from the database on warm core rings at Woods Hole Oceanographic Institution, or from personal communication with the aforementioned authors. In general, measurements were made 2-3 times per day within 15 km of the center of each ring, at 6-9 m depths within the euphotic zone. Daily flow estimates were derived by integration over 24 hours and over the euphotic zone, and 3 separate daily estimates were averaged for each month. Thus each daily estimate given in Table 8.2 is derived from 36-81 discrete measurements.

For each of the three months, carbon and nitrogen flow networks were computed by the methods described in Appendix A and in Vézina (1989, this volume). These data were entered into the SCOR file format using the SCOR data-entry program DATA. The resulting 12 files were analyzed with SCOR flow analysis software. In particular we used the program NETWRK3. The output from this program is discussed in Kay *et al.* (1989, this volume, see also Ulanowicz 1986). The results of those analyses are discussed in the following section. In this chapter we compare and discuss only the nitrogen networks although the construction of the analogous carbon networks is described.

Table 8.1. Properties of Warm Core Ring plankton systems

Month	Mixed Layer		Euphotic Zone	Primary Production	f-ratio*
	temp	meters	meters	mgC m^{-2} day^{-1}	
April	16	450	75	900	0.74
June	18	<20	30	790	0.51
October	29	80	100	360	0.12

*This is the apparent f-ratio, or (total NO_3 uptake/total observed N uptake). The true f-ratio, which considers only phytoplankton uptake, will differ.

Table 8.2. Warm Core Rings integrated mixed layer N and C flows as mg (N or C) m^{-2} h^{-1} *

Month	NO_3^- (N)	NH_4^+ + Urea (N)	Primary production (C)	Macrozooplankton ingestion (C)	Bacterial production (C)
April	5.04	1.79	75	0.46	2
June	5.49	5.26	66	2.47	4.5
October	0.58	4.32	30	1.09	2.8

* Multiply primary production by 12 and other rates by 24 for daily rates.

Table 8.3. Biomass (mgN/m^{-2}) of plankton components in Warm Core Rings.

Component	April	June	October
Phytoplankton	138	150	99
Protozoans/Microzooplankton	48	394	51
Macrozooplankton	111	56	80
Bacterioplankton	32	48	38

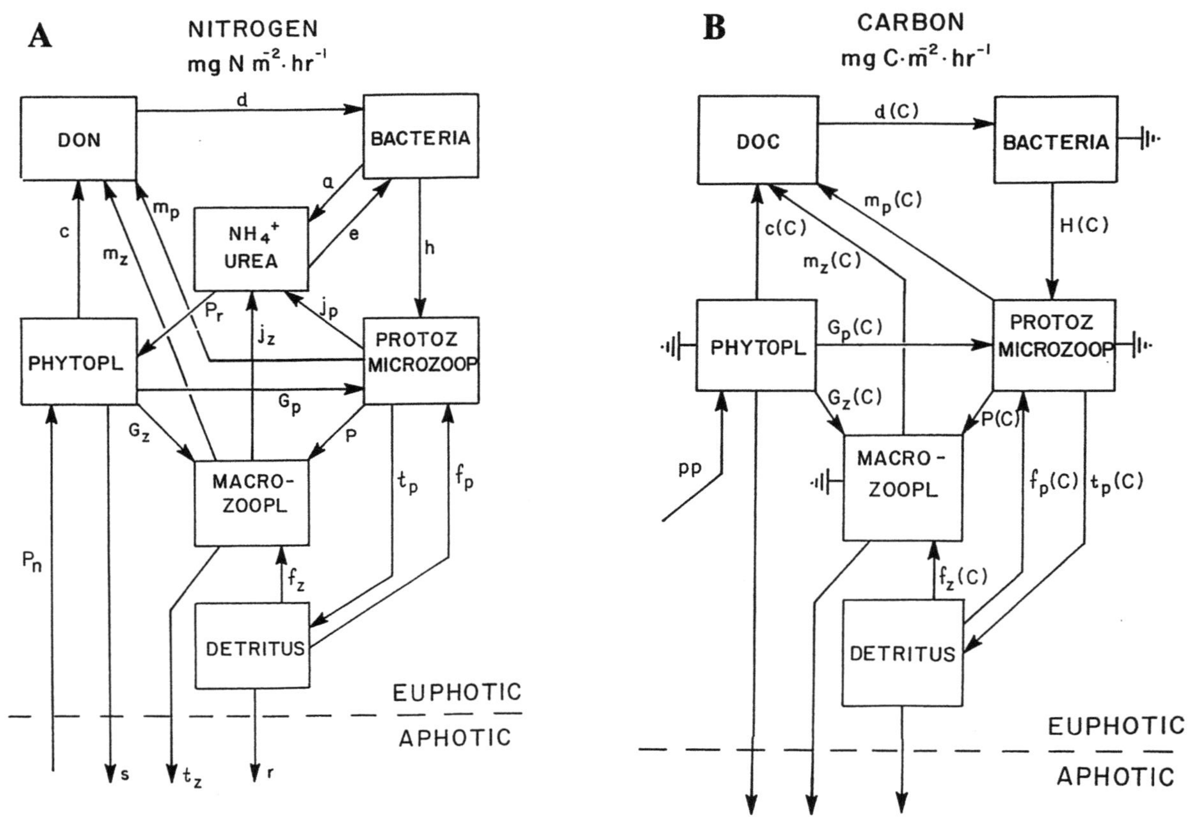

Figure 8.1. The structure and flow terminology for plankton networks. A) Nitrogen B) Carbon

RESULTS AND DISCUSSION

Overview and comparison of the networks

Steady-state solutions were obtained for nitrogen flow networks for each of the 3 months using both the a priori and inverse approaches as described in the Appendices. Figures 8.2a, 8.3a and 8.4a depict the resulting patterns of nitrogen flows determined by the a priori approach for April, June and August, respectively. Figures 8.2b, 8.3b and 8.4b show the corresponding solutions obtained by the inverse approach. We were able to obtain solutions near steady state for all 6 networks, though some slight adjustment of parameter values was necessary in some cases. Tables 8.4, 8.5 and 8.6 summarize the flow data for the 6 networks. Table 8.4 presents the rank order of the magnitudes of the flows; while in Tables 8.5 and 8.6 the flows are normalized to the total primary production (new production, P_n, plus regenerated production, P_r). These tables facilitate comparison of flows in different months when the total throughput varies.

Table 8.4. Rank orders of flows for Nitrogen Networks in a priori and inverse models. Flow symbols are defined in Fig. 8.1

		A PRIORI MODEL				INVERSE MODEL			
Flow (Donor - Recipient)	Flow (Symbol)	April	June	October	Mean	April	June	October	Mean
Input -Phytoplankton	P_n	1	1	11	4	3	4	11	6
Phytoplankton -export	S	2	2	19	8	1	19	15	12
Macrozooplankton -export	t_z	15	14	13	14	10	14	12	12
Detritus -export	r	20	20	20	20	2	6	17	8
Phytoplankton -Protozoa	G_p	4	4	1	3	17	1	2	7
Phytoplankton -Macrozoplankton	G_z	17	16	16	16	18	15	16	16
Phytoplankton -DON	c	14	15	15	15	19	8	19	15
Protozoa -Macrozooplankton	P	10	8	6	8	16	10	3	10
Protozoa -NH4	j_p	5	5	3	4	15	3	8	9
Protozoa -DON	m_p	16	17	14	16	12	18	18	16
Protozoa -Detritus	t_p	6	6	4	5	9	5	13	9
Macrozooplankton -NH4	j_z	13	9	7	10	14	13	9	12
Macrozooplankton -DON	m_z	19	19	17	18	20	17	4	14
Bacteria -Protozoa	h	11	12	10	11	13	12	7	11
Bacteria -NH4	a	9	11	9	10	4	11	6	7
NH4 -Phytoplankton	P_r	3	3	2	3	7	2	1	3
NH4 -Bacteria	e	8	10	8	9	5	16	14	12
DON -Bacteria	d	12	13	12	12	11	7	5	8
Detritus -Protozoa	f_p	7	7	5	6	6	9	10	8
Detritus -Macrozooplankton	f_z	18	18	18	18	8	20	20	16

It is apparent from examination of the networks that variations in input data contributed most to the differences between the networks, with the two modeling approaches giving similar values to most of the flows for a given month. For example, when the individual flows for each month were ranked by size, and then averaged within each model's results, the two models gave similar mean ranks for each flow (Table 8.4). The main trends in the input data are summarized in Tables 8.1 and 8.2, along with some of the physical factors (temperature, mixing) underlying those differences. The major differences between models are attributed to flows G_p, j_p, and export from the detrital compartment, particularly in April.

Some of the main features of the flow networks are discussed below. We compare months and modeling approaches with respect to the following processes: system inputs (new production), regenerated and total production, bacterial uptake and regeneration of ammonium, zooplankton grazing and detritivory and, finally, system exports. Then a more detailed flow analysis of the networks is presented in the following sections.

In all our solutions the normalized input P_n is equivalent to the f-ratio, or the fractional amount of new production in each network. This value was highest in April, with values of 0.78-0.81 which are very high for systems outside of the major upwelling provinces (Eppley & Peterson 1979). McCarthy & Nevins (1986) speculated on the magnitude of this value in Warm Core Ring 82B. They suggested that the f-ratio was very high not only because nitrate levels were high (6 mg-atoms l^{-1}) in the surface water, but perhaps also because of poor coupling between phytoplankton and grazers, resulting in low production of reduced N. We note (see below) that some of the NH_4^+ which was produced was taken up by bacterioplankton, further reducing the flux to the phytoplankton. But because our models took the fluxes of regenerated nitrogen observed by McCarthy & Nevins (1986) as a boundary condition, we cannot offer any further explanation for the apparently high f-ratio in April. We were able to obtain steady state solutions in which grazing was low, and which were consistent with the biomass fields observed (i.e., in the inverse approach, grazing was adequate to support the zooplankton biomass). Thus we concur with McCarthy & Nevins (1986) that for some unknown reason, grazing in April was very low, resulting in surprisingly high f-ratios.

The total primary production, in nitrogen mass units, was highest in June (Figure 8.5), whereas the f-ratio was highest in April (Tables 8.1, 8.5, and 8.6), when total carbon production was highest. Eppley & Peterson (1979) found a hyperbolic relationship between total (carbon) production and f-ratio, with an asymptote of $f \approx 0.5$ at production levels above about 500 g C m^{-2} y^{-1}. Although the April f-value is higher than values in the data sets available to Eppley & Peterson (1979), our f-values and carbon fixation rates for June and October (Tables 8.1, 8.5, 8.6) fall roughly on Eppley & Peterson's trend line.

Our results indicate that C:N ratios for phytoplankton uptake were also very high in April when the f-ratio was high, but were near the Redfield value of 6.7 in June and October when f-ratios were lower (Figure 8.6). We specified that the particulate production by phytoplankton be set at the Redfield ratio of 6.7. Thus respiratory losses and losses of dissolved organic carbon were highest in April, when these two losses were estimated by our two approaches to reach 55-59% of the total carbon fixation (Figure 8.6). Thus our models "explain" the high C/N assimilation ratios for April by hypothesizing high respiration and DOC release rates.

An important result of our modeling effort is a "dissection" of the uptake of regenerated nitrogen, U_r, into its components, which in these networks are phytoplankton (P_r) and bacterial uptake (e). Direct observation of P_r vs e is technically difficult and the results to date are ambiguous (Laws *et al.* 1985; Wheeler & Kirchman, 1986), though these authors suggest that e could be over 50% of U_r. The two modeling approaches yielded quite different estimates for e compared to U_r and P_r. The a priori results suggest that although bacterial uptake of regenerated nitrogen (e) increased from 9 to 22% of the total primary production (P_n + P_r; Table 8.5), the bacterial fraction of total regenerated nitrogen (i.e. e/U_r) uptake declined from 33 to 20% between April and October. Thus as the overall importance of regenerated production rose, or as the f-ratio fell from 81 to 14%, the bacterial component of U_r declined. In other words, as the supply of reduced nitrogen rose, phytoplankton obtained an increasingly large share of it apparently by outcompeting the bacteria (cf. Suttle *et al.* 1987). Over the same period, DON and reduced DIN contributed 40 and 60% of the bacterial N uptake, respectively, regardless of other conditions (Figure 8.7). Thus the results of the a priori models suggest that competition with phytoplankton for N supply was a major limiting factor for bacterial production under conditions of decreasing f-ratio. However because bacterial incorporation of regenerated nitrogen was a quantitatively important process at all times, the "apparent" f-ratio $[P_n/(P_n+U_r)]$, tended to underestimate the actual f-ratio, $[P_n/(P_n+P_r)]$, which refers only to phytoplankton uptake, as has been suggested recently by Harrison *et al.* (1987) (Table 8.7).

The inverse method gave similar qualitative, but very different quantitative estimates of bacterial N uptake. Flow e (from regenerated N to bacteria) amounted to 50% of the total primary production in April, and only 2-8% thereafter. Similarly, bacteria contributed 70% of U_r in April and only 2-8% later. Thus both models showed that the bacterial component of U_r declined in parallel with the f-ratio. But whereas the uptake of reduced inorganic nitrogen was a constant 40% of the bacterial nitrogen nutrition in the a priori results, this fraction ranged from 8-23% in the inverse models. Thus the two approaches found a very different role for DON in bacterial nutrition, especially under stratified, moderate to low f-ratio conditions: in the inverse results DON contributed 80-90% of the total bacterial N intake, but only 40% in the a priori results. Thus unlike the a priori model, the inverse model suggests that the DON flux is a major limiting factor for bacterial production. At this point we are unable to say which of these views is correct for each month. The important result of this comparison is to suggest a variety of control mechanisms for different ecosystem components, which future observational and modeling efforts may clarify. These comparisons suggest that determining the relative importance of DIN and DON in bacterial nutrition could be a major priority for marine microbial ecologists.

The two modeling approaches also differed with respect to nitrogen remineralization by bacteria. Some supply of reduced DIN from bacteria was required by both approaches in all months to meet the observed demand, U_r. However, the a priori model suggested that the uptake and release of NH_4^+ and Urea by bacteria were nearly balanced, with a slight excess of uptake in all months (Table 8.5). In contrast, the inverse model found either a slight (April) or great (June-October) excess of regeneration over immobilization. Thus in the former result, bacteria essentially supplied NH_4^+ and Urea only for their own nutrition, whereas in the latter case, bacteria made a great contribution to phytoplankton N nutrition. The a priori results are supported by the observations of Goldman *et al.* (1985) who found little or no net release of NH_4^+ by bacteria, whereas the inverse results contradict the observations of Wheeler & Kirchman (1986) who found net uptake of NH_4^+. We note here that most observational approaches to date (Goldman *et al.* 1985; Wheeler & Kirchman 1986) observe the net balance of the two processes, and have not distinguished these two complementary flows. To do so requires a modeling approach which considers the interdependence of all the components of the plankton system.

Our boundary conditions included observations of the ingestion of phytoplankton production by macrozooplankton (Table 8.2) and this flow (G_z) was always low, ranging from 1-6% of the total nitrogen-based primary production. As described in Appendix A, one of the main tasks of the flow modeling approach was to obtain estimates of the ingestion of phytoplankton by protozoan herbivores and zooplankton smaller than 333 μm. In general both approaches found that small herbivores were consistently the major source of phytoplankton removal, contributing from 8 to over 25 times as much of the total herbivory as the larger zooplankton (Tables 8.5, 8.6). The sole exception was the inverse result for April, in which each group ingested only 2-3% of the production. Both models established that in June and October, herbivory by protozoans and microzooplankton was by far the major removal mechanism for primary production. This is the most robust finding of our effort, and it is consistent with recent observations (Landry *et al.* 1984; Welschmeyer *et al.* 1985) on oceanic systems. The reason for this finding in our study is that when the models were constrained to produce the observed flows of NH_4^+ + Urea, and with realistic parameter values for their physiology, the bacteria and macrozooplankton could not serve as major sources of NH_4^+, except in April when the low regeneration rates obviated the need (in our models) for much grazing activity.

Since phytoplankton necessarily formed a small portion of the macrozooplankton diet, other sources for the macrozooplankton input were specified by the model. Both models found that predation on microzooplankton and protozoans (Rassoulzedegan & Sheldon 1986) formed an increasingly large flux between April and October (Tables 8.5, 8.6), ranging from 9 to 37% and from 6 to 69% of the total primary production in the a priori and inverse models, respectively. In general, detritus ingestion formed a small fraction of the macrozooplankton diet. This latter result may well be an artifact of our assumptions (see Appendix A: evaluating the nitrogen flow parameters). Macrozooplankton diets are discussed further in the next section.

Our models were also constrained to have exports from the system balance the input in the form of NO_3^-, or new production, P_n. But while the total export was set by the boundary conditions, the form, or source of the export, was left free to be specified by variations in foodweb dynamics, a theme introduced by Frost (1985) and elaborated by Vézina & Platt (1987). Our two models took slightly different forms; however three generic sources for the vertical export of particulate matter can be identified: the sinking of uneaten phytoplankton cells and fecal pellets produced by macrozooplankton, or of the fecal products of protozoans which were not reingested. This latter category might include marine snow formed by the aggregation or condensation of suspended particulate matter into packages large enough to sink sufficiently rapidly to escape further processing within the euphotic zone. The model results suggest a consistent seasonal variation in the composition of the vertical flux.

The sinking of uningested phytoplankton was the major source of export in April, when ca. 80% of the production was required to be exported in some form. Under these conditions in which low observed rates of N regeneration forced the models to minimize trophic transformation of the phytoplankton biomass, intact phytoplankton were 65 (inverse) to 95% (a priori) of the total export (Figure 8.8). Model results differed for June. The a priori result assigned the major share of the export to uneaten phytoplankton, whereas the inverse approach had most of the export derived from uneaten detrital output from the protozoans and microzooplankton. This difference can be attributed to the fact that the two models differed in estimates of herbivory by the protozans, with the inverse result being about double that in the June a priori result. Both models found that for October, the major export, which amounted to only 5-14% of the total production, was derived from fecal materials. Thus in general the models show a seasonal evolution in the composition of the vertical flux, progressing from a phytoplankton dominated flux in the spring (Smetacek 1985), though an intermediate summer period in which marine snow may be important, to late summer, when large, rapidly sinking fecal pellets comprise much of the exported material (Figure 8.8)

In summary, we note that while a rigorously controlled comparison between the models was not entirely possible (see Appendix B), we found that the two approaches obtained broadly similar solutions from the same boundary conditions. That is, most flows had similar magnitudes in both models, and for the most part, the structure (topology) of the models was similar. Furthermore, these results were generally consistent with our ideas about the seasonal evolution of plankton systems, and about the overall structure of oceanic foodwebs and the dependence of that structure on the balance between new and recycled production. However major differences are apparent. The relative magnitudes of some flows vary greatly depending on the derivation, as does the presence or absence of some linkages. Thus our two approaches show that it is possible to obtain a variety of network solutions, all of which generally satisfy similar constraints. This important point is treated below.

Table 8.5. Nitrogen fluxes (mgN $m^{-2} h^{-1}$) and normalized fluxes (see text) in a priori networks. R = rank order of flow magnitude.

APRIL				JUNE				OCTOBER			
R	Flow	Flux	Norm	R	Flow	Flux	Norm	R	Flow	Flux	Norm
1	P_n	5.04	0.81	1	P_n	5.49	0.59	11	P_n	0.58	0.14
2	S	4.80	0.77	2	S	4.63	0.50	19	S	0.01	0.00
15	t_z	0.24	0.04	14	t_z	0.83	0.09	13	t_z	0.57	0.14
20	r	0.00	0.00	20	r	0.02	0.00	20	r	0.00	0.00
4	G_p	1.05	0.17	4	G_p	3.78	0.41	1	G_p	3.62	0.90
17	G_z	0.08	0.01	16	G_z	0.43	0.05	16	G_z	0.19	0.05
14	c	0.31	0.05	15	c	0.47	0.05	15	c	0.20	0.05
10	P	0.57	0.09	8	P	1.81	0.19	6	P	1.51	0.37
5	j_p	0.79	0.13	5	j_p	2.51	0.27	3	j_p	2.39	0.59
16	m_p	0.09	0.01	17	m_p	0.28	0.03	14	m_p	0.25	0.06
6	t_p	0.75	0.12	6	t_p	2.37	0.25	4	t_p	2.14	0.53
13	j_z	0.43	0.07	9	j_z	1.47	0.16	7	j_z	1.13	0.28
19	m_z	0.04	0.01	19	m_z	0.15	0.02	17	m_z	0.13	0.03
11	h	0.47	0.07	12	h	1.05	0.11	10	h	0.65	0.16
9	a	0.57	0.09	11	a	1.28	0.14	9	a	0.80	0.20
3	P_r	1.20	0.19	3	P_r	3.82	0.41	2	P_r	3.45	0.86
8	e	0.59	0.09	10	e	1.44	0.15	8	e	0.87	0.22
12	d	0.44	0.07	13	d	0.89	0.10	12	d	0.58	0.14
7	f_p	0.69	0.11	7	f_p	2.13	0.23	5	f_p	2.01	0.50
18	f_z	0.06	0.01	18	f_z	0.21	0.02	18	f_z	0.13	0.03

Table 8.6. Inverse Nitrogen Flow Networks.
R = Rank order of flow magnitude.
t_h = Flow of bacterial production to detritus.

Flow	APRIL R	APRIL Flux	APRIL Norm	JUNE R	JUNE Flux	JUNE Norm	OCTOBER R	OCTOBER Flux	OCTOBER Norm
P_n	3	3.55	0.78	4	3.76	0.36	11	0.29	0.11
t_z	22	0.00	0.00	19	0.00	0.00	18	0.00	0.00
r	2	3.81	0.84	6	2.83	0.27	17	0.12	0.05
G_p	17	0.12	0.03	1	8.50	0.80	2	1.89	0.71
G_z	18	0.07	0.02	15	0.37	0.03	16	0.16	0.06
c	19	0.00'	0.00	8	1.71	0.16	21	0.00	0.00
s	1	4.38	0.96	20	0.00	0.00	15	0.21	0.08
P	16	0.29	0.06	10	1.55	0.15	3	1.83	0.69
j_p	15	0.29	0.06	3	5.87	0.55	8	0.47	0.18
m_p	12	0.63	0.14	18	0.12	0.01	19	0.00	0.00
t_p	9	0.83	0.18	5	3.60	0.34	13	0.23	0.09
j_z	14	0.42	0.09	13	0.93	0.09	9	0.36	0.14
M_z	21	0.00	0.00	17	0.17	0.02	4	1.22	0.46
t_z	10	0.77	0.17	14	0.82	0.08	12	0.27	0.10
h	20	0.00	0.00	12	1.07	0.10	7	0.62	0.23
a	4	2.54	0.56	11	1.11	0.10	6	0.65	0.25
t_h	13	0.47	0.10	21	0.00	0.00	20	0.00	0.00
p_r	7	1.00	0.22	2	6.82	0.64	1	2.37	0.89
e	5	2.28	0.50	16	0.17	0.02	14	0.22	0.08
d	11	0.67	0.15	7	2.00	0.19	5	0.94	0.35
f_p	6	1.87	0.41	9	1.58	0.15	10	0.35	0.13
f_z	8	0.84	0.18	22	0.00	0.00	2	0.00	0.00

Table 8.7. Bias in f-ratio estimates due to bacterial uptake of NH_4^+. e = bacterial uptake of NH_4^+, U_r = bacterial plus phytoplankton uptake of NH_4^+

f-ratio	APRIL A Priori	APRIL Inverse	JUNE A Priori	JUNE Inverse	OCTOBER A Priori	OCTOBER Inverse
"apparent"	0.74	0.52	0.51	0.35	0.12	0.10
"real"	0.81	0.78	0.59	0.36	0.14	0.11
e/U_r	0.33	0.69	0.27	0.02	0.20	0.09

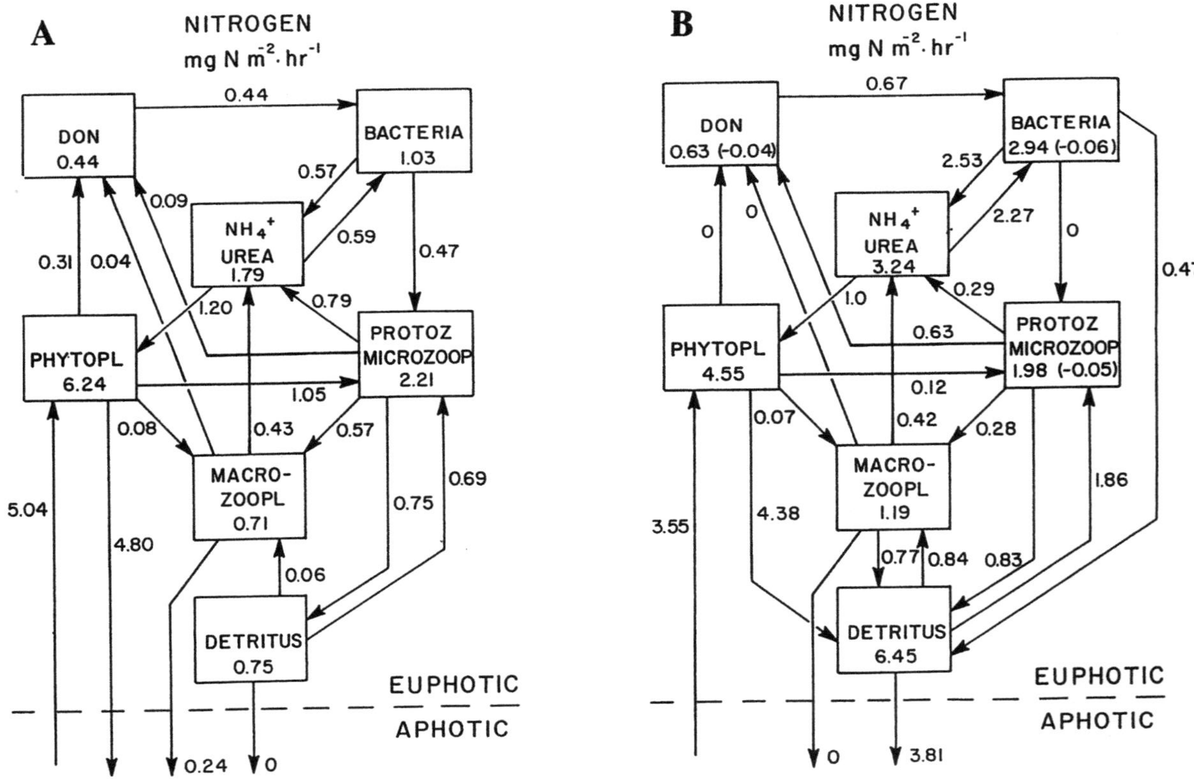

Figure 8.2. Flow networks derived for flows of Nitrogen in Warm Core Rings for April. A) A priori networks. B) Inverse networks. For inverse networks, the numbers in parantheses are the excess of output over input, for that compartment.

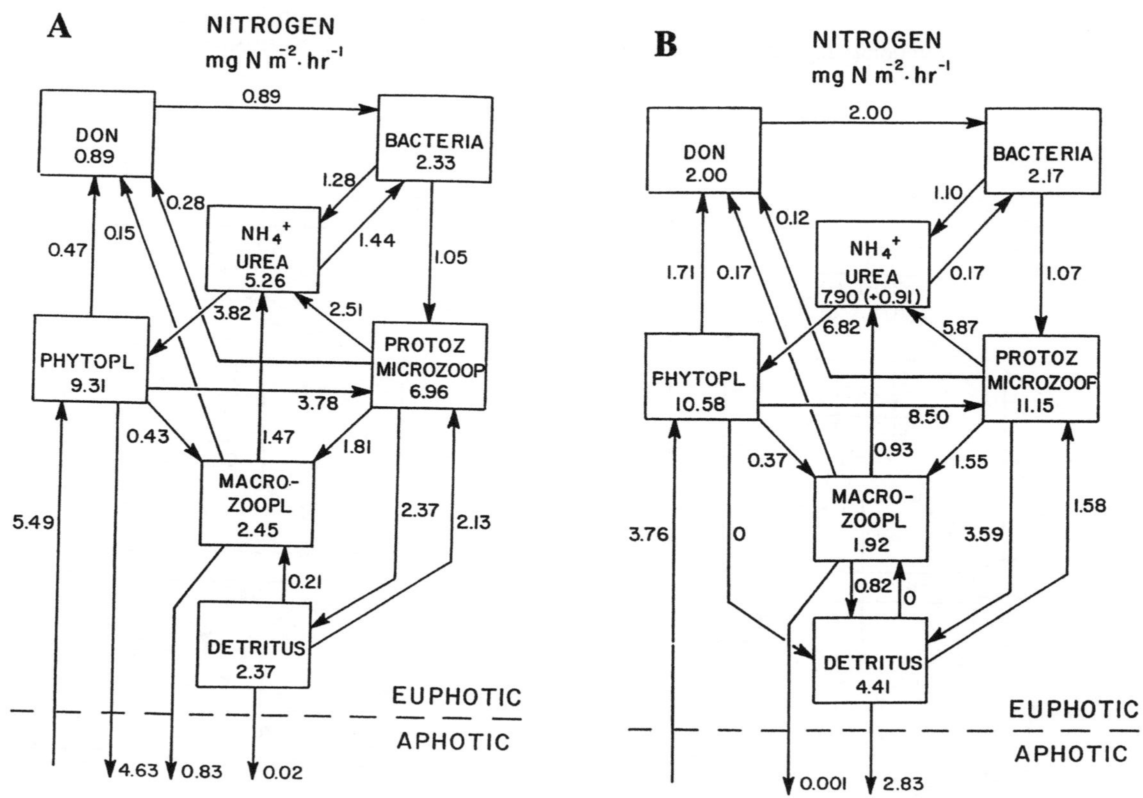

Figure 8.3. Flow networks derived for flows of Nitrogen in Warm Core Rings for June. A) A priori networks. B) Inverse networks.

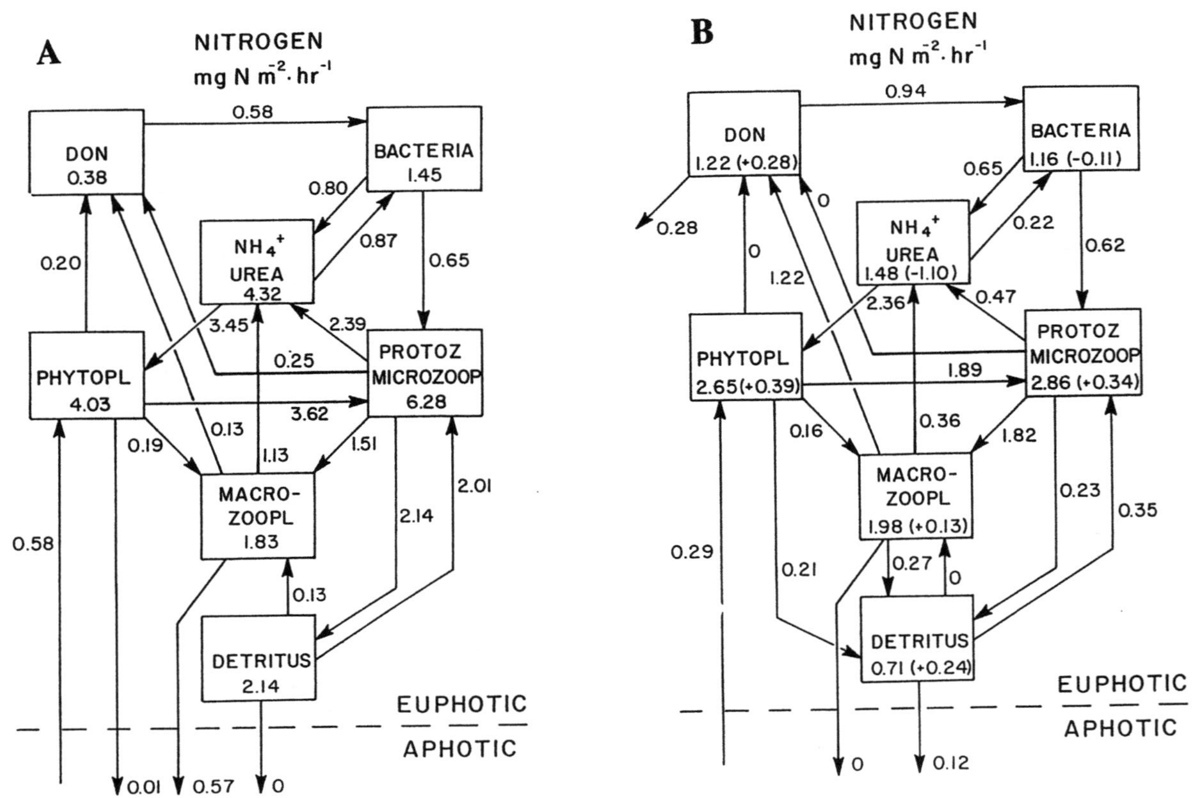

Figure 8.4. Flow networks derived for flows of Nitrogen in Warm Core Rings for October. A) A priori networks. B) Inverse networks.

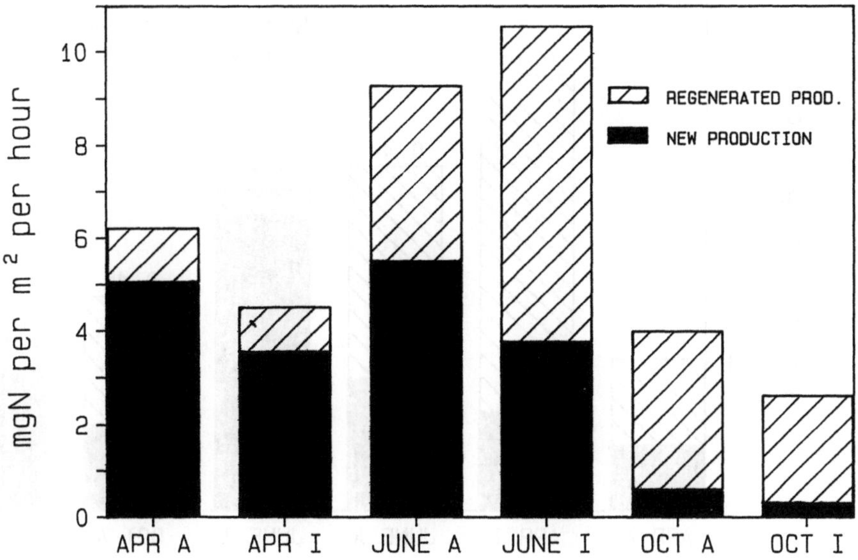

Figure 8.5. New and regenerated primary production (mgN m^{-2} hr^{-1}) as estimated for Warm Core Ring systems in each month. A = A priori networks. I = Inverse networks.

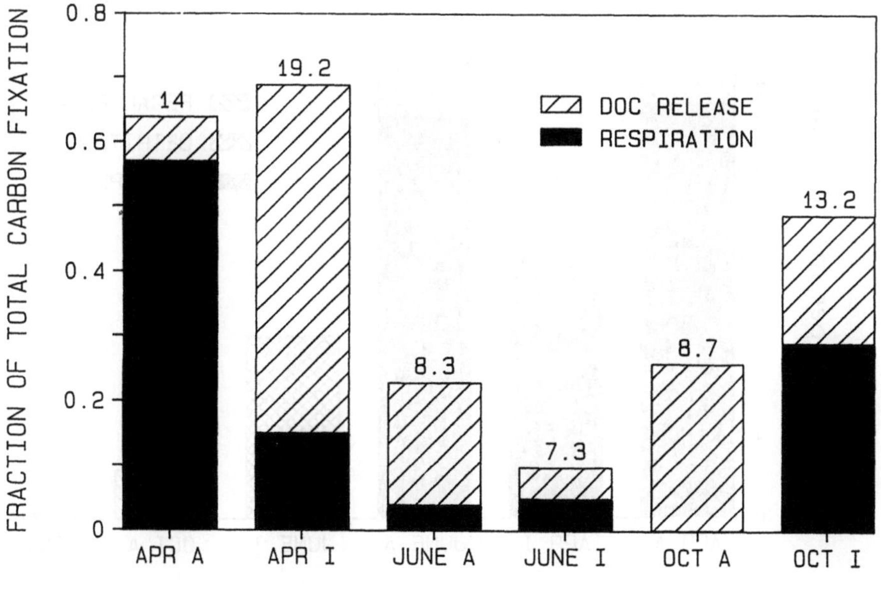

Figure 8.6. Losses from phytoplankton to respiration (////) and release of DOC (\\\\) in each month. Numbers above the bars are the C:N assimilation ratios (PP/P_n+U_r). A = A priori networks. I = Inverse networks.

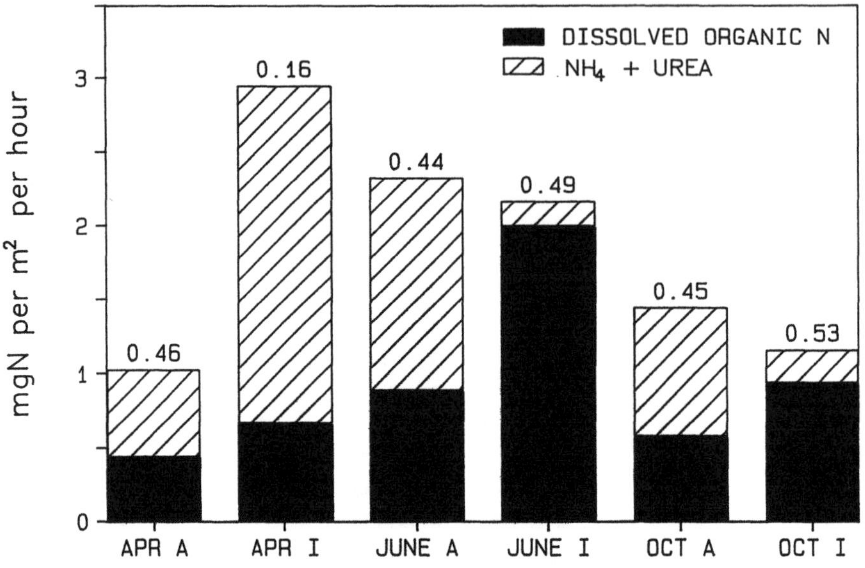

Figure 8.7. Inputs of dissolved organic Nitrogen (///) and ammonium + Urea (\\\\) to bacterioplankton in each month. Numbers above bars are the bacterial conversion efficiency for Nitrogen. A = A priori networks. I = Inverse networks.

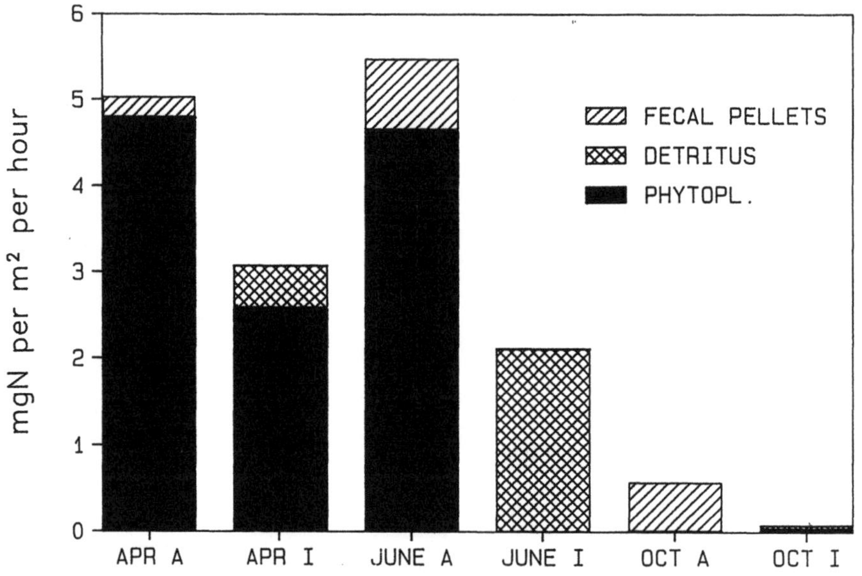

Figure 8.8. Composition of export as uningested phytoplankton (///), protozoan fecal material (\\\) and macrozooplankton fecal pellets (////) for each month. A = A priori networks. I = Inverse networks.

Analysis of the networks

We limited our analyses of these networks to the options available in the program NETWRK (Kay *et al.* 1989, this volume). The results of the analyses of network structure, Lindeman Trophic Aggregations and biogeochemical cycles are discussed in turn below. The theoretical basis and nature of each of these analyses are outlined in Ulanowicz (1984,1986).

Structure Analyses

Some of the simplest measures of community structure are matrices of the fractional inflows and outflows for each compartment. While they can be readily calculated from inspection of the flow diagrams shown in Figures 8.2-8.4, several flow analysis routines print the **G*** and **G**** matrices which summarize these data. By way of example we present in Table 8.8 some of the fractional outflows produced by FLOW for our networks. These data show the relative importance of each mechanism of removal for each component of the network: ingestion by different heterotrophs, excretion to the dissolved and detrital pools, direct export, and so on. There are variations between models and between months with some seasonal patterns evident. The fraction of NH_4^+ and Urea going to phytoplankton rose through time in both models, while the portion going to bacteria fell. Protozoan grazing grew in importance as a phytoplankton removal mechanism in the a priori networks, but fell slightly between June and October in the inverse solutions. Protozoan contributions to each pool were uniform in time in the a priori model as a consequence of our parameterizations, but varied greatly in the inverse model. It is evident from this analysis, that when the networks are investigated more closely, the two approaches obtained foodweb solutions with differing organization, which behaved differently in time. Thus although the broad outlines of the networks were similar, there were many differences in detail, as we show below.

The fractional output analysis is rather superficial, and does not take into account the effects of recycling, or the concept of "extended diets" (Ulanowicz 1986). For example, one can see from inspection of the network diagrams that bacterial nitrogen flows through the protozoans to the macrozooplankton and thus forms part of their diet. The structural analysis package available in NETWRK yields information on the total contributions from one compartment to another, over all possible direct and indirect pathways, and also the total dependency of a recipient compartment on each donor over all pathways. In this way, the effects of recycling can be assessed in a manner which is not obvious from inspection of the networks.

We have hypothesized that as the overall level of recycling increases, i.e. as the f-ratio falls, contribution and dependency should also increase. We also speculate that more complex analyses will reveal more differences among the networks. We test these ideas by examining the dependency relationships among: a) phytoplankton and regenerated DIN; b) phytoplankton and protozoans; and c) phytoplankton and macrozooplankton; d) phytoplankton and bacteria, e) bacteria and macrozooplankton and f) macrozooplankton and protozoans in each of our networks. The first four choices illustrate relationships more or less intimately associated with the support of regenerated production and with the f-ratio; while the latter two are related to the role of the microbial foodweb as a conduit or "link" between dissolved organic matter and higher trophic levels. The dependency of the phytoplankton and macrozooplankton on their own outputs are also given. The concept of dependency is defined as the fraction of the total input to a compartment which passed through a given donor compartment on its way to the recipient, over all possible pathways. The data are summarized in Table 8.9.

Table 8.8. Fractional outflows from selected compartments.
Proportion of total output from donor flowing directly to each recipient.

Donor-Recipient	APRIL A Priori	APRIL Inverse	JUNE A Priori	JUNE Inverse	OCTOBER A Priori	OCTOBER Inverse
$NH4^+$/Urea-Phytoplankton	0.67	0.31	0.73	0.86	0.80	0.92
$NH4^+$/Urea-Bacteria	0.33	0.69	0.27	0.02	0.20	0.08
Phytoplankton-Protozoa	0.17	0.03	0.41	0.80	0.90	0.71
Phytoplankton-Macrozoop	0.01	0.02	0.05	0.03	0.05	0.06
Phytoplankton-DON	0.04	0	0.05	0.16	0.05	0
Phytoplank.-Detritus*	0.88	0.96	0.49	0	0	0.08
Protozoa-Macrozoop	0.26	0.14	0.26	0.14	0.24	0.64
Protozoa-$NH4^+$/Urea	0.36	0.14	0.36	0.53	0.38	0.16
Protozoa-DON	0.04	0.31	0.06	0.01	0.04	0
Protozoa-Detritus	0.34	0.41	0.34	0.3	0.34	0.08

* In A priori model, the flow is phytoplankton export.

Table 8.9. Total dependency coefficients for selected interchanges.

Donor-Recipient	APRIL A Priori	APRIL Inverse	JUNE A Priori	JUNE Inverse	OCTOBER A Priori	OCTOBER Inverse
$NH4^+$/Urea-Phytoplankton	0.19	0.22	0.41	0.64	0.86	0.89
Protozoa-Phytoplankton	0.17	0.17	0.38	0.61	0.85	0.46
Macrozoop.-Phytoplankton	0.08	0.09	0.21	0.23	0.67	0.39
Bacteria-Phytoplankton	0.11	0.18	0.22	0.30	0.64	0.37
Phytoplankton-Phytopl.	0.19	0.19	0.41	0.64	0.85	0.45
Protozoans-Macrozoop.	0.91	0.50	0.89	0.92	0.98	0.96
Bacteria-Macrozoop.	0.35	0.25	0.36	0.36	0.69	0.51
Macrozoop.-Macrozoop.	0.14	0.23	0.26	0.25	0.68	0.50

The obvious, but still striking conclusion from this analysis is the large size of some of the dependencies compared to the direct outputs given in Table 8.8 above. For example, in April, 67 and 31% of the output from NH_4^+ and Urea go directly to phytoplankton in each model, yet 97-79% of total input to the phytoplankton passed through the NH_4^+/Urea pool on its way to the phytoplankton. An increasing amount of phytoplankton output returned to the phytoplankton, obviously as NH_4^+ or Urea, in the a priori networks, but not for all months in the inverse models. In the a priori solutions, the amount of phytoplankton output returning to its origin was equal to (1-f), but this was not the case for the inverse networks. A similar pattern was observed for the dependency of macrozooplankton on protozoans. In the a priori networks for June and October, 26-24% of the protozoan output passed directly to macrozooplankton, when 64-99% of the total input to the macrozooplankton passed through the protozoans before arrival. In the inverse solutions, the values were 14 and 64% for direct passage, and 36 and 72% for dependency. These comparisons clearly show the importance of recycling in mediating the total flow of medium to each compartment in the networks. All dependency values rose in time as f-ratios fell in the a priori solutions, but this pattern was not observed for all exchanges in the Inverse networks.

The dependency of macrozooplankton on bacteria is also striking. It is easy to calculate from the protozoan conversion efficiency (see Appendix 8A (evaluating the N flow parameters) and Table 8A.1) that 24-26% of the bacterial production ingested by the protozoans passed on to the zooplankton.

However when recycling is considered, it can be seen from Table 8.9 that up to 69% of the macrozooplankton input was mediated by the bacteria. It has been observed in mesocosm experiments that only a few per cent of labelled bacteria pass into large zooplankton (Ducklow *et al.* 1986). Since the dependency of macrozooplankton appears to rise as recycling does, it is tempting to speculate that the mesocosm results were due to low recycling. This was probably the case for the early spring experiment in a Scottish sea loch described by Ducklow *et al.* (1986).

These analyses show convincingly the growing interdependence of all components of the foodweb as recycled production increases in importance with time. The uniform increases in all dependencies in the a priori models suggest that increased recycling activity was accomplished without qualitative changes in foodweb structure in the networks. However the variable patterns of dependency with time in the inverse solutions suggest that these networks changed qualitatively to meet the boundary conditions set by observation. It is important to note that both sets of networks meet the same criteria, but do so in different ways. Either, or both sets of solutions might be correct, and in addition a nearly infinite array of other solutions could also meet these criteria. This point was made by Vézina & Platt (1987) and will be discussed further below.

Lindeman Trophic Aggregation

The results of these analyses are summarized in Tables 8.10 and 8.11. In the Lindeman analysis, the activities of the living components of the networks are aggregated into integer trophic levels reflecting the ingestion by each component of organisms below it in the food chains. The non-living compartments (in our networks, dissolved organic and regenerated nitrogen, and detritus) are pooled into an aggregated detrital component, which is contained in integer level 1. For example (Table 8.10), the Lindeman Transformation matrix for the October a priori network shows that integer level 1 contains 14% of the phytoplankton activity; level 2 contains 86% of the phytoplankton activity, 40% of the microzooplankton activity, 9% of the macrozooplankton activity, and all the bacterial activity; and so on. Reading down the columns of this matrix shows the apportioning of each network component into each trophic level. Thus 14% of the phytoplankton activity is in level 1 and 86% in is level 2.

Phytoplankton seem paradoxically to have a trophic position greater than one (Table 8.11) because they consume recycled nitrogen produced within the foodweb, as opposed to deriving all their sustenance from exogenous inputs. Thus the effective trophic level of the phytoplankton mirrors the f-ratio, being exactly equal to 1 + (1-f) (see Field *et al.* 1989, chapter 7, this volume). The results for both sets of networks are similar, with the trophic position of the phytoplankton increasing through time, as the f-ratio decreased. Dependence on recycled nitrogen leads in turn to an increase in the position of the metazoan consumers, and again the results of our two approaches are similar in that the trophic positions of the micro- and macrozooplankton increase through time. Microzooplankton feed primarily on detritus (level 1), phytoplankton (level 1 +), and bacterioplankton (level 2), and assume an effective level between 2 and 3 in all our networks. With the exception of the April inverse network, the macrozooplankton feed mostly on microzooplankton, and have a trophic position greater than 3. The exception is due to the high level of detritus ingestion by the macrozooplankton in April (Figure 8.2b).

Trophic Efficiency (Table 8.11) is a measure of the fraction of input to each of the aggregated (integer) trophic levels which is passed on to the next level. The trophic efficiency of each level increases through time, and decreases with increasing trophic level in all our networks. The temporal trend is attributable to the increasingly complete removal of phytoplankton production by grazers. By October, when all the living phytoplankton production (excluding release of DON) is grazed and none sinks, the efficiency of the first aggregated trophic level is 1, or slightly greater for the inverse network, due to an imbalance.

It is clear that the a priori and inverse methods yielded structurally similar networks in the sense that aggregation to the level of the linear Lindeman grazer chain showed the same trends and similar values for all months.

Table 8.10. Lindeman Transformation Matrix: October A Priori Network - showing proportion of activity of each living component in each level.

Level	Phytoplankton	Microzooplankton	Macrozooplankton	Bacteria
1	0.144	0	0	0
2	0.856	0.403	0.086	1
3	0	0.597	0.421	0
4	0	0	0.493	0

Table 8.11. Comparison of Lindeman Trophic Aggregations

	APRIL		JUNE		OCTOBER	
	A Priori	Inverse	A Priori	Inverse	A Priori	Inverse
			Effective Trophic Level			
Phytoplankton	1.19	1.22	1.41	1.64	1.86	1.89
Microzooplankton	2.30	2.01	2.37	2.59	2.60	2.81
Macrozooplankton	3.07	2.26	3.09	3.41	3.41	3.73
			Trophic Efficiency			
Level 1	0.50	0.65	0.80	0.79	1.00	1.11
Level 2	0.28	0.48	0.36	0.54	0.60	0.69
Level 3	0.16	0.12	0.17	0.12	0.20	0.52

Cycle Analysis.

Quite a different conclusion emerges from the cycle analysis, which serves to reveal details of the networks rather than reducing the complexity by aggregation. Cycle analysis is based on identifying and quantifying all the internal cycles inherent in a network, and classifying them by the smallest link (weak arc) in each. There can theoretically be many hundreds of cycles in even a small network, but most ecological networks (fortunately) have comparatively few. Thus Ulanowicz and Baird (1988) found only 61 cycles in a model of the Chesapeake Bay ecosystem with 36 compartments. They have suggested that identification and location of the weak arcs in network may reveal the mechanisms controlling the rates of cycling, in the sense that a rate limiting reaction controls the transformations in a system of chemical reactions (Ulanowicz 1986).

Our networks each contained 16 to 21 different cycles grouped among 9 or 10 weak arcs (Figure 8.9). All the a priori networks had the same 21 cycles and the same 10 weak arcs, though the cycles varied in which link was weak from month to month. In the inverse networks, the numbers of weak arcs and cycles, and the identity of the weak arcs varied from month to month. Our two approaches gave structurally quite different networks for each month, in the sense that in a given month the trophic exchanges controlling the rates of cycling within each system differed. Figure 8.9 shows the composite cycling for the two June networks. In each system, there are both short (e.g., microzooplankton-detritus-microzooplankton) and long (e.g. microzooplankton-macrozooplankton-DON-bacteria-NH_4^+/Urea- phytoplankton-microzooplankton) cycles, but some of the weak arcs are unique to one of the systems (e.g., NH_4^+/Urea-bacteria to the inverse), even when the link itself is in both.

In all cases, all the intercompartmental exchanges in each network were involved in cycling, showing the extensive nature of cycling mechanisms in plankton nitrogen flows, even when cycling itself is not dominant (i.e. when the f-ratio is high). This is in contrast to other networks which have been analyzed, like the Chesapeake Bay system, in which 14 of the 36 compartments are not engaged in cycling, and serve only to export or dissipate carbon (Ulanowicz & Baird 1988). The difference may be that all organisms release dissolved nitrogen which re-enters pools which are used by phytoplankton and bacteria, whereas carbon is respired and also exported and need not participate in recycling. It is also possible that our networks lack sufficient detail (i.e. have too few compartments) to reveal acyclic components.

The location of the weak (controlling) arcs differed according to the method used to derive the networks. For example, Figure 8.9 shows that in the a priori network, the phytoplankton and bacteria each had two weak arcs as outputs, suggesting that these components shared control of cycling. In contrast the two grazer components each had 2 weak arcs entering and two leaving. In the inverse network, control appeared to be shared among the phytoplankton (1 out), bacteria (1 in, 2 out) and macro- zooplankton (1 in, 2 out). The identity of the controlling group was constant over the 3 months in the a priori networks. In the inverse networks, control shifted from the phytoplankton (2 out) and microzooplankton (1 in, 4 out) in April, to a more equal sharing (phytoplankton 2 out; others 1 in 2 out) by October.

The quantitative results of the cycle analysis are presented in Tables 8.12 and 8.13. In April, there was significantly more cycling residing in 2- and 3-arc loops in the inverse networks, and about the same

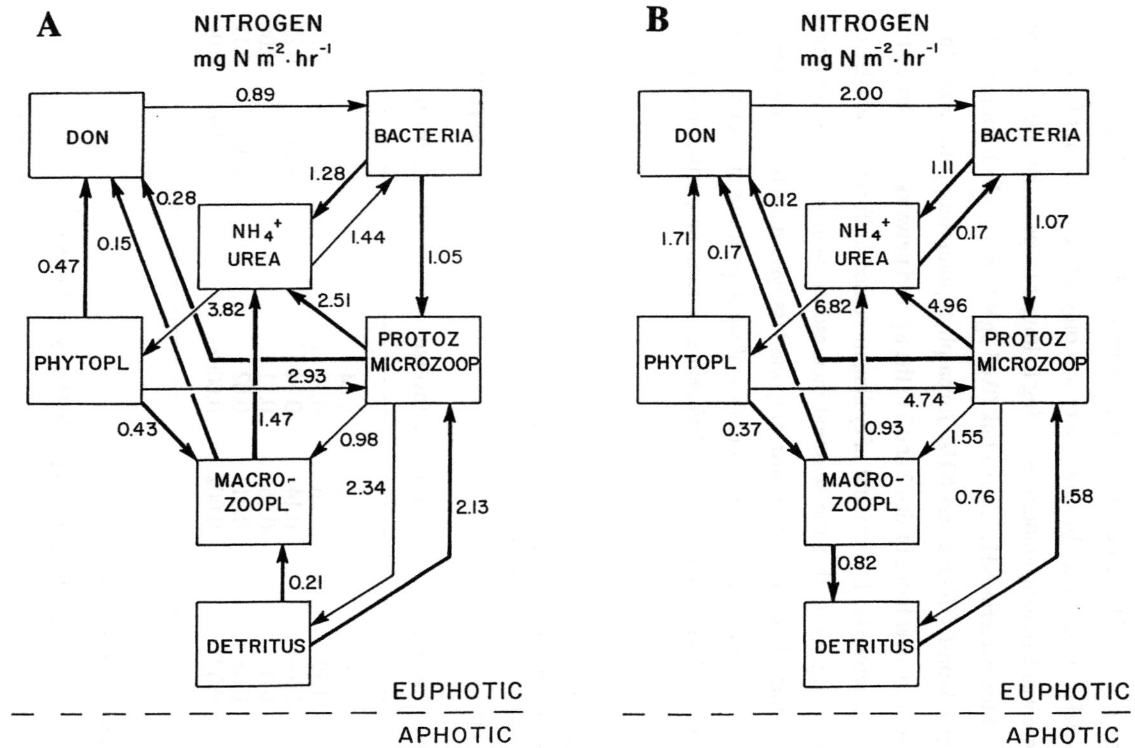

Figure 8.9. Composite cycling loops for June. Weak arcs are shown as bold arrows. Flow magnitudes are mgN $m^{-2} hr^{-1}$. A) A priori networks. B) Inverse networks.

amount in the the longer cycles. This is due to the large amount of cycling (2.28 μgN l^{-1} hr^{-1}) in the NH_4^+/Urea-bacteria-NH_4^+/Urea cycle. In June, there was only a small amount of cycling in 2-arc cycles. In October, there was a striking amount of cycling in longer loops in the inverse network, due again to the large size of the bacteria-NH_4^+/Urea link (0.65 μgN l^{-1} hr^{-1}), which controlled the cycle protozoans-macrozooplankton-DON-bacteria-NH_4^+/urea-phytoplankton-protozoans. In general, most of the cycling took place over loops of size 2 and 3 in the a priori networks, but shifted to 3-4 loop cycles in the June inverse networks. The amount of cycling increased uniformly in all size categories in the a priori networks, suggesting that the seasonal increase in cycling was not accompanied by changes in foodweb structure. In contrast, the variability was less systematic in the inverse networks. With the simultaneous variability in identity and number of weak arcs, this supports the conjecture that the inverse method produced foodwebs with different structural properties in each month (cf. Appendix B).

Table 8.12. Comparison of Cycle Analyses.
Fraction of total system throughput in loops with 2 to 7 arcs.

Loop Size	APRIL A Priori	APRIL Inverse	JUNE A Priori	JUNE Inverse	OCTOBER A Priori	OCTOBER Inverse
2	0.11	0.28	0.17	0.04	0.23	0.39
3	0.15	0.39	0.26	0.37	0.38	0.13
4	0.13	0.14	0.14	0.14	0.19	0.25
5	0.02	0.03	0.05	0.11	0.07	0.06
6	0.007	0.06	0.01	0.02	0.02	0.17
7	0.001	0.002	0.002	0.0005	0.0009	0.04

Table 8.13. Comparison of Cycle Analyses.

Cycling Efficiency of System Components[1]

	APRIL A Priori	APRIL Inverse	JUNE A Priori	JUNE Inverse	OCTOBER A Priori	OCTOBER Inverse
Phytoplankton	0.19	0.22	0.41	0.64	0.85	0.45
Microzooplankton	0.54	0.37	0.64	0.67	0.91	0.57
Macrozooplankton	0.14	0.25	0.26	0.25	0.69	0.50
Bacterioplankton	0.41	0.71	0.44	0.32	0.72	0.44
Regenerated-N	0.44	0.71	0.56	0.65	0.88	0.49
Dissolved Org. N	0.12	0.17	0.16	0.29	0.49	0.40
Detritus	0.41	0.45	0.47	0.29	0.79	0.19
Finn Cycling Index[2]	0.42	0.54	0.64	0.67	0.89	0.68

[1]The fraction of output which is recycled to that compartment.
[2]Total cycling index.

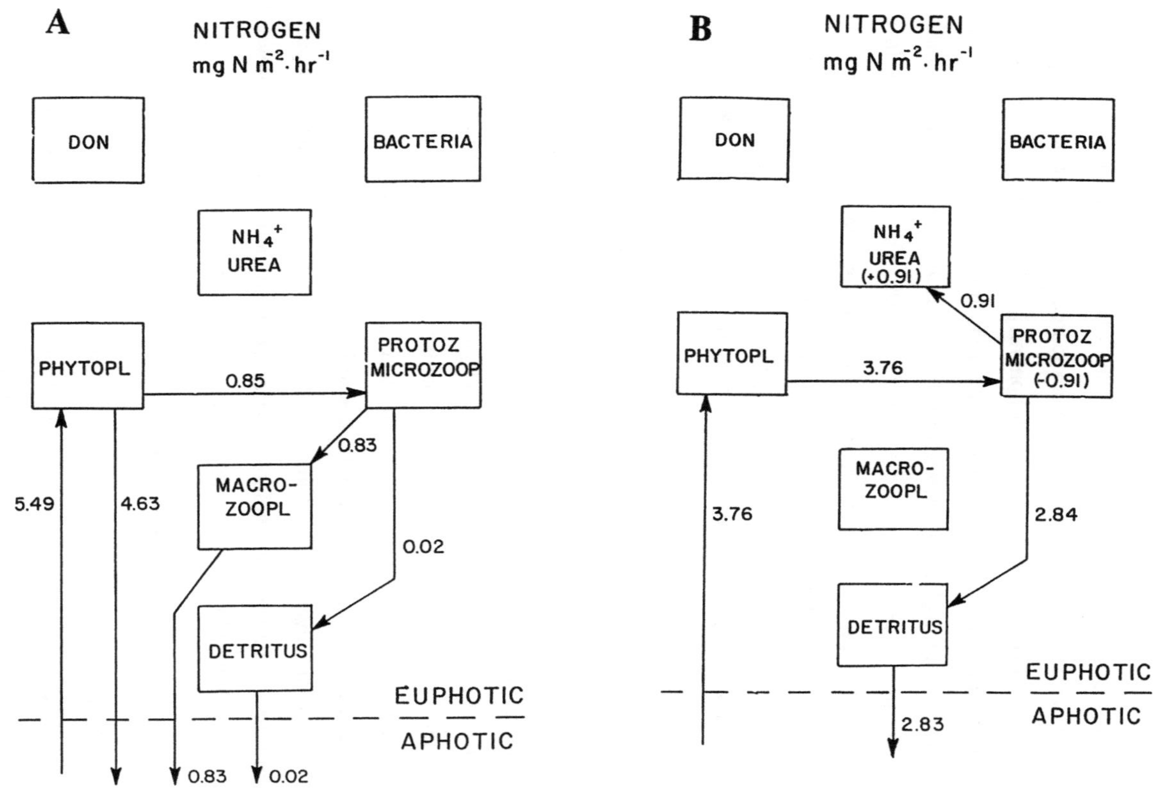

Figure 8.10. Residual, acyclic networks for June showing flow of noncycled material through and exported from the A priori (A) and Inverse (B) networks.

The cycling efficiency index measures the fraction of the throughput in a compartment which eventually returns to that compartment, over all pathways. Table 8.13 shows that in the a priori networks, the efficiency of each component, and the total (Finn) cycling index increased with time, again suggesting that the overall seasonal increase in cycling was achieved without major changes in foodweb structure. The microzooplankton had the highest efficiency in all 3 months, indicating their importance in cycling even though they do not appear to control it. The October values are similar to those found by Fasham (1985) in an analysis of nitrogen flows in a generic plankton network. The results for the inverse network are less uniform, in spite of monthly decreases in the f-ratio. For example, the efficiency for regenerated-N decreased with time, while it increased for DON, and showed no pattern for other components. The Finn index did not change between June and October, even though the f-ratio decreased 3-fold. These data also suggest that the inverse method obtained network solutions with different trophic organization for the 3 months.

Once the total amount of cycled flow is removed from the total system of flows in each network, the matrix of residual acyclic flows remains. These are the direct flows from each input to the system (new production in our networks) through the network and out again via export or respiration. The residual acyclic networks produced by the a priori model were structurally similar for all months, revealing that acyclic flow was primarily split between direct output from phytoplankton, and export from macrozooplankton, which was derived mostly from protozoan herbivory (Figure 8.10a). Because the inverse networks were not always in exact steady state, the residual networks show some dead end flows (Figure 8.10b). In general however, they showed variations on the same theme: export via uningested phytoplankton or through the phytoplankton-protozoan conduit. The acyclic flows also suggest that there was more structural foodweb variation in the Inverse networks.

CONCLUSION

The f-ratio provides an easily-understood and simple-to-calculate index of community nutrient status. However variations in f-ratio do not appear to be linked to variations in community structure in any simple way. Community structure and cycling patterns changed greatly between April and June, while the f-ratio declined only slightly. Flow analysis offers a formal method for analyzing community structure. The present results suggest that much of the variability in more complex indices of foodweb organization revealed by flow analysis can be related to variations in the f-ratio. Thus our results point to the importance of the f-ratio as a conceptual tool for categorization of plankton communities, and reveal some of the mechanisms regulating the balance between new and regenerated production. Our models also help to suggest which variables (flows and internal parameters like growth efficiency) need to be measured for better characterization of plankton systems. The key measurements appear to be herbivory, detritovory, and fecal production by microzooplankton and protozoans, NH_4^+/Urea cycling by bacteria, and the sinking of uningested phytoplankton and detritus.

Our networks can be viewed as hypotheses about the general structure of plankton systems as they evolve from well-mixed, nutrient-sufficient conditions in spring to nutrient-limited conditions in stratified waters in mid- and late summer. The f-ratio provides a simple index of the overall pattern of nutrient cycling and export as the plankton system evolves through the year. Our analyses have shown that changes in the f-ratio are accompanied or perhaps caused by changes in the relative magnitude and organization of the system of intercompartmental flows. We have presented two different approaches to constructing plankton flow networks which meet the criteria set by existing data. These two approaches in effect present two alternative hypotheses about the mechanisms underlying the seasonal decrease in the f-ratio: the a priori solutions suggest that the shift from an inefficient system which exports most of its throughput (high f) to more efficient (low f) systems can be accomplished by an uniform increase in the cycling efficiency of all system components, and without any changes in the linkages among compartments (topology). The inverse solutions support the idea that this trend is achieved from changes in foodweb structure including more complex changes in the cycling efficiency of individual components, and in the location of rate-controlling linkages in the trophic network.

We offer no proof that one hypothesis is more likely or preferable to the other. Both may be true in different regions or when the system is viewed at different scales. Further sampling in diverse regions (subpolar vs subtropical gyres, upwelling provinces, etc.), and more intensive measurement efforts will help to resolve the dilemma. Until much more data is available to permit closure of each system of flows, we cannot specify which class of solutions is most likely. Flow analysis can help to plan and guide these efforts, and will reveal patterns in the data not easily obtained by other means.

ACKNOWLEDGEMENTS

This research was sponsored in part by NSF grants OCE 84-06524 to H.W. Ducklow, who wishes to thank in particular his colleagues J. McCarthy, M. Roman and P. Wiebe for lending unpublished data and for discussions about Warm Core Rings; J. Fuhrman and G. McManus for reading a draft of the manuscript, and R. Ulanowicz for discussions on flow analysis.

REFERENCES

Azam, F. & R.E. Hodson 1977. Size distribution and activity of marine microheterotrophs, Limnol. Oceanogr. 22: 492-501.

Banse, K. 1984. Review of "Marine planktonic protozoa and microzooplankton ecology," Limnol.Oceanogr. 29: 445-446

Bjornsen, P.K. 1986. Bacterioplankton growth yield in continuous seawater cultures, Mar.Ecol. Prog.Ser. 30: 191-96.

Chervin, M.B. 1978. Assimilation of particulate organic carbon by estuarine and coastal copepods, Marine Biol., 49: 265-275.

Ducklow, H.W. 1983. The production and fate of bacteria in the ocean, BioScience 33:494-501.

Ducklow, H.W. 1986. Bacterial biomass in warm core Gulf Stream ring 82B: Mesoscale distributions, temporal changes, and production, Deep Sea Res. 33: 1789-1812.

Ducklow, H.W., D.A. Purdie, P.J. LeB. Williams & J.M. Davies 1986, Bacterioplankton: A sink for carbon in a coastal plankton community, Science 232: 865-67.

Dugdale, R.C. & Goering, J.J. 1967. Uptake of new and regenerated forms of nitrogen in primary productivity, Limnol. Oceangr. 12: 196-206.

Eppley, R.W. & B.J. Peterson 1979. Particulate organic matter flux and planktonic new production in the deep ocean, Nature 282: 677-680.

Fasham, M.J.R.(ed.) 1984. *Flows Of Energy And Materials In Marine Ecosystems; Theory And Practice,* New York: Plenum, 733 pp.

Fasham, M.J.R. 1985. Flow analysis of materials in the marine euphotic zone, Can. Bull. Fish. Aquat. Sci. 213: 139-162.

Field, J.G., F. Wulff & K.H. Mann 1989. The need to analyze ecological networks, In: Wulff, F., J.G. Field & K. H. Mann(eds) *Network Analysis in Marine Ecology. Methods and Applications,* Lecture Notes in Coastal and Estuarine Studies, Springer-Verlag, New York.

Field, J.G., C.L. Moloney & C.G. Attwood 1989. Network analysis of simulated succession after an upwelling event. In: Wulff, F., J.G. Field & K. H. Mann(eds) *Network Analysis in Marine Ecology. Methods and Applications,* Lecture Notes in Coastal and Estuarine Studies, Springer-Verlag, New York.

Franks, P.J.S., J.S. Wroblewski, & G.R. Flierl 1986. Prediction of phytoplankton growth in response to the frictional decay of a warm-core ring, J.Geophys.Res. 91C: 7603-7610.

Frost, B. W. 1984. Utilization of phytoplankton production in the surface layer, In Global Ocean Flux Study: Proceedings of a workshop. Washington, D.C.: Natl Academy Press, pp 125-135.

Goldman, J.C., D.A. Caron, O.K. Andersen & M.R. Dennett 1985. Nutrient cycling in a microflagellate food chain: I. Nitrogen dynamics, Mar. Ecol. Prog. Ser., 24: 231-242.

Goldman, J.C., D.A. Caron & M.R. Dennett 1987. Regulation of gross growth efficiency and ammonium regeneration in bacteria by substrate C:N ratio, Limnol. Oceanogr. 32: 1239-1252.

Harrison, W.G., T. Platt & M.R. Lewis 1987. f-Ratio and its relationship to ambient nitrate concentration in coastal waters, J. Plankton Res. 9: 235-248.

Hitchcock, G.L., C. Langdon & T.J. Smayda 1985. Seasonal variations in the phytoplankton biomass and productivity of a warm core Gulf Stream Ring, Deep Sea Res. 32: 1287-1300.

Hobbie, J.H., R.J. Daley & S. Jasper 1977. Use of Nuclepore filters for counting bacteria by epifluorescence microscopy, Appl. Environ. Microbiol. 33: 1225-1228.

Johnson, P.W. & J. McN. Sieburth 1979. Chroococcoid cyanobacteria in the sea: A ubiquitous and diverse phototrophic biomass, Limnol. Oceanogr. 24: 928-35.

Joint, I.R. & R.J. Morris 1982. The role of bacteria in the turnover of organic matter in the sea, Oceanogr. Mar. Biol. Annu. Rev. 20: 65-118.

Joyce, T. & P. Wiebe 1983. Warm Core Rings of the Gulf Stream, Oceanus 26: 34-44.

Joyce, T.M. 1985. Gulf Stream warm-core ring collection: An introduction, J. Geophys. Res., 90: 8801-8802.

Kay, J.J., L. Graham & R.E. Ulanowicz 1989. A detailed guide to network analysis, In: Wulff, F., J.G. Field & K. H. Mann (eds). *Network Analysis in Marine Ecology. Methods and Applications.* Lecture Notes in Coastal and Estuarine Studies, Springer-Verlag, New York.

Landry, M.R. 1977. A review of important concepts in the trophic organization of pelagic ecosystems, Helgolander wiss. Meeresunters 30: 8-17.

Landry, M.R., L.W. Haas & V.L. Fagerness 1984. Dynamics of microbial plankton communities: experiments in Kaneohe Bay, Hawaii, Mar. Ecol. Prog. Ser. 16: 127-133.

Laws, E.A., W.G. Harrison & G.R. DiTullio 1985. A comparison of nitrogen assimilation rates based on ^{15}N uptake and autotrophic protein synthesis, Deep-Sea Res. 32: 85-95.

McCarthy, J.J. & J.L. Nevins 1986. Utilization of nitrogen and phosphorus by primary producers in warm-core ring 82-B following deep convective mixing, Deep Sea Res., 33: 1773-1788.

Michaels, A.F. & M.W. Silver 1988. Primary production, sinking fluxes and the microbial foodweb, Deep-Sea Res. 35: 473-490.

Pace, M.L., J.E. Glasser & L.R. Pomeroy 1984. A simulation analysis of continental shelf food webs, Mar. Biol. 82:47-63.

Paffenhofer, G.A. & J.D.H.Strickland 1970. A note on the feeding of Calanus helgolandicus on detritus, Marine Biol. 5:97-99.

Paffenhofer, G.A. & S.C. Knowles 1979. Ecological implications of fecal pellet size, production and consumption by copepods, J. Mar. Res. 37: 35-49.

Parker, R.L. 1977. Understanding inverse theory, Ann. Rev. Earth Planet. Sci. 5: 35-64.

Patten, B.C., R.W. Bosserman, J.T. Finn & W.G. Gale 1976. Propagation of cause in ecosystems, pp. 458-579 In: B.C. Patten, Ed., Systems Analysis and Simulation in Ecology, vol. IV. Academic Press, New York.

Payne, W.J. & W.J. Wiebe 1978. Growth yield and efficiency in chemosynthetic microorganisms, Ann. Rev. Microbiol. 32: 155-183.

Platt, T., K.H. Mann & R.E. Ulanowicz (eds) 1981. *Mathematical Models In Biological Oceanography,* Monogr. Oceanographic Methodology 7, Paris:UNESCO, 157 pp.

Pomeroy, L.R. 1974. The ocean's food web: A changing Paradigm, BioScience 24: 499-504.

Rassoulzadegan, F. & R.W. Sheldon 1986. Predator-prey interactions of nanozooplankton and bacteria in an oligotrophic marine environment, Limnol. Oceanogr., 31: 1010-1021.

Roman, M.R. 1984. Utilization of detritus by the copepod, *Acartia tonsa*, Limnol.Oceanogr. 29: 949-959.

Roman, M.R., C.S. Yentsch, A.L. Gauzens & D.A. Phinney 1986. Grazer control of the fine-scale distribution of phytoplankton in warm-core Gulf Stream rings, J. Mar. Res. 44: 795-813.

Smetacek, V. 1985. Role of sinking in diatom life history cycles: ecological, evolutionary, and geological significance, Mar. Biol. 84: 239-251.

Steele, J.H. 1974. *The Structure of Marine Ecosystems,* Cambridge, Harvard, 128 pp.

Steele, J.H. & B.W. Frost 1977. The structure of plankton communities, Phil. Trans. Royal Soc. London B. Biol. Sci., 280: 485-534.

Suttle, C.A., J.A. Fuhrman & D.G. Capone 1987. Rapid ammonium turnover times and concentration dependent resource partitioning in planktonic communites measured using ^{13}N, EOS 68: 1760.

Turley, C.M. 1985. Biological studies in the vicinity of a shallow-sea tidal mixing front. IV. Seasonal and spatial distribution of urea and its uptake by phytoplankton, Phil. Trans. R. Soc. Lond. B310:471-500.

Ulanowicz, R.E. 1984. Community measures of food networks and their possible applications, pp 23-48 In: In M. Fasham (ed.), *Flows of energy and materials in marine ecosystems: Theory and practice*, Plenum Press.

Ulanowicz, R.E. & T. Platt (eds) 1985. *Ecosystem Theory for Biological Oceanography,* Can. Bull. Fish. Aquat. Sci., 213: 260 p.

Ulanowicz, R.E. 1986. *Growth and Development: Ecosystems Phenomenology,* Springer-Verlag, New York. 232 pp.

Ulanowicz, R.E., & D. Baird 1989. The seasonal dynamics of the Chesapeake Bay ecosystem: III. Structure of recycling. Ecology, in press.

Vézina, A.F. 1989 Construction of flow networks using inverse methods, In: Wulff, F., J.G. Field & K. H. Mann(eds) *Network Analysis in Marine Ecology. Methods and Applications,* Lecture Notes in Coastal and Estuarine Studies, Springer-Verlag, New York.

Vézina, A.F. & T. Platt. 1987. Small-scale variability of new production and particulate fluxes in the ocean, Can. J. Fish. Aquat. Sci., 44: 198-205.

Vézina, A.F. & T. Platt 1988. Foodweb dynamics in the ocean. Part 1. Best-estimates of flow networks using inverse methods. Mar. Ecol. Prog. Ser. 42: 269-287.

Waterbury, J.B., S.W. Watson, R.R. Guillard, & L.E. Brand 1979. Widespread occurrence of a unicellular, marine, planktonic cyanobacterium, Nature 277:392-94.

Welschmeyer, N.A., and C.J. Lorenzen 1985. Chlorophyll budgets: Zooplankton grazing and phytoplankton growth in a temperate fjord and the Central Pacific Gyres, Limnol. Oceanogr. 30: 1-21.

Wheeler, P.A. & D.L. Kirchman 1986. Utilization of inorganic and organic nitrogen by bacteria in marine systems, Limnol. Oceanogr. 31: 998-1009.

Wiebe, P.H. & T.J. McDougall (Eds) 1986. Warm Core Rings: Studies of their Physics, Chemistry, and Biology, Deep-Sea Res, 33: 1455-1922.

Williams, P.J. leB. 1981. Incorporation of microheterotrophic processees into the classical paradigm of the planktonic food web, Kieler Meeresforsch. 5: 1-28.

Wunsch, K. 1978. The North Atlantic general circulation west of 50° West determined by inverse methods, Rev. Geophys. Space Phys. 16: 583-590.

APPENDIX 8A : DERIVATION OF FLOW NETWORKS (A PRIORI APPROACH)

Estimating the nitrogen flows

Model structure

The model structure (Fig. 8.1) is designed to represent the main nitrogen flows in the euphotic zone and consists of seven compartments viz., 1) phytoplankton, 2) protozoans and microzooplankton less than 333 μm in size, 3) macrozooplankton greater than 333 μm in size, 4) free-living bacterioplankton, 5) labile dissolved organic nitrogen (DON), 6) regenerated ammonium and urea, and 7) suspended (or slowly sinking) euphotic zone detritus, derived from compartment 2.

It is assumed that ammonium and urea are produced by the excretion processes of zooplankton (compartments 2,3) and bacteria and are taken up by both phytoplankton and bacteria. Zooplankton and phytoplankton are assumed to excrete DON which is (in our model) the main substrate utilized by the free bacteria. In this simple model nitrate nitrogen is only utilized by phytoplankton, while urea is considered to be regenerated nitrogen available to phytoplankton and bacteria. Turley (1985) has speculated that urea is utilized primarily or solely by phytoplankton. This hypothesis remains untested in natural waters, and for the sake of simplicity, we include it with NH_4^+ in the regenerated DIN pool, rather than adding yet another compartment to the model. Goldman *et al.* (1987) have shown that cultured bacteria enriched from natural waters do not take up urea. Thus our model may require elaboration to simulate microbial nitrogen dynamics in nature. In our model, there is no export of DON and the entire pool is assumed to cycle (i.e. we did not add a pool of refractory DON).

Both size classes of zooplankton produce fecal products, but it is assumed that the macrozooplankton pellets have a fast sinking rate and so sediment out of the euphotic zone before any significant re-ingestion can take place. In contrast, the protozoan/microzooplankton fecal pellets are assumed to sink slowly (i.e. less than 10-50 meters per day) and so form part of the euphotic zone detritus capable of being reingested by both size classes of zooplankton. Any ungrazed detritus or phytoplankton is considered to have sedimented out of the euphotic zone. This loss, together with the sedimenting phytoplankton and macrozooplankton fecal pellets, represents the net particulate export from the euphotic zone and will, at steady state, be equal to the new, or nitrate based, primary production (Eppley & Peterson 1979).

The protozoans/microzooplankton are assumed to graze bacteria, phytoplankton and detritus, while the macrozooplankton are assumed to graze phytoplankton, protozoans/microzooplankton and detritus (see section: Estimating the carbon flows below).

Calculation of protozoan/microzooplankton grazing on phytoplankton

The problem to be solved is as follows: we have experimental measurements of bacterioplankton production, macrozooplankton ingestion of phytoplankton, and particulate uptake of NO_3^- and NH_4^+ by ^{15}N tracer techniques and we wish to estimate the ingestion of phytoplankton by the protozoa/ microzooplankton and as many other of the fluxes shown in Fig. 8.1 as possible. The problem is obviously grossly under-determined in that there are 20 fluxes to be calculated from only four measurements. We attempt to solve this problem using the technique of input-output flow analysis, which assumes that the system is in steady-state and that any unmeasured output flows from a compartment are proportional to the total compartmental throughput (Patten *et al.* 1976). The latter assumption is commonly made by ecologists when employing the concepts of assimilation efficiency and gross growth efficiency. The flow nomenclature used is defined in Fig. 8.1. and Table 8A.1. If it is assumed that heterotrophs excrete regenerated nitrogen in proportion to their ingestion then, at steady-state, the uptake of regenerated nitrogen by phytoplankton and bacteria must be related, in some way, to total zooplankton grazing. If this relation could be determined then G_p, the ingestion of phytoplankton by protozoans and microzooplankton, could be calculated.

The first problem concerns the interpretation of the measured uptake of NH_4^+ plus urea, U_r. The ^{15}N technique used does not distinguish between uptake by phytoplankton or by bacteria, and so

$$U_r = P_r + e \qquad (1)$$

(P_r is the phytoplankton production based on regenerated nitrogen, in contrast to new production, P_n, based on nitrate uptake (Eppley & Peterson 1979)). If the balance of flows through the U_r compartment is now considered, then

$$U_r = P_r + e = j_p + j_z + a \qquad (2)$$

The flow a can be obtained from the measured bacterial carbon production (H) by defining a gross growth efficiency k_N and the bacterial C:N ratio R_B. Thus, the bacterial nitrogen production h is given by

$$h = H/R_B = k_N\,(e + d) \qquad (3)$$

Equating the nitrogen flows through the bacterial compartment gives

$$e + d = h + a, \qquad (4)$$

Whence using equation (3)

$$a = [(1 - k_N)/k_N]h = \omega h, \qquad (5)$$

where we have defined $\omega = (1 - k_N)/k_N$.

The next step is to obtain expressions for j_p and j_z. This is done by defining the following constants:

α_z = fraction of macrozooplankton ingestion excreted as NH_4^+ and/or urea

α_p = fraction of protozoans/microzooplankton ingestion excreted as NH_4^+/urea

k_p = gross growth efficiency of protozoans/microzooplankton

Using these definitions gives

$$j_p = \alpha_p(G_p + h + f_p) \tag{6}$$
$$j_z = \alpha_z(G_z + P + f_z) \tag{7}$$

where P, the protozoan production, is given by

$$P = k_p(G_p + h + f_p) \tag{8}$$

It only remains now to determine the detrital ingestion rates f_p and f_z and the problem of determining G_p will be solved. It is first necessary to obtain an expression for the detrital compartmental throughput T_D. Equating T_D to the sum of the inputs to the detrital compartment gives

$$T_D = t_p \tag{9}$$

To proceed we define the following parameters;

β_z = fraction of detrital throughput ingested by macrozooplankton,

β_p = fraction of detrital throughput ingested by protozoa/microzooplankton,

ζ_z = fraction of macrozooplankton ingestion excreted as DON,

ζ_p = fraction of protozoa/microzooplankton ingestion excreted as DON.

Using these definitions we can now write

$$T_D = (1 - \alpha_p - \zeta_p - k_p)(G_p + h + f_p) \tag{10}$$

Substituting $f_p = \beta_p T_D$ and rearranging gives

$$T_D = C_1(h + G_p) \tag{11}$$

where

$$C_1 = (1 - \alpha_p - \zeta_p - k_p)/(1 - \beta_p(1 - \alpha_p - \zeta_p - k_p))$$

Using the equation for T_D and the definitions of β_p and β_z we can now substitute for f_p and f_z in equations (6) and (7) to give explicit expressions of j_p and j_z containing only measured flows, the unknown G_p and the parameters α_z, ζ_z, β_z, α_p, ζ_p, β_p, k_p. Returning now to equation (2) and using equation (5) we can, after some rearranging, obtain an equation for G_p. To simplify this expression we first define the quantity

$$C_2 = \alpha_p \beta_p + \alpha_z \beta_z + \alpha_z k_p \beta_p$$

and then

$$G_p = [U_r - (\omega + \alpha_p + \alpha_z k_p + C_1 C_2) h - \alpha_z G_z]/[\alpha_p + \alpha_z k_p + C_1 C_2]$$

Assuming that we are able to assign reasonable values to the eight parameters, α_z, β_z, α_p, β_p, ζ_p, k_p, k_N and R_B, it is now possible to calculate G_p from the measured flows H, U_r and G_z. (If the macrozooplankton grazing G_z was measured in carbon units then a ninth parameter, the carbon-to-nitrogen ratio of phytoplankton will be required to convert this flow to nitrogen units).

Calculation of remaining flows

It now remains to calculate the remaining flows in Fig. 8.1. As we have assumed that nitrate is taken up only by phytoplankton then it follows that P_n is equal to the measured uptake of ^{15}N nitrate. A number of other flows can be derived using previous definitions and the calculated value of G_p. Thus,

$$m_z = \zeta_z (G_z + p + f_z) \qquad (13)$$

$$m_p = \zeta_p (G_p + h + f_p) \qquad (14)$$

and

$$t_z = (1 - \alpha_z - \zeta_z)(G_z + p + f_z) \qquad (15)$$

When considering the flows through the macrozooplankton compartment the fate of the macrozooplankton production must be discussed. In the ocean this production will be consumed by the next trophic level, carnivorous zooplankton and fish. This trophic level has not been included in the model but its effect can be allowed for by distributing the production (in the range of, say, 10-20% of ingestion) amongst the outflow to DON, ammonium and detritus. This point should be remembered when choosing values for α_z and ζ_z.

The sedimenting detritus is that fraction of the detrital throughput not grazed by zooplankton, and so

$$r = (1 - \beta_z - \beta_p) T_D \qquad (16)$$

The flow s can be obtained by equating flows across the euphotic-aphotic boundary, giving

$$s = P_n - t_z - r \qquad (17)$$

It now only remains to calculate the flows c, d and e associated with bacterial utilization of DON and for this it is necessary to introduce another parameter, γ_N, the fraction of total phytoplankton production that is excreted as DON. The flow c is therefore given by

$$c = \gamma_N (P_n + P_r)$$

$$= \gamma_N (P_n + U_r - e) \qquad (18)$$

From the balance of flow s through the DON compartment we have

$$d = c + m_p + m_z \tag{19}$$

Using equations (3) and (18) to substitute for d and c and rearranging gives

$$e = [h/k_N - \gamma_N(P_n + U_r) - m_z - m_p] / (1 - \gamma_N) \tag{20}$$

Having calculated e it is then a simple matter to calculate c and d.

Evaluating the Nitrogen flow parameters

The a priori approach required the explicit evaluation of 11 parameters (including the C:N ratios for phytoplankton and bacterial biomass production) for the nitrogen networks (see above) and an additional 4 parameters (including the detrital C:N ratio) for the carbon networks (Table 8A.1). These parameters were evaluated using values from the literature, as reviewed in Fasham (1985) and Vezina & Platt (1987). However because the known ranges for some parameter values are large (e.g. the bacterial conversion efficiency for carbon is thought to range from ca. 0.2 to 0.9; Bjornsen 1986), we also performed sensitivity analyses in order to examine the dependence of model solutions upon particular choices for parameter values. The results of these analyses are shown in Figures 8A.1 and 8A.2. Eight cases are presented. Two choices of bacterial production (i.e. $k_N = 0.3$ or 0.6), two choices of protozoan production ($k_p = 0.2$ or 0.4) and two choices of excretion ($\alpha_p = 0.2/0.4, \alpha_z = 0.7/0.4$; ζ_p and $\zeta_z =$ 10% of the corresponding alpha value) were analyzed using the June input values, and the results can be summarized as follows:

1. G_p was inversely proportional to β_p and β_z with the change being fairly linear except in the low excretion cases. This is because if the herbivores ingest significant amounts of detrital N, their intake of phytoplankton (G_p) need not be as high to satisfy the observed levels of U_r.

2. Increasing k_N increases the magnitude and range of G_p for the other parameter choices.

3. Increasing the protozoan production (k_p) decreases the G_p values for low beta values, but this is not a very sensitive parameter.

4. Increasing the excretion parameters reduces the magnitude and range of G_p. These are sensitive parameters: a doubling of these parameter values produces a halving of G_p, because for the same amount of ingestion, larger excretion parameters produce the required U_r flows.

With these insights into the behavior of the model, we picked final parameter values by considering two other issues: ingestion of detritus, and flows through the microbial foodweb. The microbial foodweb is schematized in our model primarily by the flows into and out of the protozoan compartment. The nature of this compartment, and its relationship to a more realistically complex microbial foodweb are discussed in the next section. The detrital parameters were evaluated by adopting a simple criterion relating microzooplankton and macrozooplankton grazing to phytoplankton (G_p, G_z) and detritus (f_p, f_z) as described here.

Table 8A.1. A Priori Model variables and parameter values.

Measured variable	Symbol	Element	Apr	June	Oct
Macrozooplankton ingestion of phytoplankton	$G_z(C)$	C	0.46	2.47	1.09
Total uptake of regenerated NH_4^+ + urea	U_r	N	1.79	5.26	4.32
Bacterial production	H(C)	C	2	4.5	2.8
Total NO_3^- uptake (new production)	P_n	N	5.04	5.49	0.58
Primary Production	PP	C	75	66	30
Parameter	**Symbol**	**Element**	**Apr**	**June**	**Oct**
Phytoplankton C:N ratio (atoms)	R	--	6.7	6.7	6.7
Bacterial C:N ratio (atoms)	R_B	--	5	5	5
Detritus C:N ratio (atoms)	R_{det}	--	9	9	9
Bacterial conversion efficiency	k_C	C	0.30	0.30	0.30
Bacterial conversion efficiency	k_N	N	0.45	0.45	0.45
Protozoan conversion efficiency	k_p	N	0.26	0.26	0.24
Protozoan conversion efficiency	$k_p(C)$	C	0.14	0.15	0.14
Protozoan respiration (% input)	$\alpha_p(C)$	C	0.40	0.40	0.43
Protozoan NH_4^+ excretion (% input)	α_p	N	0.36	0.36	0.38
Protozoan DON excretion (% input)	ζ_p	N	0.04	0.04	0.04
Protozoan detritus ingestion (%tot)	β_p	N	0.92	0.90	0.94
Macrozoop NH_4^+ excretion (% input)	α_z	N	0.60	0.60	0.62
Macrozoop DON excretion (% input)	ζ_z	N	0.06	0.06	0.07
Macrozoop detritus ingestion (%tot)	β_z	N	0.08	0.09	0.06
Phytoplankton DON excretion (% input)	γ_N	N	0.05	0.05	0.05

It is a well established observation in the field (Paffenhofer & Strickland 1970; Chervin 1978) and laboratory (Roman 1984) that zooplankton will ingest and can assimilate non-living particulate matter including fecal matter (Paffenhofer & Knowles 1979). Our model required evaluation of two parameters, β_p and β_z, which specify the fraction of the detrital production which would be ingested by the micro- and macrozooplankton. These values can clearly range from $\beta_p = \beta_z = 0$ to $\beta_p + \beta_z = 1.0$, for at steady state, the ingestion cannot exceed the input. However the total ingestion need not reach its maximum values, and if the sum is less than 1.0, the remaining detrital input is exported. It is also clear that a large number of choices for the two parameters will yield solutions to the networks

(Figures 8A.1-2). Some further criterion was needed to constrain the range of possible choices. We chose the simple criterion that β_p/β_z must $= G_p/G_z$. In biological terms, this means that we required that each group would eat the same proportion of the phytoplankton and detrital production. In order to meet this criterion, the beta values were adjusted to produce the proper ratio G_p/G_z for each month. For each month, it was necessary to have set β_p above 0.9, in order to avoid negative flows for the export of phytoplankton-N (Figures 8A.1-2) or for carbon flow m_p (both calculated by difference, see equations 17 above and 30 below). As a consequence, detritus is not exported in the a priori model results.

Another consequence of these criteria was that the specific removal rates for phytoplankton and detritus by each group differed from month to month, and thus so did the degree to which detritus was selected against as a food source (Table 8A.2). Detritus was most attractive as a food source in April, when it was chosen by each group about 70% as often as phytoplankton, and least favored in June, when it was only selected 1/5 as often as phytoplankton. In each month, micro- and macrozooplankton selected phytoplankton over detritus to the same extent. Had we chosen a different criterion, for example, $\beta_p = \beta_z = 0.5$, the food selection patterns for the two groups would have been very different.

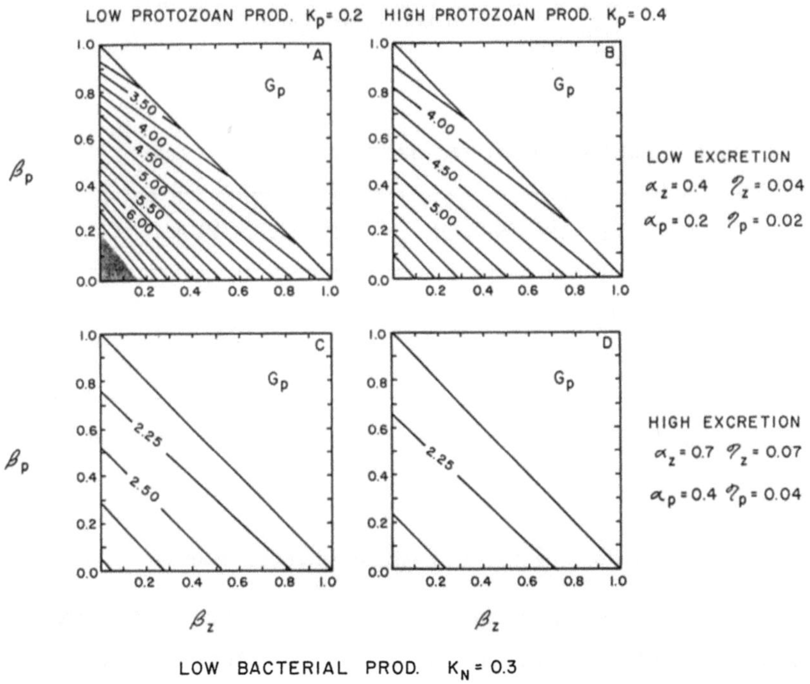

Figure 8A.1. Sensitivity analyses of A priori networks for low bacterial production ($K_n = 0.3$) with: (A,C) low ($K_p = 0.2$) or B,D) high ($K_p = 0.4$) protozoan production and with: A,B) low excretion ($\alpha_z = 0.4; \alpha_p = 0.2; \zeta_z = 0.04; \zeta_p = 0.02$) or with C,D) high excretion ($\alpha_z = 0.7; \alpha_p = 0.4; \zeta_z = 0.07; \zeta_p = 0.04$).

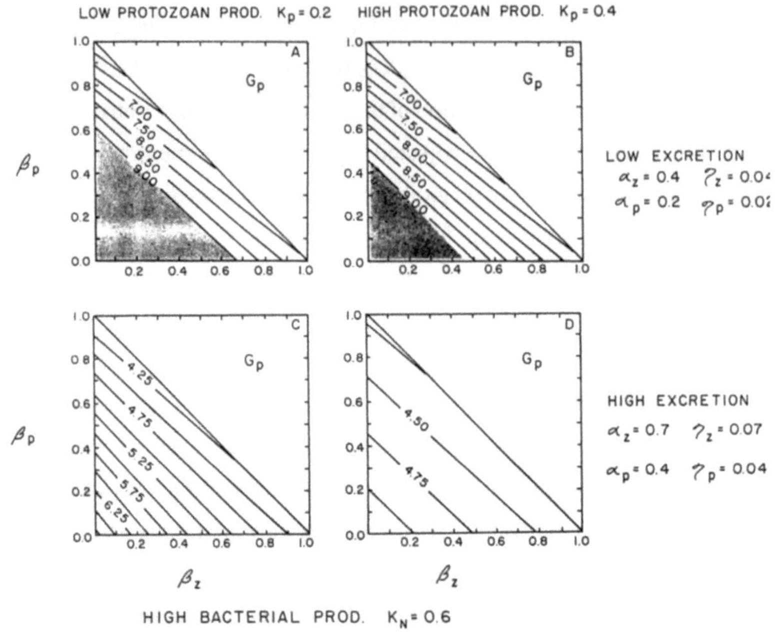

Figure 8A.2. Sensitivity analyses of A priori networks for high bacterial production (K_n = 0.6) with: (A,C) low (K_p = 0.2) or B,D) high (K_p = 0.4) protozoan production and with: A,B) low excretion (a_{ζ} = 0.4;α_p = 0.2;ζ_z = 0.04;ζ_p = 0.02) or with C,D) high excretion (α_z = 0.7;α_p = 0.4;ζ_z = 0.07;ζ_p = 0.04).

Table 8A.2. Selection of detritus by micro- and macroconsumers[1].

MONTH	MACROZOOPLANKTON	MICROZOOPLANKTON
APRIL	0.75	0.66
JUNE	0.20	0.23
OCTOBER	0.61	0.50

[1]Selectivity defined as (detrital ingestion/detrital biomass)/(Phytoplankton ingestion/Phytoplankton biomass).

Estimating flows through the microbial foodweb

Some further elaboration of the functioning of the protozoan and macrozooplankton compartments is necessary. As mentioned above, the macrozooplankton parameters were evaluated by taking into account the effects of predation by carnivorous zooplankton and fish. The effect of that predation is to dissipate the macrozooplankton production, or in model parlance, to distribute it amongst DOC and DON, NH_4 and respired CO_2, and fecal pellets. If a three-step food chain (copepods ⇒ carnivores

⇒ fish) is postulated, with 40% respiration, 20% growth efficiency, 10% DOM excretion, and 30% fecal production at each stage, the aggregated output from the macrozooplankton compartment is 50% to respiration, 37% to fecal pellets, and 13% to DOC. The values we adopted for were slightly different, but the outcome was similar in that 60-70% of the input went into dissolved excretory products, and 30-40% to fecal pellets (Table 8A.1). A similar approach was adopted to evaluate the parameters for the protozoan/microzooplankton compartment.

Our protozoan/microzooplankton compartment is also an aggregated compartment, in which the combined outputs from the microbial foodweb (Figure 8A.3) are condensed into a single unit for simplicity. To evaluate the aggregated outputs from the protozoan compartment we assumed the following:

1. The microbial foodweb contains three principal components, heterotrophic microflagellates (HNAN) capable of ingesting small phytoplankton and bacteria, small ciliates (CILIATES) capable of ingesting phytoplankton, bacteria and HNAN, and microzooplankton (loricate ciliates, juvenile zooplankton, etc. all small enough to pass through 333 μm mesh), which ingest phytoplankton, smaller ciliates, HNAN, and detritus, but not bacteria.

2. The entire production by HNAN and CILIATES is consumed. The only production available to higher trophic levels (macrozooplankton) is that of the microzooplankton (P in our main model). In reality, HNAN and CILIATES are probably consumed by copepods but we exclude these flows for simplicity's sake.

3. The proportions of phytoplankton production eaten by each of the three herbivorous microbial groups were left as free parameters. In the standard case, of the total protozoan ingestion, G_P, 10% was eaten by HNAN, 30% by CILIATES, and 60% by other larger microzooplankton. These values could be changed if data on size-fractionated phytoplankton biomass or size- specific primary production or removal were available (Michaels & Silver 1988). Clearly more consumption by HNAN would lower the apparent efficiency of the protozoan component, whereas more consumption by microzooplankton would yield greater apparent efficiencies.

4. Similarly, the proportions of bacterial production eaten by HNAN and CILIATES, and of HNAN production eaten by CILIATES and microzooplankton were left freely adjustable. In the standard case, 70% of the bacteria were eaten by HNAN, and 30% by CILIATES. The HNAN production was apportioned equally between the two HNANivorous groups.

5. Gross growth efficiencies for each group, NH_4 production rates, and DON production rates were left as parameters to be set, though for simplicity, we chose to have each be equal for the three groups. Detrital production was left to be calculated by difference. A further parameter, β_p, was used to specify the amount of detrital reingestion by the microzooplankton (see above). Again for simplicity's sake, we chose to have only the microzooplankton eat the aggregated detrital output from all groups.

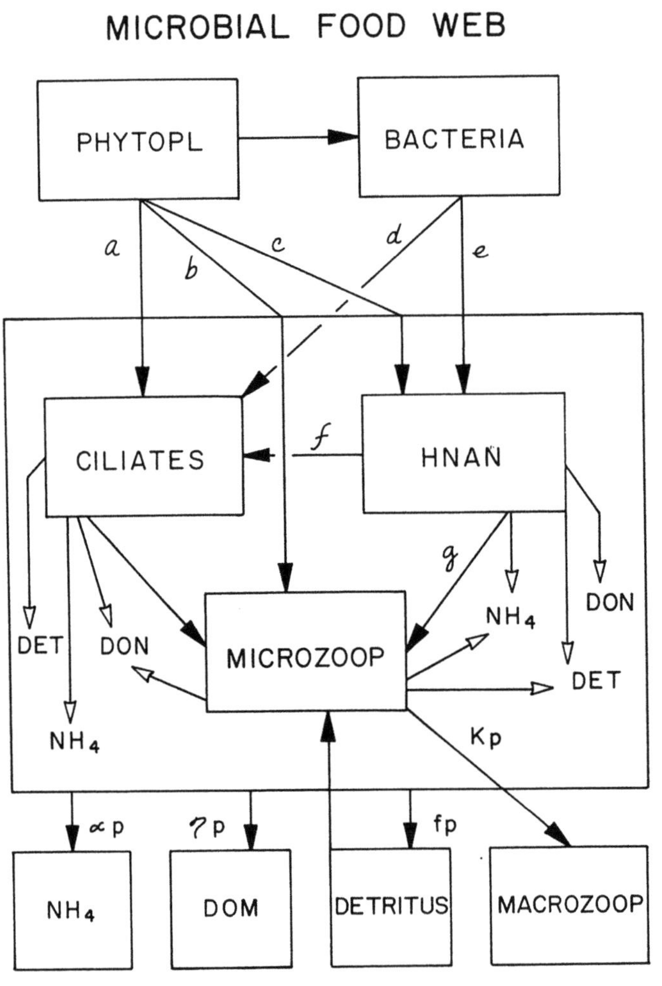

Figure 8A.3. Flow structure for a microbial food web network used to compute outputs from the protozoan/microzooplankton compartment (A priori network) of Figures 1-4. In this submodel, terms a-g symbolize the freely adjustable proportions of prey production ingested by each predator, with the constraint that $a+b+c = d+e = f+g = 1.0$.

With the feeding parameters set as given in points 3 and 4 above, and with individual growth efficiency, respiration, DOM production and detrital ingestion coefficients for all the microbial consumers set at ca. 0.35, 0.3, 0.03, and 0.9, respectively, we derived the parameter values for the microzooplankton outputs α_p, ζ_p, f_p and k_p (Figure 8A.3) as given in Table 8A.1. As noted in Section 2, some further adjustment of these values was necessary to balance all flows in the carbon and nitrogen networks.

Our model is theoretically free to have all of G_p consumed by the microzooplankton and all of the bacterial production consumed by ciliates (i.e., $b=d=1.0$) thus providing a scenario for efficient linkage between the microbial and macrozooplankton compartments. Alternatively, all of G_p and the bacterial production could be ingested at the base of a linear food chain (i.e., $c=e=f=1.0$), which simulates an inefficient coupling of picoplankton production to the macrozooplankton level. The few data available (Ducklow *et al.* 1986) support this "microbial sink" scenario. It is easy to show that these extreme scenarios give k_p values ranging from 0.16 to 0.30. Unless there is physiological variability associated with such mutually exclusive trophic configurations, these values can be used to define the limits of the "link vs sink" question regarding the role of microbial foodwebs in the sea (Payne & Wiebe 1978; Banse 1984; Ducklow *et al.* 1986). The values of k_p we have derived for our main model (0.25, cf Table 8A.1) are near the high end of this range. Therefore our model foodwebs behave more like "links" than "sinks".

Estimating the carbon flows

The measured carbon flows are primary production, bacterial production, and macrozooplankton ingestion. The N flows corresponding to the measurements of macrozooplankton carbon ingestion and bacterial carbon production were converted using assumed C:N ratios for phytoplankton and bacteria, respectively, as described earlier. In the following discussion, the same symbols used to denote nitrogen flows will be used for the corresponding carbon flows, with the addition of the suffix, (C), for carbon. Several of the carbon flows were converted from the calculated N flows using C:N ratios:

$$G_p(C) = G_p\ R \tag{21}$$

$$t_p(C) = t_p\ R_{det} \tag{22}$$

$$f^P(C) = f^P\ R_{det} \tag{23}$$

$$f_z(C) = f_z\ R_{det} \tag{24}$$

$$s(C) = s\ R \tag{25}$$

Taking each compartment in turn, the remaining flows were calculated as follows:

The DOC output, d(C), was calculated from the measured bacterial production by assuming that all DOC was consumed by bacteria:

$$d(C) = H(C)/K_c \tag{26}$$

$$B_{resp} = d(C) - H(C) = H(C)/(1-K_c) \tag{27}$$

Protozoan flows were calculated after obtaining the total protozoan ingestion, P_{in}:

$$P_{in} = G_p(C) + H(C) + f_p(C) \tag{28}$$

$$P(C) = K_p(C)\ P_{in} \tag{29}$$

$$P_{resp} = P_{in}\ \alpha_p(C) \tag{30}$$

$$m_p(C) = P_{in} - P(C) - P_{resp} - t_p(C) \tag{31}$$

The C:N ratio of the detritus was estimated from data on standing stocks of microbial plankton, POC and PON, as the difference between the living and total C and N pools (Wiebe *et al.* in prep.). This value (R_{det}) was then used as a constraint in the model. Since the protozoans are assumed to be the sole source of detritus, the export of that material is calculated as the excess of production (Eqn. 22) over consumption (Eqns. 23, 24):

$$r(C) = t_p(C) - f_p(C) - f_z(C) \tag{32}$$

The evaluation of $K_p(C)$ and $\alpha_p(C)$ is discussed in the previous section. Macrozooplankton flows were calculated exactly as the protozoan flows after obtaining the total macrozooplankton ingestion, MZ_{in}:

$$MZ_{in} = G_z(C) + P(C) + f_Z(C) \tag{33}$$

$$MZ_{resp} = 0.5\ MZ_{in} \tag{34}$$

$$t_z(C) = 0.37\ MZ_{in} \tag{35}$$

$$m_z(C) = MZ_{in} - t_z - MZ_{resp} \tag{36}$$

The choice of the respiration and egestion parameters is discussed in the previous section. Note that while the protozoan fecal output ("detritus") was calculated via an estimated C:N ratio (see below), the C and N fluxes embodied in macrozooplankton fecal pellets are calculated independently, allowing fecal pellet C:N ratios to be estimated by the model. In addition, the production of dissolved organic carbon by microzooplankton and macrozooplankton was calculated by difference. Some adjustment of the nitrogen parameters was necessary to ensure that $m_p(C)$ and $m_z(C)$ were positive.

Finally, the remaining phytoplankton flows were calculated by difference from the previously calculated flows as shown:

$$c(C) = d(C) - m_p(C) - m_z(C) \tag{37}$$

$$PHY_{resp} = PP - G_z(C) - G_p(C) - s(C) - c(C) \tag{38}$$

APPENDIX 8B: DERIVATION OF FLOW NETWORKS (INVERSE APPROACH)

The application of geophysical inverse methods (Parker 1977; Wunsch 1978) to the derivation of flow networks for marine food webs is described in detail elsewhere (Vézina 1989, this volume; Vézina & Platt 1988) and will not be repeated here. Here we highlight the features of the inverse approach, both general and specific for this application to the Warm Core Ring data sets, that result in the differences between the a priori and inverse flow networks.

First, it might be useful to summarize these differences. They are: (1) the structure of the trophic linkages (topology) differs between the a priori and inverse models; (2) the topology varies between the inverse flow networks for the April, June and October data sets whereas it remains the same for the a priori networks; (3) in contrast with the a priori networks, the inverse networks are not always in steady state, that is, inflows and outflows through compartments do not always balance (e.g. Figure 8.4b); and (4) the apparent f-ratios differ in the a priori and inverse solutions. We first discuss general properties of inverse solutions followed by their implications for the differences between the a priori and inverse syntheses of the Warm Core Ring data.

General considerations

The particular properties of the inverse solutions stem from the fact that the unknown flows are estimated directly from the available data on food web structure and dynamics. The estimates are calculated to satisfy conservation of mass and general biological constraints. The latter constraints simply specify the bounds that the flow estimates cannot cross if they are to remain biologically realistic in the light of current knowledge. Many of the transport pathways postulated a priori to operate in oceanic food webs are left unconstrained due to the scarcity or absence of quantitative information. Some of these pathways may be removed through the inverse solution process if they are not required to conserve mass and satisfy basic biological constraints in a particular situation. The site-specific data used by the inverse method will include information on the standing stocks and elemental composition (e.g. C:N) of the compartments as well as the flux measurements used by the a priori technique. Thus, the inverse solution must not only balance the inflows and outflows through the various compartments, but also preserve the size and elemental composition of the various compartments. There is no guarantee that this goal can be met: the observations may be inaccurate or reflect transient events, the steady state assumption may be wrong or the model may neglect crucial transport processes.A corollary is that different models may vary in how well they reconcile the various observations.

In the inverse procedure, the observations on standing stocks and elemental composition are treated as independent variables. The expected mass balance for each of the compartments in the system (0 at steady state) and the measured flows are the dependent variables. Since the independent variables are fixed, an inverse solution that cannot preserve the structure and the dynamics of the food web simultaneously will result in a lack of fit to the mass balances and measured flows. In fact, given inaccuracies in the data, a perfect fit cannot be expected. The inverse solution is always approximate, leaving some residual for each of the compartmental mass balances and measured flows.

In most cases, these residuals are small, implying a change of less than 1% per day in the standing stocks or elemental ratios. The average magnitude of the residuals may vary depending on the model fitted to a particular data set. Sometimes, the residuals for some compartments or fluxes may become very large, indicating a problem with the data, the model, or both.

In summary, the inverse method works backward from the data to estimates of the unknown flows. As a result, inverse solutions will behave very differently from a priori solutions. The topology of the inverse solutions can respond to observed variations in the structure and dynamics of food webs. Also, there is a statistical component to inverse techniques that does not exist in the a priori techniques: some error is associated with all inverse solutions and solutions may fail in specific situations.

Implications for the Warm Core Ring analyses

Our initial intention was to begin the a priori and inverse derivations of the WCR flow networks from the same starting point: same initial model structure, assumptions and constraints. This goal was not entirely met; basically, we could not obtain reasonably balanced inverse solutions under the conditions posed for the a priori derivations. The consequences of this are discussed below.

The inverse solutions were initially derived using the generic food web model presented by Vézina & Platt (1988) and Vézina (1989, this volume). This model differs slightly from the model used in the a priori derivations in that the exports from the phytoplankton and large zooplankton pools (dead cells, fecal pellets) are routed to a general detrital pool instead of directly out of the euphotic zone as in the a priori model. This allows for increased recycling through the reingestion of dead phytoplankton and fecal matter and for the formation of heterogeneous aggregates (e.g. marine snow).

The losses from the surface layer through sinking are strictly from that detrital pool. Both models are caricatures of the true processes, but the generic inverse model is more conservative in view of our current lack of knowledge on detritus recycling and particle-particle interactions.

In addition to this difference in flow structure between the two methods, there is also a difference in the way the site-specific information on N fluxes was handled. In the inverse approach, the *total flow* of N from the inorganic pools to the phytoplankton and bacteria was constrained by the sum of the measured uptake rates of new and regenerated N. The observed range in apparent f-ratios for each month was used as a constraint on the partitioning between new and regenerated flows. In contrast, the a priori solutions are constrained by the *separate* estimates of new and regenerated N uptake, thereby fixing the f-ratio for that month. Thus, the apparent f-ratio of the inverse solution for a particular month was free to differ from the average f-ratio for that month.

In the interest of minimizing as much as possible the differences between the inverse and a priori approaches, we tried inversions using the a priori model structure and method of constraining the N fluxes.

Table 8B.1 lists the residual sums of squares of the inverse solutions for the three WCR data sets and the two models (generic inverse model and a priori structure). For April and June, the solutions based on the generic model gave orders of magnitude better fits to the mass balance conditions and measured flows than those based on the a priori structure. Both models provided roughly equivalent fits for the October data, except that the apparent f-ratio calculated from the inverse solution using the a priori model structure violated the observed range (0.10-0.14). However, neither solution fitted the October data: both left large residuals in the flows of regenerated N (Figure 8.4b). The data characterising the structure and dynamics of the WCR food web in October was inconsistent with balanced N flows, according to the inverse methodology.

From these results, it appears that both (1) the assumption that the large, rapidly sinking detrital material is not recycled in the euphotic zone and (2) the procedure of using the separate estimates of new and regenerated N uptake to constrain the N fluxes cause problems in attempting to construct internally consistent inverse solutions for the WCR data sets. We have not yet investigated in detail how these problems arise.

Table 8B.1. Residual sums of squares of the inverse solutions to the WCR data.

Inverse solution	April	June	October
A priori structure	7.77	3.53	0.47*
Generic structure	0.10	0.023	1.006

*f-ratio constraint violated: f = 0.28 compared to the observed range of 0.10-0.14.

Chapter 9

NETWORK ANALYSIS OF THE NORTH INLET SALT MARSH ECOSYSTEM

Milton L. Asmus[1] *& Henry N. McKellar, Jr.*[2]
Belle W. Baruch Institute for Marine Biology and Coastal Research,
University of South Carolina, Columbia, SC 29208 USA

[1]Also *Departamento de Oceanografia, Fundação Universidade do Rio Grande, RS, Brazil*
[2]Also *Department of Environmental Health Science, University of South Carolina, SC, U.S.A.*

INTRODUCTION

Salt marshes occur along the east coast of North America from the arctic marshes of Canada, through New England to the semi-tropical marshes of Florida (Teal & Teal, 1969). These systems reach their most extensive development behind the barrier beaches of South Carolina and Georgia. Considerable interest in the functioning of salt marsh ecosystems derives from their role in maintaining high biological productivity and in providing nutrient-rich nursery areas for valuable aquatic and wetland biota. The importance of salt marshes as a source of dissolved and particulate organic matter to the estuarine area and surrounding ocean was pointed out by Odum & de la Cruz (1967). Teal (1962) quantified the potential for large exports of organic matter from salt marshes to adjacent estuarine waters despite the high respiration rate of the dominant producer (*Spartina alterniflora*). On the other hand, Nixon (1980) reviewed the functioning of salt marshes and concluded that while many marshes do export carbon, it is doubtful that they make a significant difference to the production of finfish and shellfish in the adjacent waters. Similar conclusions were reached by Mann (1986). However, recent data from the extensive salt marshes of the Southeast US continue to indicate the importance of marsh-estuary exchanges in terms of organic export and trophic relationships (Kjerfve & McKellar 1980; Pomeroy & Wiegert 1981; Chrzanowski *et al.* 1982, 1983; Mitsch & Gosselink 1986).

The salt marshes near North Inlet in Georgetown County, South Carolina are relatively pristine and have been the focus of much ecological research over the past decade. The marshes are bounded partly by scrub oak and long-leaf pine forests on the landward edges, by a meso-haline embayment on the southwestern edges and by barrier beaches on the eastern border separating the marsh from the Atlantic ocean. The 3200 hectare marsh is dominated by intertidal stands of cord grass (*Spartina alterniflora*) interlaced with numerous tidal creeks overlying subtidal benthic communities. Dense oyster reefs also occur in patches along the intertidal areas between the vegetated marsh and the low tide level in the creeks.

Earlier attempts to describe the holistic structure and functioning of the North Inlet salt marsh resulted in a set of deterministic simulation models examining the role of subsystem interactions in controlling carbon exchange with the sea (Summers & McKellar 1979, 1981a, 1981b; Summers *et al.* 1980). Simulation results indicated the importance of tidal driving forces coupled with seasonal changes in sea level and the areal distributions of subsystems in controlling total system carbon exchange. We suggest that these variables may be critical in explaining the large variability in system "outwelling" that is observed for different marsh-estuarine systems (see review by Nixon 1980). More recent simulation analyses have focused on more mechanistic interactions within the tidal creek water columns (Childers & McKellar 1987) emphasizing the coupling of carbon and nitrogen exchanges.

Input-output analyses have been used by several researchers in quantifying structure of system networks, the degree of cycling, and the magnitudes of interdependency among components (Finn 1976; Hirata & Ulanowicz 1984; Ulanowicz 1986). It can be a useful way to quantify basic elements of trophic structure. The path length (number of pathways of exchange among biotic components) for example, can quantify the number of trophic levels (Finn 1976). Ulanowicz & Kemp (1979) demonstrated that a complex network of ecosystems flows could be reduced to a conceptual trophic chain configuration to quantify aspects of the system's trophic structure. They also suggested a systematic trophic aggregation scheme to simplify the conceptualization of the system structure. The number of pathways can still be used to classify an effective trophic position of any compartment and help define the degree of trophic specialization in the system (Levine 1980). Hannon (1973) commented that network analysis defines the relative dependence of each compartment on the range of energy sources, and the interdependence among all compartment.

In this Chapter, we apply several basic elements of network analysis to the Summers & McKellar model for the North Inlet ecosystem. We use this analysis to quantify the degrees of dependence (direct and indirect) between major components and subsystems, to define key pathways of organic matter cycling, and to identify the dominant structure of trophic exchanges.

THE NORTH INLET NETWORK

The network used in this analysis was derived directly from the model developed by Summers *et al.* (1980) considering organic matter exchanges among four major subsystems; intertidal marsh, oyster reef, tidal creek water column and the subtidal benthos. The model includes a spectrum of inter subsystem exchanges among 19 compartment of energy storage including primary producers, meiofauna, macro-invertebrates, nekton and birds (Fig. 9.1). A recent, independent review of the model is provided by Hopkinson & Day (1988).

Data for the model (Appendix 9.A) came from a variety of sources including direct measurements, literature values from other marsh systems, and assumptions considering a total energy balance. Although the model uses energy units for organic matter exchanges, values were derived largely using a ratio of 10 kcal per 1 g carbon. System inputs represent primary production (by *Spartina alterniflora*, macroalgae in the marsh and oyster reef, and edaphic microalgae in the marsh) and immigration of nekton and birds (Table 9.A1). System outputs are respiration, emigration, and tidal export to the sea. Table 9.A2 tabulates the direct exchanges among all 19 compartments.

In the original model (Summers *et al.* 1980) all values were normalized in area units (m^2) for each subsystem. Here, we have prorated area units over the entire system to enhance the network analysis for the total system. We made a few corrections in the original data set based on more recent information; however, all except one of the adjustments were less that 10% of the total exchange for the compartments involved. The largest change (+25%) was an increase in marsh sediment respiration to adjust an energy balance for sediment organic matter.

NETWORK ANALYSIS

For the present analysis we used a package of computer algorithms described in Kay *et al.* (1989, this volume) to standardize matrix manipulations developed by several earlier works in network analysis (Hannon 1973, Finn 1976, Patten *et al.* 1976, Ulanowicz & Kemp 1979). Specific elements of this network analysis included (1) the development of contribution and dependency coefficients for each donor-recipient pair of organic matter exchange, (2) the identification of all pathways and path lengths of cycling within the network, and finally, the reduction of the complex network (Fig. 9.1) into a linear chain of trophic interactions where each component is assigned an effective trophic level.

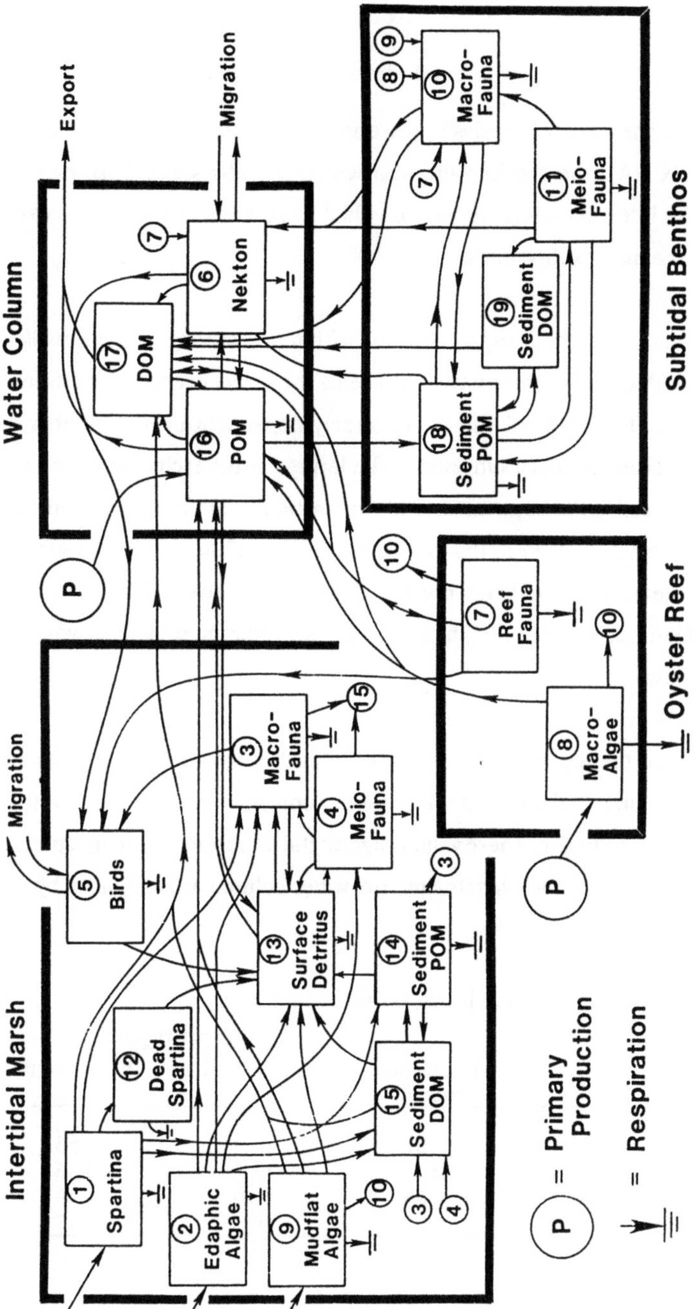

Figure 9.1. Energy flow in North Inlet salt marsh ecosystem.

RESULTS AND DISCUSSION

Total Flows and Subsystem Dependencies.

Although the network analysis was done using information from all 19 compartments (Fig. 9.1, Appendix 9.A), the results were aggregated by subsystem. This approach retains the information related to internal cycling among individual components but emphasizes exchanges and dependencies among the major subsystems.

The direct exchange of organic matter within and among subsystems is tabulated in Table 9.1. The principal diagonal of this matrix represents the direct exchanges between compartments within each subsystem. The diagonal values are largely dependent on the number of individual compartments conceptualized for each subsystem. The largest direct exchange is for exchanges within the intertidal marsh reflecting the large number of compartments in the marsh submodel as well as the relative size of the marsh area (80% of the total area). Clearly, if the oyster reef submodel had been articulated in greater detail with reef fauna portioned into filter feeders, deposit feeders, predators and microbiota (as in Dame & Patten 1981), higher values for internal exchanges would be reflected in the matrix of direct flows.

Other than the internal exchanges within each subsystem, the dominant direct flows are for export from the marsh to the tidal creek water column, and flows from the water column to subtidal benthos and back to the marsh. These values suggest the importance of marsh productivity for the whole system. However, some of the exports from the water column are enhanced by plankton productivity as well.

Table 9.1. Matrix of Direct Flows (kcal/m^2·day)

	RECIPIENT			
DONOR	Intertidal marsh	Oyster reef	Water column	Subtidal benthos
Intertidal marsh	7.07	0	3.56	0.05
Oyster reef	0.01	0	0.36	0.03
Water column	0.88	0.02	0.99	0.88
Subtidal benthos	0	0	0.15	1.52

The coefficients of total contribution, provided in network analysis for each pair of compartments, represent the fraction of the total throughput of each donor which eventually enters each recipient along direct as well as indirect pathways. We multiplied these coefficients by the total throughput of the donor compartments to derive a matrix of total contribution flows (Table 9.2) as described by

Szyrmer & Ulanowicz (1987). The difference between the direct flows (Table 9.1) and the total contribution flows (Table 9.2) reflects the magnitude of indirect pathways of exchange. Whereas, the model shows about 7 kcal/m^2· day of direct exchanges taking place within the marsh subsystem (Table 9.1), there is almost four times additional flow through indirect pathways (Table 9.2). The effect of indirect exchanges is clear between those subsystems which have no direct exchanges, such as between the marsh and the oyster reef, or between the subtidal benthos and the marsh. Szyrmer & Ulanowicz (1987) pointed out the utility of total contribution flows in ranking the total importance of each compartment in a network and in assessing bilateral trophic relationship between any two compartments.

Network analysis also provides coefficients of total dependency which represent the fraction of the total input to each recipient compartment which has passed through each of the donor compartments. We multiplied the dependency coefficients for each donor-recipient pair by the total inputs to each recipient to obtain total dependency flows, aggregated by subsystem (Table 9.3). Szyrmer & Ulanowicz (1987) called these values the "total flows from the input perspective". The difference between the total contribution flows (Table 9.2) and the total dependency flows (Table 9.3) indicates the degree of cycling among compartments along the pathway between donor and recipient.

Table 9.2. Total Contribution Flows (kcal/m^2·day)

	RECIPIENT			
DONOR	Intertidal marsh	Oyster reef	Water column	Subtidal benthos
Intertidal marsh	31.82	1.73	10.56	1.93
Oyster reef	0.68	0.37	1.59	0.46
Water column	1.69	2.18	2.80	2.34
Subtidal benthos	0.12	0.05	0.57	2.75

Table 9.3. Total Dependency Flows (kcal/m^2·day)

	RECIPIENT			
DONOR	Intertidal marsh	Oyster reef	Water column	Subtidal benthos
Intertidal marsh	39.44	2.23	12.98	2.94
Oyster reef	0.30	0.38	1.70	0.55
Water column	1.61	2.10	2.87	2.70
Subtidal benthos	0.10	0.05	0.46	2.67

The largest difference for the North Inlet model was for flows from the marsh to the subtidal benthos. With a total input (including indirect flows) from the marsh to the subtidal benthos of 1.9 kcal/m^2/day (Table 9.2), the total dependency of the benthos on the marsh (including the inherent cycling en route) is about 50% higher (2.9 kcal/m^2·day, Table 9.3). This value indicates that, of the total input to the subtidal benthos, an amount of 2.9 kcal/m^2·day passes through the marsh and is involved in intermediate cycles between the marsh, water column, and oyster reef. Although Table 9.3 indicates the presence of cycling among some of the other subsystems, the overall similarity with values in Table 9.2 reflects the dominance of once- through flows among the subsystems in this system.

Table 9.4 represents the matrix of total dependency coefficients among the aggregated subsystems. Szyrmer & Ulanowicz (1987) refer to these coefficients as "total intermediate input coefficients", which reflect the normalized interdependencies among components. Subsystem dependencies indicate again the importance of the intertidal marsh as an exporter of organic matter in the salt marsh ecosystem. All recipients showed considerable dependency on the marsh, reflecting its role as a major donor component. Table 9.4 also indicates high dependencies of the oyster reef and the subtidal benthos on the water column suggesting the additional importance of plankton productivity as an input of organic matter for the system.

Table 9.4. Total Dependency Coefficients

	RECIPIENT			
DONOR	Intertidal marsh	Oyster reef	Water column	Subtidal benthos
Intertidal marsh	0.342	0.348	0.502	0.521
Oyster reef	0.003	0.124	0.179	0.197
Water column	0.016	0.693	0.237	0.941
Subtidal benthos	0.001	0.015	0.021	0.603

The Structure of Cycling

In addition to quantifying the degree of cycling within a system in terms of total flows (as discussed above) the Ulanowicz network analysis also provides algorithms for identifying and enumerating the cycles in the network. The North Inlet model, as expressed in Fig. 9.1, contains 198 different possible cycles with path lengths involving 2 to 12 compartments. Cycles of 2 path lengths dominate 69% of the cycle flows (Fig.9.2). Cycles of path length 5, 3, and 4 are next in importance, accounting for most of the remaining cyclical flow. All other loop sizes represent less than 1% of the cycled flow.

Despite the large number of possible cycles in the system, total energy flow as described by this model is largely non-cyclic. The Finn Cycling Index (Finn, 1976; Ulanowicz & Kemp, 1979) is the ratio of cyclical to non- cyclical flow. This index for the North Inlet model (Fig.9.2) indicates that only 8% of the total energy flow in the system is represented by cyclical pathways. Although there are some important functional cycles involved, the dominant tendency for organic matter exchange is unidirectional flow from primary production, through adjacent subsystems, leaving the system via respiration, export, or emigration.

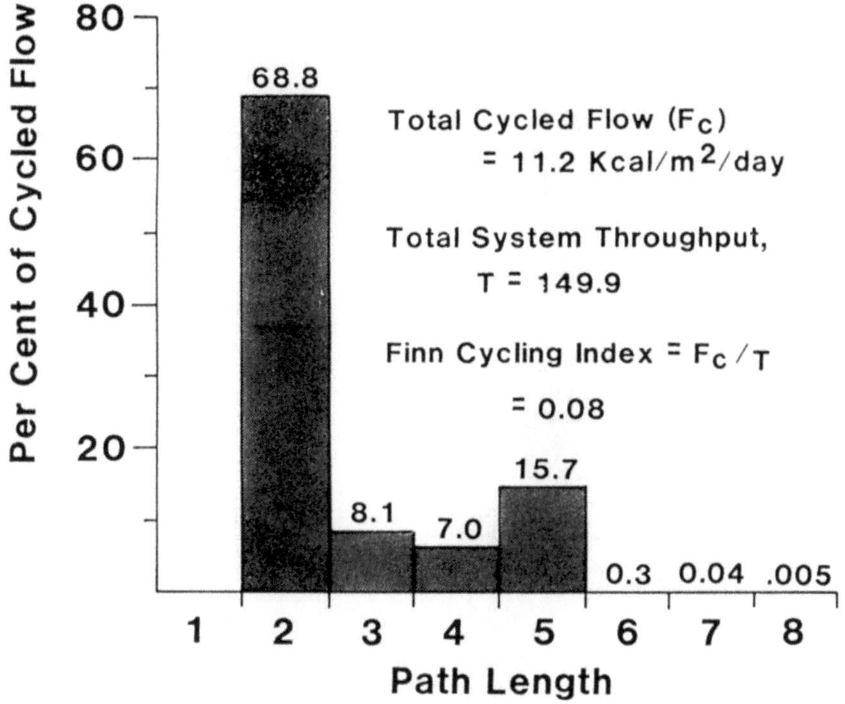

Figure 9.2. Distribution of cyclical pathways.

Trophic Configuration

Finally, we examine the trophic level of each heterotrophic compartment by considering the total number of pathways of organic matter transfer for each pathway of food intake. The analysis package provides a "Lindeman transformation matrix" which partitions the energy input for each consumer into discrete trophic positions and then weights the total inputs in order to evaluate the effective trophic level for each aggregated consumer compartment (Fig. 9.3). Counting primary producers and detritus as trophic level 1, the trophic behavior of the major consumers ranged from 2 to 5. Meiofauna and macrofauna dominated the 2nd trophic position whereas nekton and birds dominated the higher levels.

Using this information, the complex network of trophic transfers was reduced to a linear food chain representation as described by Ulanowicz & Kemp (1979). The energy budget of each trophic level (Fig. 9.4) indicates the total input from the lower trophic level as well as export, respiration, detrital recycle, and transfer to the next higher trophic level. Such a representation demonstrates the "pyramid" structure of food webs decreasing energy at each level. Quantification of this structure for total ecosystems can be useful in evaluating ratios of energy quality and the increasing amount of total embodied energy in valuable species at the top of the food web (Ulanowicz & Kemp 1979, Odum 1983).

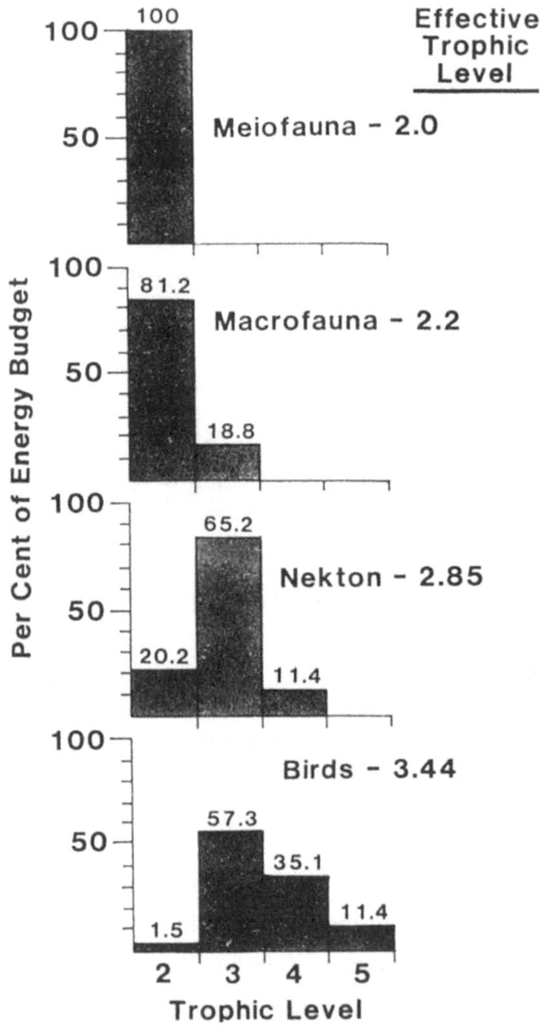

Figure 9.3. Trophic spectrum and effective trophic level of heterotrophic compartments.

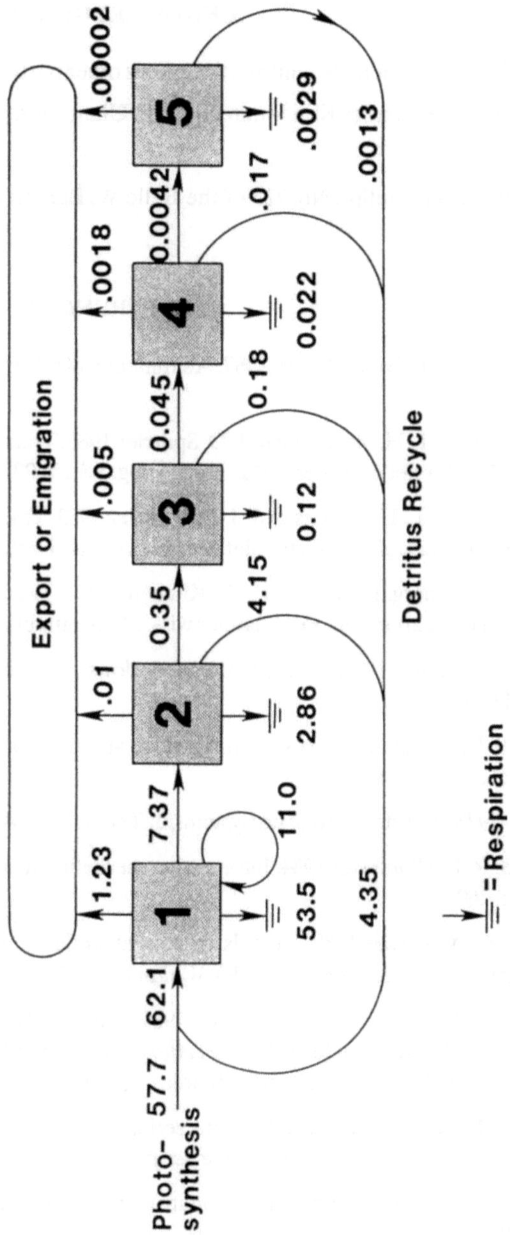

Figure 9.4. North Inlet linear food chain (kcal/m^{-2}·day) .

ACKNOWLEDGEMENTS

We acknowledge considerable contributions from other participants at the workshop and are especially grateful for comments from Ken Mann and Bob Ulanowicz.

This chapter is Contribution No. 723 of the Belle W. Baruch Institute for Marine Biology and coastal research.

REFERENCES

Childers, D. L. & H. N. McKellar, Jr. 1987. A simulation of saltmarsh water column dynamics, Ecol. Model., 36: 211-238.

Chrzanowski, T. H., L. H. Stevenson & J. D. Spurrier 1982. Transport of particulate organic carbon through the North Inlet ecosystem, Mar. Ecol. Prog. Ser., 7: 231-245.

Chrzanowski, T. H., L. H. Stevenson & J. D. Spurrier 1983. Transport of dissolved organic carbon through a major creek of the North Inlet ecosystem, Mar. Ecol. Prog. Ser., 13: 167-174.

Dame, R., F. Vernberg, R. Bonnell & W. Kitchens 1977. The North Inlet marsh-estuarine ecosystem: A conceptual approach, Helgolander wiss. Meeresunters., 30: 343-356.

Dame, R. & B. Patten 1981. Analysis of energy flows in an intertidal oyster reef, Mar. Ecol. Prog. Ser. 5:115-124.

Finn, J. T. 1976. Measures of ecosystem structure and function derived from analysis of flows, J. Theor. Biol., 56: 363-380.

Hannon, B. 1973. The structure of ecosystems, J. Theor. Biol., 41: 535- 546.

Hirata, H. & R. E. Ulanowicz 1984. Information theoretical analysis of ecological networks, Int. J. Sys. Sci. 15: 260-270.

Hopkinson, C. & J. Day 1988. Models of coastal wetland and estuarine systems, In: Mitsch, W., M. Straskraba, & S. Jorgensen (eds), *Wetland modeling*, Elsevier, NY.

Kay, J. J., L. A. Graham & R. E. Ulanowicz 1989. A detailed guide to network analysis, in: Wulff, F., J. G. Field & K. H. Mann (eds) *Network Analysis in Marine Ecology: Methods and Applications*, Lecture Notes in Coastal and Estuarine Studies, Springer-Verlag, New York.

Kjerfve, B. & H. N. McKellar, Jr. 1980. Time series measurements of estuarine material fluxes, pp.341-358, In: Kennedy, V.S. (ed.), *Estuarine perspectives*, Academic Press, New York.

Levine, S. 1980. Several measures of trophic structure applicable to complex food webs, J. Theor. Biol., 83: 195-207.

Mann, K. 1986. The role of detritus at the land-sea boundary, pp.123-140, In: P. Lassere and J.M. Martin (eds), *Biogeochemical processes at the land-sea boundary*, Elsevier, Amsterdam.

Mitsch, W. J. & J. G. Gosselink 1986. *Wetlands*, Van Nostrand Reinhold Company, New York, 539 pp.

Nixon, S. W. 1980. Between coastal marshes and coastal waters- A review of twenty years of speculation and research on the role of salt marshes in estuarine productivity and water chemistry, pp 261-290, in: Hamilton, P. & K. McDonald (eds), *Estuarine and Wetlands Process*, Plenum Press, New York.

Odum, E. P. & A. A. de la Cruz 1967. Particulate organic detritus in a Georgia saltmarsh estuarine ecosystem, pp. 383-389, In: Lauff G.H. (ed.), *Estuaries*, Am. Assoc. Adv. Sci., Publ., 83.

Odum, H. T. 1983. *Systems ecology, John Wiley and Sons, NY, 644 pp.*

Patten, B. C., R. W. Bosserman, J. T. Finn & W. G. Cale 1976. Propagation of cause in ecosystems, pp. 457-479, In: Patten B.C. (ed.), *System analysis and simulation in ecology, Vol.4*, Academic Press, New York.

Pomeroy, L. R. & R.G. Wiegert (ed.) 1981. *The ecology of a salt marsh*, Springer-Verlag, New York, 539pp.

Summers, J. K. & H. N. McKellar, Jr. 1979. A simulation model of estuarine subsystem coupling and carbon exchange with the sea. I. Model structure, pp. 323-366, In: Jorgensen, S.E. (ed.), *State-of-the-art in ecological modelling*. Int. Soc. Ecol. Modelling, Copenhagen.

Summers, J. K. & H. N. McKellar, Jr. 1981a. The role of physical forcing functions in an estuarine model of carbon exchange with the sea, ISEM Journal, 3: 71-101.

Summers, J.K. & H.N. McKellar, Jr. 1981b. A sensitivity analysis of an ecosystem model of estuarine carbon flow, Ecol. Model. 13: 282-301.

Summers, J. K., H. N. McKellar, Jr. R. F. Dame & W. M. Kitchens 1980. A simulation model of estuarine subsystem coupling and carbon exchange with the sea. II. North Inlet model structure, output and validation, Ecol. Model.11: 101-140.

Szyrmer, J. & R. E. Ulanowicz 1987. Total flows in ecosystems, Ecol. Model., 35: 123-136.

Teal, J. M. 1962. Energy flow in a salt marsh ecosystem of Georgia, Ecology, 43: 614-624.

Teal, J. & M. Teal 1969. *Life and death of a salt marsh*, New York, N.Y., Ballantine Books, Inc., 278 pp.

Ulanowicz, R. E. & W. M. Kemp 1979. Toward canonical trophic aggregations, Am. Nat. 114: 871-883.

Ulanowicz, R. E. 1986. *Growth and development: ecosystems phenomenology*, Springer-Verlag, New York, 203 pp.

APPENDIX 9A

Table 9A.1 Total System Inflow and Outflow Vectors (kcal/m^2/day).
P = Photosynthesis, I = Immigration or import from external sources, R = Respiration, E = Emigration or export. Compartment numbers are the same as in Fig. 9.1.

COMPART-	INPUTS		OUTPUTS	
MENT	P	I	R	E
1	49.9		40.4	
2	2.82		0.28	
3			0.974	
4			0.83	
5		0.0003	0.0517	0.0004
6		0.003	0.038	0.009
7			0.75	
8	0.83		0.48	
9	1.07		0.625	
10			0.039	
11			0.392	
12			1.49	
13			3.78	
14			2.62	
15				
16		3.09	3.29	0.106
17				1.10
18			0.515	
19				

RECIPIENT (columns) / **DONOR** (rows)

DONOR	1	2	3	4	5	6	7	8	9	10	11	12	13	14	15	16	17	18	19
1			0.52									5.16		3.1	0.62		0.09		
2			0.72	0.55									0.97		0.19	0.11			
3					0.031								0.88		0.555				
4			0.15										1.1		0.1				
5													0.0222						
6					0.037											0.069	0.01		
7					0.006					0.014						1.08	0.004		
8										0.02						0.27	0.06		
9										0.051			0.24			0.08	0.076		
10						0.099											0.022	0.116	
11						0.009				0.038								0.352	0.045
12													3.67						
13			1.02	1.63												1.97			
14			0.028										0.635		0.237				
15													0.044	0.424			0.123		
16						0.014	1.87						0.843				0.249	0.951	
17							0.00004									0.652			
18						0.019				0.153	0.773								0.002
19																	0.004	0.043	

Table 9A.2. Exchange matrix for the North Inlet ecosystem (kcal/m^2·day). The numbers identify compartments as in Fig 9.1

Chapter 10

ANALYSIS OF THE FLOW NETWORK IN AN ESTUARINE BENTHIC COMMUNITY

Richard M. Warwick and Philip J. Radford
Plymouth Marine Laboratory
Prospect Place, The Hoe
Plymouth PL1 3DH
U.K.

INTRODUCTION

An intertidal mudflat in the Lynher estuary, Cornwall, U.K. has been the site of intensive investigations of energy flow through its biological compartments, and of interactions between its compartments. Joint (1978) quantified autotrophic and heterotrophic microbial production both in the water column and sediments. Several detailed studies have been made of energy flow within the macrobenthos and meiobenthos (Price & Warwick, 1980a,b; Tear & Price, 1979; Warwick, 1981; Warwick & Price, 1975, 1979). Field manipulations (caging experiments) have been conducted to investigate trophic relationships between the faunal and floral components (Warwick *et al.*, 1982; Gee *et al.*, 1985). The results of these and other unpublished studies have been assembled into a steady state energy flow diagram and a dynamic simulation model of the benthic system by Warwick *et al.* (1979) in which most of the flows between compartments have been actually measured rather than assumed, and these models have now found their way into a number of authoritative texts (e.g. Longhurst, 1981; Mann, 1982).

These models are only descriptive of part of the estuarine ecosystem, i.e. the benthos, and therefore an analysis of this ecological network cannot be usefully compared with more extensive (but less detailed) studies of whole systems which appear in this volume (Asmus & McKellar, 1989, Wulff & Ulanowicz, 1989). We have therefore capitalised on the level of detail and realism of this network to focus on two specific topics: an empirical analysis of the effects of perceived detail in the system (i.e. levels of functional/taxonomic aggregation of species) on measured flow indices, and an analysis of seasonal changes in these indices.

All flow indices are, of course, dependent on the investigator's perception of the system being studied, and it is clearly essential that all important flows be identified and quantified, particularly if the output is to be used to compare one system with others. Inevitably some level of aggregation of the species populations into compartments has to be made. The methods of doing this are discussed by Briand (1983). In practice, aggregation may be based on function (e.g. suspension-feeders, deposit-feeders), taxonomic identity (e.g. polychaetes, molluscs) or size (e.g. macrobenthos, meiobenthos), and is often done rather arbitrarily using a mixture of these criteria, as the inspection of many of the flow diagrams in this volume will testify. Compartments aggregated by one of these criteria will inevitably be heterogeneous with respect to the other two. Aggregation is common at lower trophic levels because the species tend to be smaller, less well known ecologically and more diverse, and is less frequently used at higher trophic levels (Pimm, 1982). The influences of these varying types of aggregation on flow indices is unknown, and we have used the Lynher data to test them.

Whilst most flow analyses are based on seasonally averaged steady-state situations, there is increasing interest in changes in these indices through time. Repeated measurements of a network of flows over time are generally impracticable, and using the output of simulation models of various kinds seems to be a promising approach to the problem (Field *et al.*, 1989, Chapter 7 this volume; Kremer, 1989, Chapter 6 this volume). In this paper we have also adopted this approach.

MODELS AND METHODS

Flows among the model compartments have been analysed following Kay *et al.* (1989; Chapter 2, this volume) using Ulanowicz' program NETWRK and the program of Kay (1984) ECOSYS. The main measures of whole system network properties derived from NETWRK which are considered here are Total System Throughput, Development Capacity, Ascendency and Redundancy, and from ECOSYS Resource Niche Breadth (S) and Flow diversity (D).

Experiments with various levels of aggregation are based on the steady state carbon flow diagram given by Warwick *et al.* (1979), reproduced in Fig 10.1. The individual species data are already aggregated to some extent. For example, in calculating the throughput for the compartment Nematoda, Warwick & Price (1979) determined respiration and production for each of the 40 species individually. However, it is not possible to conduct network analyses on the data in their most dissected form, due to computational limitations on the total number of compartments. The initial aggregation has been based on the amalgamation of species which are both taxonomically and functionally similar, e.g. the suspension-feeding bivalves *Cardium* and *Mya* or the deposit-feeding spionid polychaetes *Pygospio* and *Streblospio*. There is a total of 19 compartments, as follows:

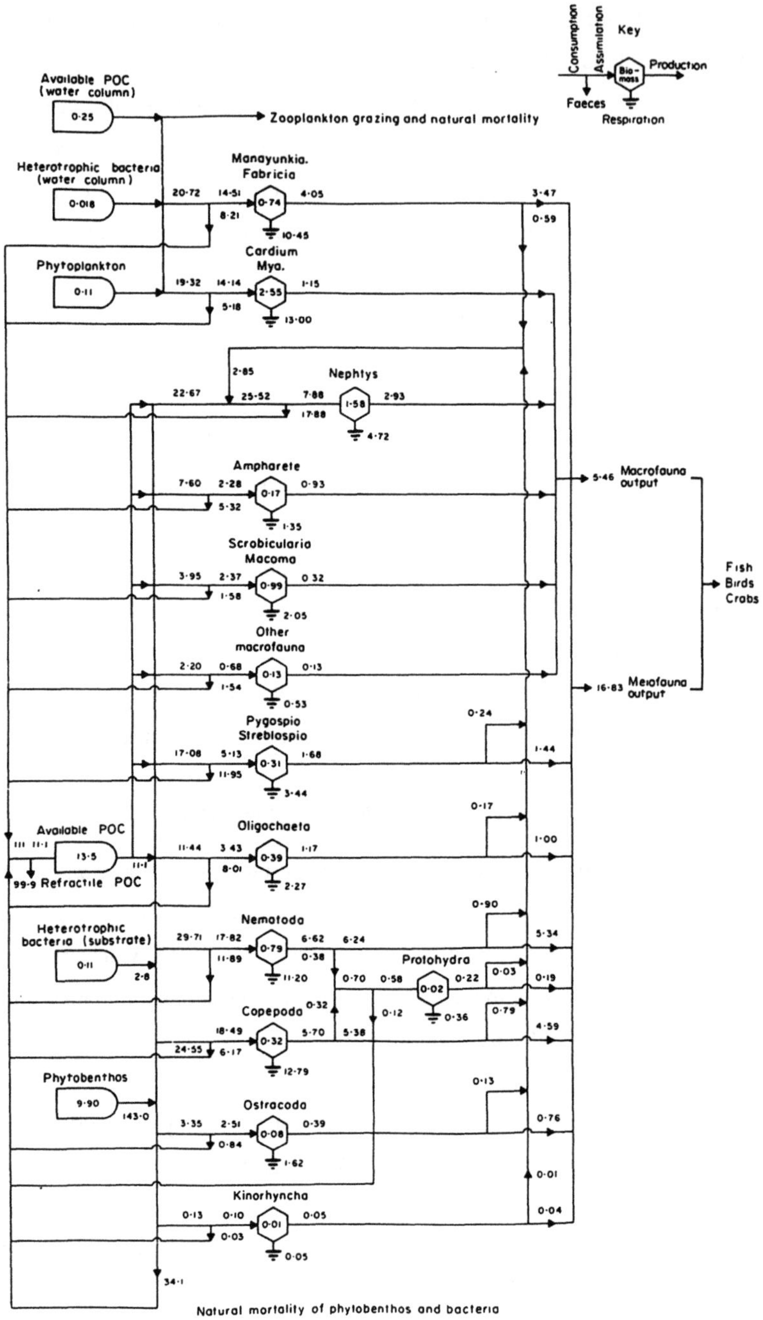

Fig 10.1. Steady state carbon flow network for the Lynher mudflat. Standing stocks in gC m^{-2}, flow rates in $gCm^{-2}y^{-1}$. From Warwick *et al.* (1979).

1. Heterotrophic bacteria (water column)

2. Phytoplankton

3. Heterotrophic bacteria (sediment)

4. Phytobenthos

5. *Manayunkia* and *Fabricia* (small sabellid polychaetes: meiobenthos)

6. Cardium and *Mya* (suspension-feeding bivalves: macrobenthos)

7. *Nephtys hombergi* (omnivorous polychaete: macrobenthos)

8. *Ampharete acutifrons* (deposit-feeding polychaete: macrobenthos)

9. *Scrobicularia* and *Macoma* (deposit-feeding bivalves: macrobenthos)

10. Other macrobenthos (six species)

11. *Pygospio* and *Streblospio* (deposit-feeding polychaetes: meiobenthos)

12. Oligochaeta (two species: meiobenthos)

13. Nematoda (40 species: meiobenthos)

14. *Protohydra leuckarti* (predatory hydroid: meiobenthos)

15. Copepoda (8 species: meiobenthos)

16. Ostracoda (2 species: meiobenthos)

17. Kinorhyncha (2 species: meiobenthos)

18. Available Particulate Organic Carbon (water column)

19. Available Particulate Organic Carbon (sediment)

Four separate aggregations of the data were made (using the program AGG):

1. Aggregation of primary food sources in water column (1,2,18) and sediment (3,4,19).

2. Taxonomic aggregation of consumers: Annelida (5,7,8,10,11,12), Mollusca (6,9), "Aschelminthes" (13,17), Coelenterata (14), Arthropoda (15,16).

3. Functional (trophic) aggregation of consumers: suspension feeders (5,6), deposit feeders (8,9,10,11,12,13,15,16,17), carnivores (7,14).

4. Size aggregation of consumers: macrobenthos (6,7,8,9,10,11,12), meiobenthos (5,13,14,15,16,17).

The simplified process flow diagram on which the simulation model is based is given in Fig 10.2. The assumptions and structure of the model, written in CSMP, are given in Warwick *et al.* (1979). Suffice it to say that those state variables which are simulated are in square boxes, whilst those in circles are exogenous forcing functions measured as mean monthly values. Flows among the model compartments were saved at 30 day intervals (starting on January 1) as matrices for analysis using NETWORK and ECOSYS .

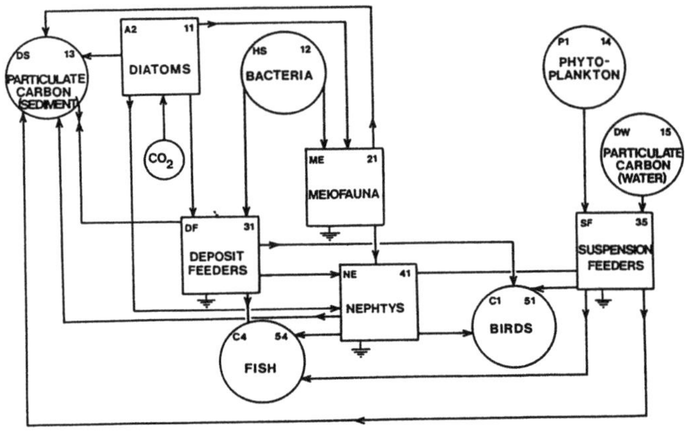

Fig 10.2 Process flow diagram for the dynamic simulation model of the Lynher estuary.

Table 10.1. Network flow indices for the Lynher estuary benthos using various levels of aggregation. UNAG = unaggregated data, i.e. compartments as in Fig. 10.1; Food = aggregation of primary food sources in water column and sediment; TAX = taxonomic aggregation of consumers; TROPH = trophic aggregation of consumers; SIZE = size aggregation of consumers. Units for Total System Throughput are $gCm^{-2}y^{-1}$ and for Development Capacity, Ascendency and Redundancy gC-bits.

	UNAG	FOOD	TAX	TROPH	SIZE
Total System Throughput	699	699	699	699	699
Development Capacity (DC)	2125	1955	1754	1488	1463
Ascendency (A)	1286	1253	894	972	696
A/DC (%)	60.5	64.1	51.0	65.3	47.6
Redundancy (R)	344	196	571	306	612
R/DC (%)	16.2	10.0	32.6	20.6	41.8
Flow Diversity (D)	1.15	0.98	1.13	1.10	1.00
Resource Niche Breadth (S)	0.47	0.25	0.83	0.44	0.90

RESULTS AND DISCUSSION

Aggregation

The effects of various types of aggregation of compartments on the system network properties are shown in Table 10.1. Some of these were anticipated intuitively, others not. A Total System Throughput of 699 gC $m^{-2}y^{-1}$ was unaffected by any form of aggregation because of the way the Program AGG pools specified categories, with "self-loops" between pooled categories in the new pooled categories (see Kay *et al.* 1989, Chapter 2 this volume). The Development Capacity is the product of the size (i.e. T) and the complexity of the network, assuming a fully developed system with maximum evenness and number of flows, i.e. when the flows connecting all compartments are equal in magnitude and unique (no parallel paths, or redundancy). Its value not unexpectedly declined in proportion to the number of compartments in the aggregated network, since the T was constant. Ascendency is a measure of the degree to which the real system complies with this "fully developed" condition. In absolute terms its value varied almost twofold, but as a proportion of the development capacity there was less variation. The system appeared to be most highly developed when the species were aggregated on a trophic basis (65.3% of full Development Capacity), and least well developed when aggregated on a taxonomic (51.0%) or size (47.6%) basis. The shortfall caused by Redundancy (i.e. the occurrence of parallel paths) varied considerably both in absolute and relative terms according to the type of aggregation. When food sources in the water column and sediment were aggregated, redundancy accounted for only 10.0% of the shortfall from full Development Capacity, but when consumers were aggregated on the basis of size (i.e. into macrobenthos and meiobenthos) this value increased to 41.8%.

The Diversity of Flows (Kay's D) was rather constant with respect to the type of aggregation, but the flow Resource Niche Breadth (Kay's S) varied considerably and predictably. Its value was lowest when the various primary food resources were not separately defined but simply input as particulate organic carbon in the water column and sediment. Its value was highest when the consumer compartments were trophically heterogeneous, i.e. when aggregated taxonomically or by size.

Time series

Seasonal changes in the biomass of the major components of the benthic system are shown in Fig 10.3. The greatest seasonal variation is shown in the biomass of phytobenthos and suspension feeders, whereas deposit-feeders, *Nephtys* and meiofauna fluctuate within rather narrow limits.

Daily Total System Throughput showed almost a threefold increase between winter (0.331 gC.m^{-2}.d^{-1} at day 60, i.e. the end of February) and summer (1.175 gC $m^{-2}d^{-1}$ at day 210, i.e. the end of July) (Fig

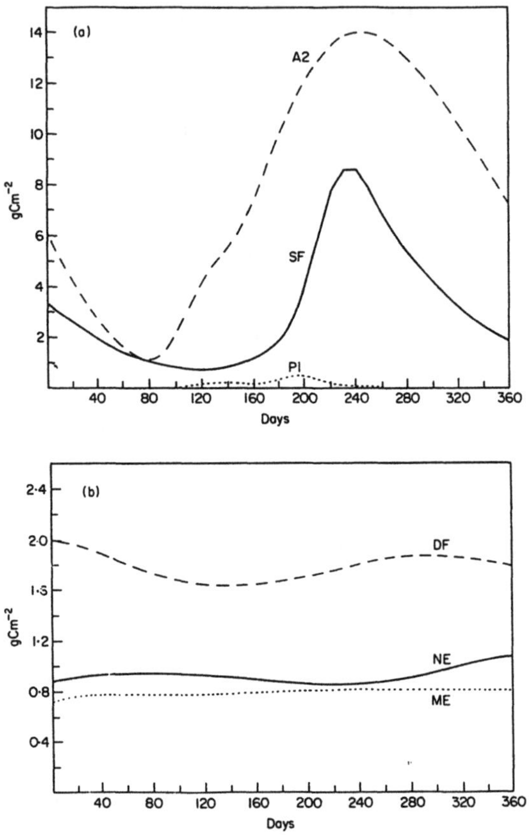

Fig. 10.3. Seasonal changes in standing stocks of the main biological components of the Lynher estuary, derived from the simulation model. A2 = phytobenthos, SF = suspension feeders, P1 = phytoplankton, DF = deposit feeding macrobenthos, NE = Nephtys hombergi, ME = meiobenthos.

10.4). Since the structure of the network remains constant, this is paralleled by a similar trend in the Development Capacity. However, there is slightly more than a threefold increase in Development Capacity between winter (0.882 gC-bits at day 60) and summer (3.476 gC.bits at day 210), indicating that although Total System Throughput accounted for most of the pattern, slightly increased Flow Diversity in the summer was also partly responsible. Similar results were found with Kremer's (1989) estuarine model. Another similarity to Kremer's model is that the relative contribution of Ascendency to Development Capacity was rather constant, but was lower in the winter than the summer, ranging from 0.52 on day 0 to 0.68 on day 210 (Fig 10.5). Variations in the relative importance of redundancy were also quite small, but showed the reverse trend with a maximum in winter (0.35 on day 360) and minimum in summer (0.22 on day 210). Flow Diversity (Kay's D) does not show a smooth seasonal cycle of variation, but is highest in the summer (day 240), and Resource Niche Breadth (Kay's S) follows a similar pattern (Fig 10.6). This is somewhat counter-intuitive, since deposit feeding benthos, with a

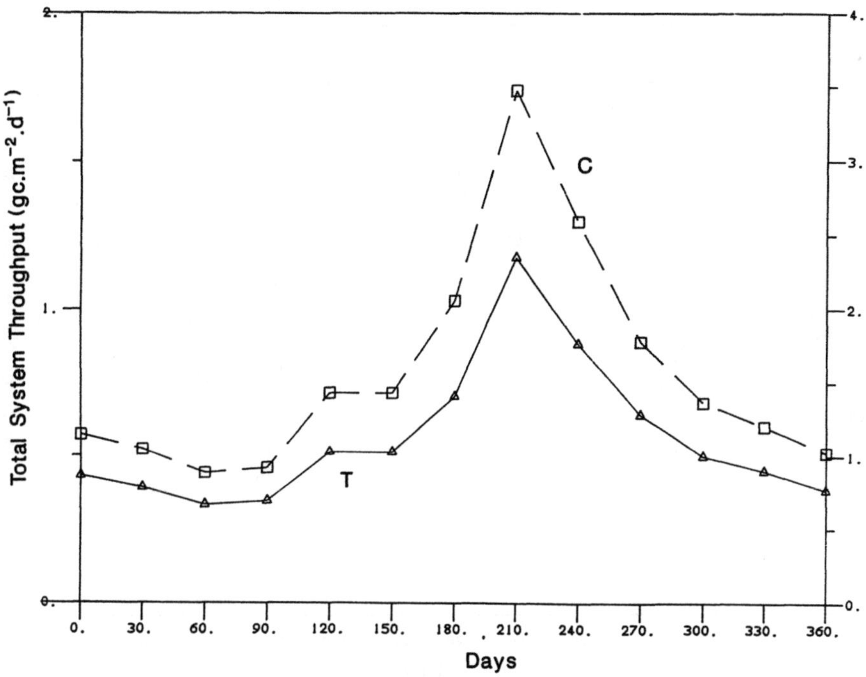

Fig 10.4. Seasonal changes in Total System Throughput and Development Capacity.

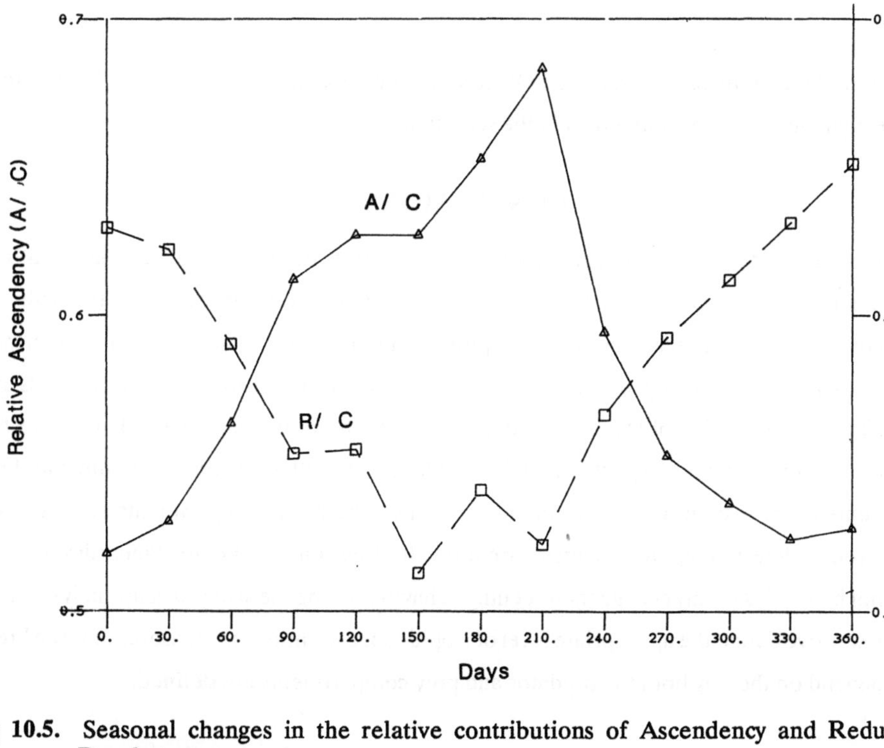

Fig 10.5. Seasonal changes in the relative contributions of Ascendency and Redundancy to Development

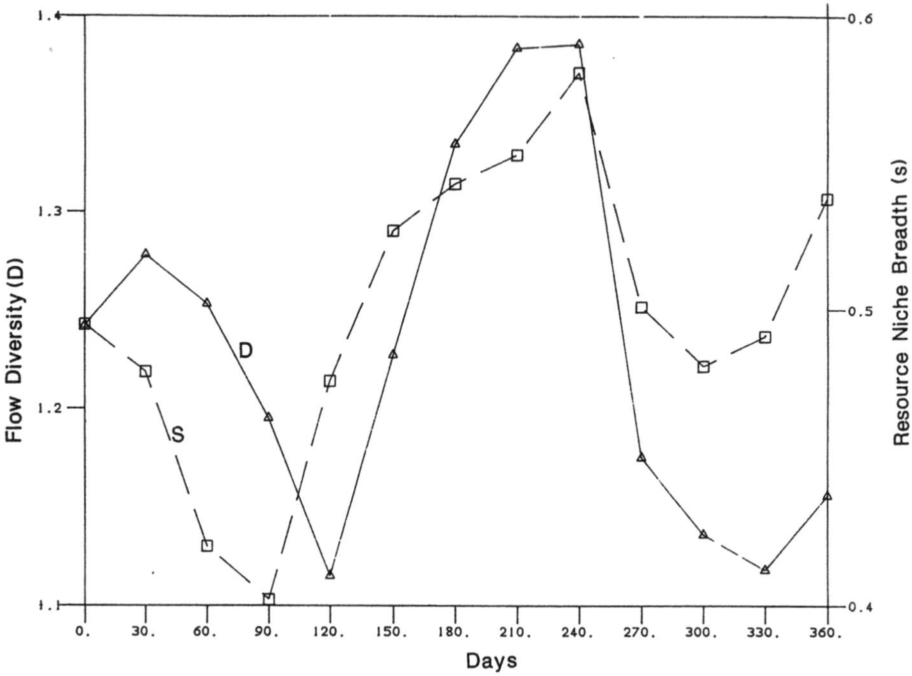

Fig. 10.6. Seasonal changes in Diversity of Flows (Kay's D) and Resource Niche Breadth (Kay's S).

mixed diet of phytobenthos, bacteria and POC have a much less heterogeneous diet in the summer due to the high biomass of phytobenthos in the sediment.

CONCLUSIONS

Experiments with varying types of aggregation tend to confirm the opinion of Kremer (1989), that comparisons of different systems in terms of their flow indices are difficult to make without an unusually high degree of standardisation (perception of important pathways, method of determining flows, degrees and type of aggregation). Most the measures are properties of network diagrams drawn on paper. The absolute value of Development Capacity and its components Ascendency, Overheads (on inputs, respiration and exports) and Redundancy are all dependent on the number of compartments in the system. Unlike taxonomic diversity, where the compartments are species, the units involved in determining flow diversity are difficult, if not impossible, to standardise or define. They cannot be species (many change their feeding behaviour ontogenetically or seasonally) or trophic groups (how do we define the appropriate level of trophic specialisation?). Measures of niche breadth will also depend on the way both the predator and prey compartments are defined.

Since the starting point for flow analysis is a perceived food web, it is relevant to note here that leading proponents of food web theory have recently been going through an exercise of critical self-examination which has a strong bearing on the above discussion (see the Special Feature on food Web Theory, Ecology, vol. 69, No. 6). Many of the concepts of food web theory are similar to those of network flow theory; for example the connectance of food webs (the proportion of total possible trophic connections between compartments that is actually observed in nature) clearly has a strong bearing on the Development Capacity. Paine (1988), whilst accepting the value and necessity of the theory, has criticised the way in which theory has been linked to the data on natural communities. He has argued that the manner in which field biologists draw linkages is usually informal and idiosyncratic. Hull (1968) had already pointed out that no two workers confronted with the same biological system would construct similar food webs. Paine (1988) further argued that the aggregation of species into larger compartments was unwarranted, and that the concept of connectance depended on so many arbitrary decisions that it was not a sound basis on which to build theories about natural communities. What network flow analysis tells us about the functioning of real ecological systems must, therefore, similarly be viewed with some caution.

Relative values of indices such as Ascendency, Overheads and Redundancy as a proportion of Development Capacity do however seem to be more robust to variations in the definition of compartments, and some generalities are beginning to emerge which may reflect conservation characteristics of ecological flow networks. On the other hand, it could equally be argued that, since these values are so robust to aggregation, they are products of the analysis and are insensitive to the biology (Cohen, 1988).

Ascendency ranged between 47.6 and 65.3% of Development Capacity with different levels of compartment aggregation in the steady state model, and varied seasonally between 51.9 and 68.3% in the dynamic simulation model (with a mean value of 58%). Kremer (1989) also noted that a value of about 60% was representative of the Narragansett Bay model, but the significance of this apparent generality is not known. Another similarity with the Narragansett Bay model is the divergence of Development Capacity from Total System Throughput during the summer due to increased Flow Diversity. As noted above, this phenomenon is not readily explainable for the Lynher.

ACKNOWLEDGEMENTS

We are grateful to John Field and Fred Wulff for advice, and to Andrea McEvoy for computing assistance.

REFERENCES

Asmus, M .L. & H. N. McKellar 1989. Network analysis of the North Inlet salt marsh ecosystem, in: Wulff, F., J.G. Field & K. H. Mann(eds) *Network analysis in Marine Ecology: Theory and Applications*, Lecture Notes in Coastal and Estuarine Studies, Springer-Verlag, New York.

Briand, F. 1983. Environmental control of food web structure, Ecology 64: 253-263.

Cohen, J. E. 1988. Food webs and community structure, pp x-x in: Roughgarden, J., R.M. May & S.A. Levin (eds) *Perspectives in Theoretical Ecology*, Princeton University Press, Princeton, New Jersey.

Field, J. G., C. L. Moloney & C. G. Attwood 1989. Network analysis of simulated succession after an upwelling event, in: Wulff, F., J. G. Field & K. H. Mann (eds) Network Analysis in Marine Ecology. Lecture Notes in Coastal and Estuarine Studies, Springer-Verlag, New York.

Gee, J. M, R. M. Warwick, J. T. Davey and C. L. George 1985. Field experiments on the role of epibenthic predators in determining prey densities of an estuarine mudflat. Est. cstl. shelf. Sci. 21: 429-448.

Kay, J. J., L. A. Graham, & R.E. Ulanowicz 1989. A detailed guide to network analysis, in: Wulff, F., J.G. Field & K.H. Mann (eds) *Network Analysis in Marine Ecology: Theory and Applications*, Lecture Notes in Coastal and Estuarine Studies, Springer-Verlag, New York.

Hull, D. L. 1968. The operational imperative: sense and nonsense in operationism, Syst. Zool. 17: 438-457.

Joint, I. R. 1978. Microbial production of an estuarine mudflat, East. cstl. mar. Sci. 7: 185-195.

Kay, J. 1984. *Self-organization in living systems*, PhD. Thesis, Department of Systems Design Engineering, University of Waterloo, Waterloo, Ontario, Canada .

Kremer, J. N. 1989. Network information indices with an estuarine model, in: Wulff, F., J.G. Field & K.H. Mann (eds) *Network Analysis in Marine Ecology: Theory and Applications*, Lecture Notes in Coastal and Estuarine Studies, Springer Verlag, New York.

Longhurst, A. R. 1981. *Analysis of Marine Ecosystems*, Academic Press, London 741 p.

Mann, K.H. 1982. *Ecology of Coastal Waters. A Systems Approach*. Blackwell, Oxford. 322 p.

Paine, R.T. 1988. Food webs: road maps of interactions or grist for theoretical development? Ecology 69: 1648-1654.

Pimm, S.L. 1982. *Food Webs*, Chapman & Hall, London, 219 p.

Price, R. & R. M. Warwick, 1980a. Temporal variations in annual production and biomass in estuarine populations of two polychaetes, *Nephtys hombergi* and *Ampharete acutifrons*, J. mar. biol. Ass. U.K. 60: 481-487.

Price, R. & R. M. Warwick. 1980b. The effects of temperature on the respiration rate of meiofauna. Oecologia (Berl.) 44: 145-148.

Teare, M. & R. Price 1969. Respiration of the meiobenthic harpacticoid copepod *Tachidius discipes* Giesbrecht, from an estuarine mudflat. J. exp. mar. Biol. Ecol. 41: 1-8.

Warwick, R. M. 1981. The influence of temperature and salinity on energy partitioning in the marine nematode *Diplolaimelloides bruciei*. Oecologia (Berl.) 51: 318-325.

Warwick, R. M., J. T. Davey, J. M. Gee & C.L. George 1982. Faunistic control of Enteromorpha blooms: a field experiment. J. exp. mar. Biol. Ecol. 56: 23-31.

Warwick, R. M., I. R. Joint & P. J. Radford 1979. Secondary production of the benthos in an estuarine *environment*, pp 429-450. In: Jefferies, R. L. & A. J. Davey (eds) *Ecological Processes in Coastal* Environments. Blackwell, Oxford.

Warwick, R .M. & R. Price 1975. Macrofauna production in an estuarine mud- flat. J. mar. biol. Ass. U.K. 55: 1-18.

Warwick, R. M. & R. Price 1979. Ecological and metabolic studies on free-living nematodes from an estuarine mud-flat. Est. cst. mar. Sci. 9: 257-271.

Wulff, F., & R. E. Ulanowicz 1989. A comparative anatomy of the Baltic Sea and Chesapeake Bay Ecosystems, in: Wulff, F., J. G. Field & K. H. Mann (eds) *Network Analysis in Marine Ecology: Theory and Applications*. Lecture Notes in Coastal and Estuarine Studies, Springer-Verlag, New York.

Chapter 11

A COMPARATIVE ANATOMY OF THE BALTIC SEA AND CHESAPEAKE BAY ECOSYSTEMS

Fredrik Wulff
Askö Laboratory
University of Stockholm
S-10691 Stockholm
SWEDEN

Robert E. Ulanowicz
Chesapeake Biological Laboratory
University of Maryland
Solomons, Maryland 20688-0038
USA

INTRODUCTION

The theoretical basis for describing and explaining ecological processes has been historically deficient at higher hierarchical levels, such as at the level of whole ecosystems. The manifest tendency for modern technological society to perturb large natural systems has, to a large extent, motivated studies at the ecosystem level, because experience has shown that we cannot explain these changes by studying organisms and populations in isolation (Mann 1982).

The development of ecosystem theory has been hampered by the difficulties in collecting data at this higher level. However, exploited ecosystems have recently begun to offer unique opportunities for testing ecosystem level hypotheses (Leggett *et al.* 1985). The ever increasing availability of data from these systems is beginning to make comparative studies attractive. At the same time analytical tools have emerged that now make it possible to describe system-level structural and functional relationships using these new data sets. In this exercise we apply a suite of the techniques that were described earlier in this book (Kay *et al.* 1989, Chapter 2, this volume) to two exploited marine systems, the Baltic Sea and Chesapeake Bay.

DESCRIPTION OF THE SYSTEMS

The Baltic Sea

The Baltic Sea is a large non-tidal estuary (373,000 km^2) with narrow and shallow connections to the Kattegat and the North Sea (Fig. 11.1). The system is composed of three major basins: the Baltic proper, south of the Åland archipelago (257,000 km^2), the Bothnian Sea (80,000 km^2) and the Bothnian Bay (36,000 km^2). The salinity of the upper layer of the Baltic proper is about 7 ‰, due to the positive freshwater balance and the narrow and shallow connections to the ocean limiting salt water inflow. The salinity below the permanent halocline, situated at about 65 meters, is between 10 and 15 ‰. The deep water is replenished by dense water from the Kattegatt. The intermittent inflows are driven primarily by barotropic flows due to the level differences between the Baltic and the Kattegatt. A seasonal thermocline develops in May and is reinforced to a depth of about 20 meters during summer. Surface water cooling and wind driven entrainment deepens the thermocline to the depth of the halocline in the fall. The upper part of the deep water (about 10 meters) is then mixed into the surface layer by the fall-winter storms counteracting the freshwater inputs and brackish surface water outflows through the Danish sounds, keeping the salinity constant. Maximum summer temperature is about 20°C above the thermocline and 2.5 - 6°C below. Ice is regularly formed in the coastal zones during late winter but rarely in the open Baltic proper.

The Baltic ecosystem is comprised of a low diversity assemblage of euryhaline marine and brackish estuarine species, fresh water species and glacial marine or fresh water relicts (Segerstråle 1957). Many species tend to occupy broader ecological niches than in their original habitats.

The original oligotrophic Baltic proper has undergone measurable eutrophication during the last 50 years, due to increased nutrient inputs (Larsson *et al.* 1985). Primary production is nitrogen limited in the Baltic proper, as in most coastal marine systems, whereas the Bothnian Bay is phosphorus limited and still ultra-oligotrophic. Basin-wide nutrient budgets (Larsson *et al.* 1985) indicate that nitrogen inputs from land and precipitation and by bluegreen algal blooms are to a large extent counteracted by denitrification. Phosphorus inputs are accumulating in bottom sediments. There is much evidence that the biogeochemical state of the Baltic has changed. Anoxic conditions in the deep water are more common and the concentrations of both nitrogen and phosphorus in the water have increased during the last several decades (Nehring *et al.* 1987). Increased production and sedimentation of organic matter has resulted in anoxia and an almost complete elimination of the benthic macrofauna in the deep basins but a considerable biomass increase above the halocline (Cederwall & Elmgren 1980). Fish catches have also increased (Ojaveer *et al.* 1981).

This study covers conditions in the Baltic proper. The peripheral northern subsystems have lower salinities and a less pronounced stratification.

Chesapeake Bay

Chesapeake Bay, the largest drowned river valley estuary in the United States (12,500 km^2) is situated along the Atlantic Ocean coast from latitude 35°50'N at its mouth to 39°40'N at its head (Fig.11.1). About 50 % of freshwater input to the system enters via the Susquehanna River, with most of the balance coming from the Potomac River. The tidal range throughout the estuary is small, averaging about 0.6 meters. The resultant flow structure is that of a moderately mixed estuary with residual flows seaward near the surface and landward at depth. The bay is shallow with an average depth of about 9 m and exceeding 60 m only in a very few locations along the channel. The input of fresh water to the bay from the Susquehanna is seasonal, peaking at about 5,700 $m^3 sec^{-1}$ in late February and early March. During the summer months there is moderately strong stratification with the combined halocline-thermocline residing at about 8 - 10 m depth. There are frequent periods of anoxia below the halocline during summer, and winds and storms will occasionally drive the anoxic waters onto the littoral shelf. Stratification tends to diminish in late September. The region of the Chesapeake covered in this study is the mesohaline middle portion from about 6 ‰ to 18 ‰. This domain covers almost half the total area of the estuary. The flushing time of the mesohaline zone is about 42 days. The mesohaline zone remains relatively ice-free during most years. Summer temperatures in this region average 25-28°C.

The mesohaline Chesapeake is characterized by a small number of species which are euryhaline but the Baltic has an even smaller number, especially of marine species.

As far as can be determined, the Chesapeake has always been slightly eutrophic, typical of the natural state of most east coast estuaries of the United States. Sediment loads increased with the clearing of land in the watershed during the late 17th century and continued until the 1970's when sediment controls were put in place. Over the past half century, nutrient loads from domestic sewage and non-point source agricultural runoff have increased appreciably. Three significant changes in the mesohaline system evidence the increased loadings: (1) The increased frequency and duration of dinoflagellate blooms (primarily *Gymnodynium splendens*) beginning in late spring and extending into the fall (2) The virtual disappearance of submerged aquatic grasses, apparently due to suffocation by heavy periphyton overgrowth (3) The increased frequency and duration of anoxia below the thermocline. The mesohaline region is nitrogen-limited during the summer. Although there is reason to believe that the oligohaline region is occasionally phosphorus-limited during the cooler months, such limitation probably does not appreciably impinge upon the mesohaline region.

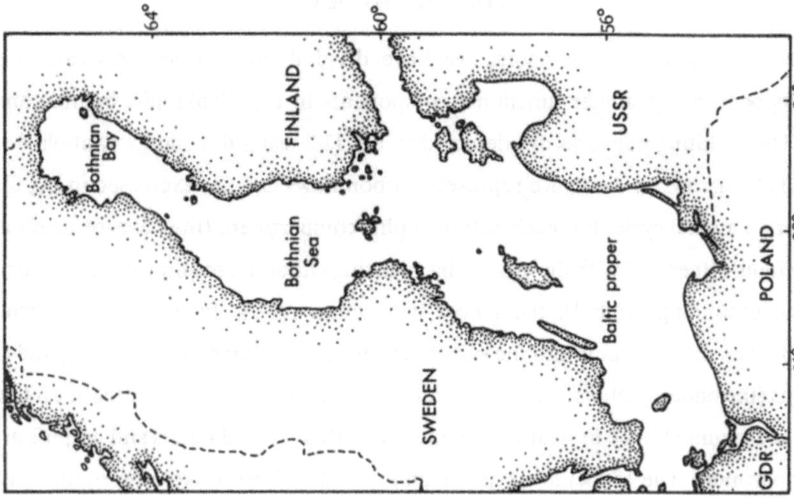

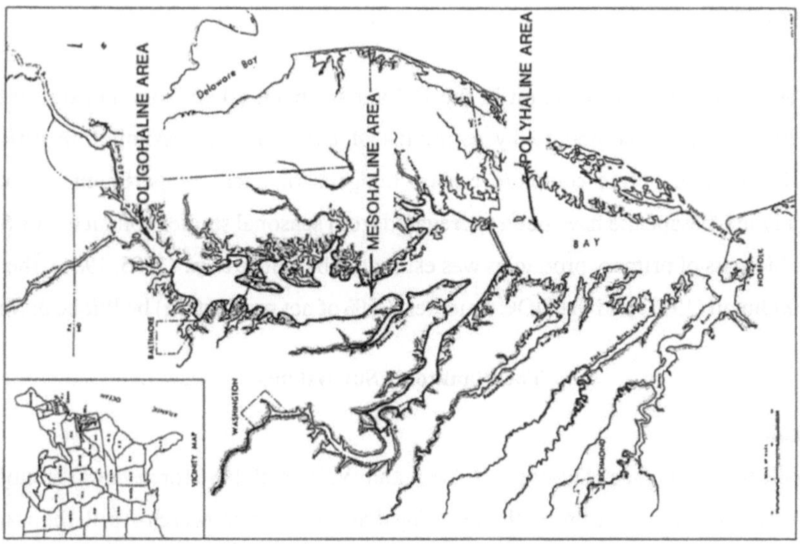

Figure 11.1. Maps of the Baltic Sea and the Chesapeake Bay.

THE NETWORKS

To facilitate cross-system comparisons, we have divided both ecosystems into the same 13 compartments comprising similar functional components in the planktonic, benthic and nektonic subsystems. The resulting networks are depicted in Fig 11.2 using the energy symbols developed by Odum (1971). All arrows in the figure represent carbon flows and are expressed in mg C m^{-2} d^{-1}, as averaged over an annual cycle. For each heterotrophic compartment (the hexagon symbols) we have balanced the inputs (feeding) with the exports by adjusting either the respiratory losses (arrows ending in an electric grounding symbol), the faeces production or the feeding by organisms in other compartments. The overall inputs into the system are the gross primary production by the autotrophic compartments (the bullet symbols) and the advection of allochotonous organic matter into the system. The systems are assumed to be in steady state, i.e., annually averaged inputs and outputs are balanced for each compartment. The 15 member representation of the Chesapeake ecosystem is an aggregated version of a 36 compartment network of flows among the ecosystem of the mesohaline reach of the Bay. Those readers desiring more details about how this system was estimated are referred to Baird & Ulanowicz (1986).

All estimates of annual net production in the Baltic have been taken from an extensive review by Elmgren (1984) and supplemented by estimates of consumption, assimilation efficiency, and respiration in order to obtain complete carbon budgets for each compartment. Annual rates of production in the Chesapeake have been aggregated from seasonal studies conducted by Sellner *et al.* (1985). The biomass of primary producers was estimated by Tuttle *et al.* (1985, 1986), the respiration by Flemer & Olman (1971) and the DOC exudates (25% of net production) by Wiebe & Smith (1977).

The Planktonic Subsystems

Phytoplankton

The *Baltic* pelagic system (see Hällfors *et al.* 1981, and Wulff *et al.* 1986 for reviews) is subject to large seasonal production pulses, starting with the spring diatom bloom in March-April, then followed by a minimum in June and a second pulse during July-August that is dominated by small flagellates. Sometimes, very conspicuous blooms of nitrogen- fixing bluegreen algae occur during late summer. Autumn blooms are sporadic, except in the southern parts, where they occur with regularity. The annual net primary production has been estimated at 160 g C m^{-2} y^{-1} by Elmgren (1984). (Gross production was derived from this figure by assuming a 30 % respiratory loss). Phytoplankton biomass in the *Chesapeake* is dominated by flagellates and small centric diatoms during the warm season and by larger

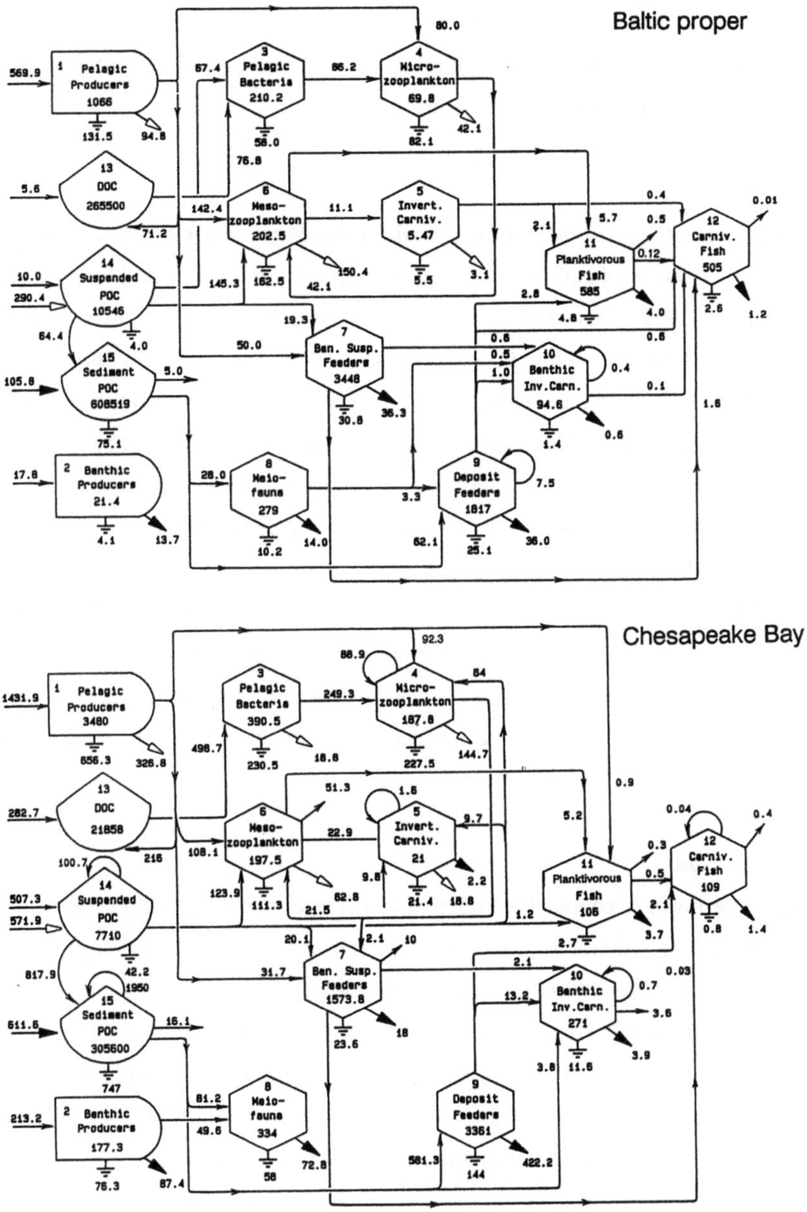

Figure 11.2. Carbon flow diagrams of the Baltic proper and the Chesapeake Bay , units are in mg C m^{-2} (biomasses) and mg C m^{-2} day^{-1} (flows). For simplicity, POC flows between compartments are not connected. Instead, open arrows indicate flows to suspended POC, filled arrows flows to sediment POC.

diatoms during the winter (Sellner *et al.* 1985). On a seasonal basis the greatest productivity occurs throughout the summer (39% of annual production), followed in order of magnitude by the spring (27%), fall (20%) and winter (14%) seasons.

Microbial loop

The importance of the microbial food chain, wherein phytoplankton exudate is utilized by bacteria and then channeled into the flagellate-ciliate-mesozooplankton food chain (Azam *et al.* 1982) has been established for the *Baltic*. To arrive at the figures shown we began with measurements of phytoplankton exudate and bacterial production (Larsson & Hagström 1979). We then assumed that bacteria grow with an carbon conversion efficiency of 60% and that the remaining requirements of the bacteria are met by the suspended POC (14). All bacteria are considered to be eaten by heterotrophic flagellates and ciliates, lumped here into the microzooplankton compartment (4). An annual biomass production of 15 g C m^{-2} and an assimilation efficiency of 50 % have been assigned to the microzooplankton. All net production by this compartment is consumed by mesozooplankton. The levels and role of bacterioplankton in *Chesapeake Bay* have been the subject of much research (Tuttle *et al.* 1985, 1986; Ducklow 1983, 1986) and some controversy (Sherr *et al.* 1987), so that a sufficient amount of actual data was available. It is worth noting that stocks of free living bacteria in the Chesapeake are generally higher than those observed in any comparable estuary. At times the pelagic bacterial production exceeds primary production (Malone *et al.* 1986).

Mesozooplankton

The overwintering mesozooplankton population (6) is very small in the *Baltic*, and maximum biomasses are observed during late summer. This compartment is dominated by the copepods (*Acartia, Temora, Pseudocalanus*), but rapidly reproducing (parthenogenetic) populations of rotifers (*Synchaeta*) and cladocerans (*Bosmina, Podon, Evadne*) are sometimes important during the late spring and summer. The carbon flows relating to the mesozooplankton are calculated by applying a C/P ratio of 4 and a P/R ratio of 0.5 to an estimated annual biomass production of 15 g C m^{-2}. The mesozooplankton community in the *Chesapeake* is dominated by the calanoid copepod, *Acartia tonsa*, during the summer and fall and by *Eurytemora affinis* in the late winter and spring (Heinle, 1966). They feed mainly on phytoplankton and suspended detrital particles (14) (Heinle & Flemer 1975; Roman 1984 a,b), but recent evidence suggests they also garner substantial sustenance from microzooplankton, such as tintinnid ciliates (Stoecker & Saunders 1985). A most interesting aspect of the Chesapeake ecosystem is that these important grazers are least abundant during that time of the year (summer) when primary production is greatest.

Carnivores

Planktonic invertebrate carnivores (5) in the *Baltic* consist primarily of medusae (*Aurelia aurita*) in the southern parts and mysids in coastal areas. A C/P ratio of 0.22 and a R/B ratio of 1 are used in conjunction with an annual biomass production of 2.4 g C m^{-2} to estimate the flows associated with this element. Their only food source is mesozooplankton and all their net production is consumed by fish. In the *Chesapeake* this compartment is dominated by the ctenophore, *Mnemiopsis leidyi*, which becomes very abundant during summer and fall. It feeds mostly on mesozooplankton and to a lesser extent upon suspended detritus and microzooplankton (Kremer 1979). Their high stocks during the summer exert heavy grazing pressure upon the mesozooplankton, and this consumption is thought to be the reason for the depressed levels of copepods during the summer (Baird & Ulanowicz 1986).

Allochtonous inputs

Regarding allochtonous organic carbon input to the *Baltic* (taken from Larsson *et al.* 1985), one third is channeled into the pool of dissolved organic matter, DOC (13), and the rest to the pool of suspended organic matter, POC (14). The POC pool receives additional inputs issuing from the faeces of pelagic organisms and from phytoplankton, however we have disregarded as insignificant any excretion of DOC by heterotrophs. Small fractions of these two pools are lost in the flow of water out of the Baltic. The concentrations of suspended detrital carbon in the *Chesapeake* waters and the rates of its deposition on to the sediment have been measured by Boynton *et al.* (1985, 1986). The input to this compartment has been calculated by difference and is one of the least reliable values appearing in the network.

The Benthic Subsystems

Shallow sandy littorals are typical in the eastern and southern parts of the *Baltic*. The shallow soft bottom vegetation is dominated by fresh water plants (*Phragmites, Potamogeton, Chara, Ruppia*), although dense beds of the marine eel grass (*Zostera*) may occur in some of the southern parts. Dense algal vegetation covers the shallow rocky shores of the barren archipelagos found in the western and northern parts of the sea. The typical zonation starts with a filamentous green algal belt (*Cladophora*) occupying the first meter, followed by a brown algae (*Fucus*) down to about 8 meters. Below that depth one finds a scattering of red algae (see Jansson & Kautsky 1977, Hällfors & Niemi 1981). Although this benthic vegetation plays an important role as a spawning and nursery ground, few herbivores utilize these plants directly, and the net production (5 g C m^{-2} y^{-1} as an average for the whole Baltic proper) is channeled into the sediment detrital pool (15).

In the Baltic, two bivalves (*Mytilus edulis, Macoma balthica*) overwhelmingly dominate the benthic filter feeding community (7), and two amphipods (*Pontoporeia affinis* and *P. femorata*) comprise the deposit feeding community (9). The polychaete, *Harmothoe sarsi*, the isopod, *Mesidotea entomon*, and the priapuloid, *Halicryptus spinulosus*, are the important benthic invertebrate carnivores (10) (see Ankar 1977 and Elmgren 1978). Echinoderms and decapod crustaceans are totally absent, except in the southernmost part. One compartment defined entirely by the size of its member organisms is the meiofauna (8), a heterogeneous mixture of predominantly deposit feeders, but including some carnivores. In this network we assumed that their food sources are primarily bacteria and sediment POC and that they are consumed by deposit feeders and by benthic invertebrate carnivores.

Rooted aquatic vegetation (*Potamogeton, Ruppia, Myriophyllum*) in the *Chesapeake* has declined drastically over the past 10-12 years and now accounts for a very small fraction of overall production. However, the biomass of subtidal, benthic diatoms is not negligible and has been estimated from live chlorophyll *a* measurements on the sediment surface (Boynton *et al.* 1986). Biomass and production are highest in the summer and lowest in winter. About 35% of the benthic microflora production is consumed by the meiofauna and the remainder becomes sediment POC.

The Chesapeake benthic community is somewhat more diverse than that of the Baltic. There are three main categories of benthic invertebrates: the suspension feeders, (including *Crassostrea virginica* and *Mya arenaria*), the deposit feeders (polychaetes such as *Nereis succinea*, bivalves like *Macoma balthica* and various amphipod crustaceans), and the commercially important predator/scavenger, *Callinectes sapidus*. The benthic-feeding carnivores are dominated by the fish *Leiostomeus xanthurus*.

The Nektonic Subsystems

Like the rest of the fauna, the fishes found in the *Baltic* are a mixture of fresh and salt water species especially in the littoral communities. The fishery is however totally dominated by three marine species; the sprat (*Sprattus sprattus*), herring (*Clupea harengus*) and cod (*Gadus morhua*). The two former species constitute the filter feeding fish compartment (11) and cod dominate the carnivorous fish compartment (12) of the Baltic network. The only other fishes for which data are available are also exploited fishes (species like flounder, turbot, salmon, pike and perch) and are lumped into carnivorous fishes. Again, the synthesis by Elmgren (1984) on available data on fish stocks and food composition has been used. Very little quantitative information on smaller and abundant species like gobiids and sticklebacks are available except in some coastal regions (Jansson *et al.* 1985) .

In the *Chesapeake Bay*, filter feeders (11) like Bay anchovy (*Anchoea mitchelli*, a major forage fish) and Atlantic menhaden (*Brevoortia tyrannus*) comprise over half the finfish biomass, and they constitute a significant forage base for the nektonic carnivores (12). The carnivorous fishes are of two types: (1) the benthic predators Atlantic croaker, hogchoker, spot, white perch and white catfish

(*Micropogonias undulatus, Trinectes maculata, Leiostomus xanthurus, Morone americana*, and *Ictalurus catus*) and (2) the pelagic feeders bluefish, weakfish, summer flounder and striped bass (*Pomatomus saltatrix, Cynoscion regalis, Paralichthys dentatus*, and *Morone saxatilis*). The latter group reaches its maximum biomass during the summer and its members emigrate from the mesohaline area during the winter. There is a notable lack of hard data on the stocks of Chesapeake fishes, although reasonable quantitative information on their diets does exist (Homer & Boynton 1978). The standing stocks of the commercial species were estimated approximately from the annual catch, divided by the published P/B ratios for each species (Winberg 1956) .

RESULTS

Given the description of the two systems discussed earlier, it is not surprising that certain differences in the key features of the networks are obvious (from figures 11.2a and b). The large difference in the magnitude of the inputs to both systems is immediately obvious. Primary productivity of the Chesapeake is almost threefold greater than that of the Baltic. Furthermore, allochtonous inflows to the Chesapeake are almost 50% of the magnitude of its primary production, while such inputs to the Baltic are less than 3% of its primary production. There are natural factors which contribute to these contrasting figures: The mean temperature in the Chesapeake is considerably higher and the ratio of its drainage basin to water area is much larger (20:1) than in the Baltic (4:1).

One other structural difference obvious from Figure 11.2 is that in the Baltic the activities of the pelagic components strongly dominate those associated with the benthos. This becomes even more obvious when one calculates the throughput of each compartment, i.e., the amounts flowing through them, as shown below in Table 11.1.

One of the purposes of the "input-output" analysis (see Kay *et al.* 1989, Chapter 2, this volume) is to trace the fate of inputs through the system. For example, Figure 11.3 depicts the percentage of primary production that leaves the system at each compartment. It is immediately obvious that the most important "sinks" for net primary production differ in the two systems. The respiration of the mesozooplankton (6) dominates the Baltic sinks, whereas bacterial respiration in the sediments (15) eventually removes most of the autochthonous production from the Chesapeake. This observation serves to underscore the predominance of the pelagic system in the Baltic versus the key role of the benthos in the more shallow Chesapeake. The picture for the eventual fates of the allochtonous inputs is similar to that for the in situ production.

Table 11.1. Total carbon throughput (mg C m^{-2} day^{-1}) for each compartment in the networks.

Compartment	Baltic	Chesapeake
Pelagic Producers (1)	570	1430
Benthic Producers (2)	178	213
Pelagic Bacteria (3)	144	499
Microzooplankton (4)	166	495
Inv. Carnivores (5)	11.1	44
Mesozooplankton (6)	330	253
Benthic Susp. Feeders (7)	69.3	53.9
Meiofauna (8)	28	131
Deposit Feeders (9)	73	581
Benthic Inv. Carnivores (10)	2.5	19.8
Planktivorous Fish (11)	10.5	7.3
Carnivorous Fish (12)	3.8	2.6
Dissolved Org. Matter, DOC (13)	76.8	499
Suspended POC (14)	300	1180
Sediment POC (15)	170	3380

Another purpose to which input-output analysis can be put is to estimate how much of the various inputs actually reaches those compartments that provide "**useful commercial products**" from the systems. In Table 11.2 are listed the percentages of inputs from the three major sources reaching four of the commercially important compartments.

It is clear that the Baltic is more "efficient" in producing commercially valuable resources. For example, a unit of phytoplankton production produces about four times as much planktivorous fish as does its counterpart in the Chesapeake.

Input-output analysis can be modified to focus on transfers occurring within the system (Szyrmer & Ulanowicz, 1987; Kay *et al.* 1989, Chapter 2, this volume). More specifically, it is possible to calculate the fraction of the total intake by a given compartment that at one time or another passed through each of the other compartments. The list of such fractions pertaining to a given species can legitimately be called the "**indirect diet**" for that component. Because the same material will have passed through several compartments on its way to the target species, the sum of such fractions will exceed unity, and in fact will be related to the trophic status of that particular compartment. Table 11.3 lists indirect diets (in %) of the carnivorous and the plankton-feeding fish in both systems.

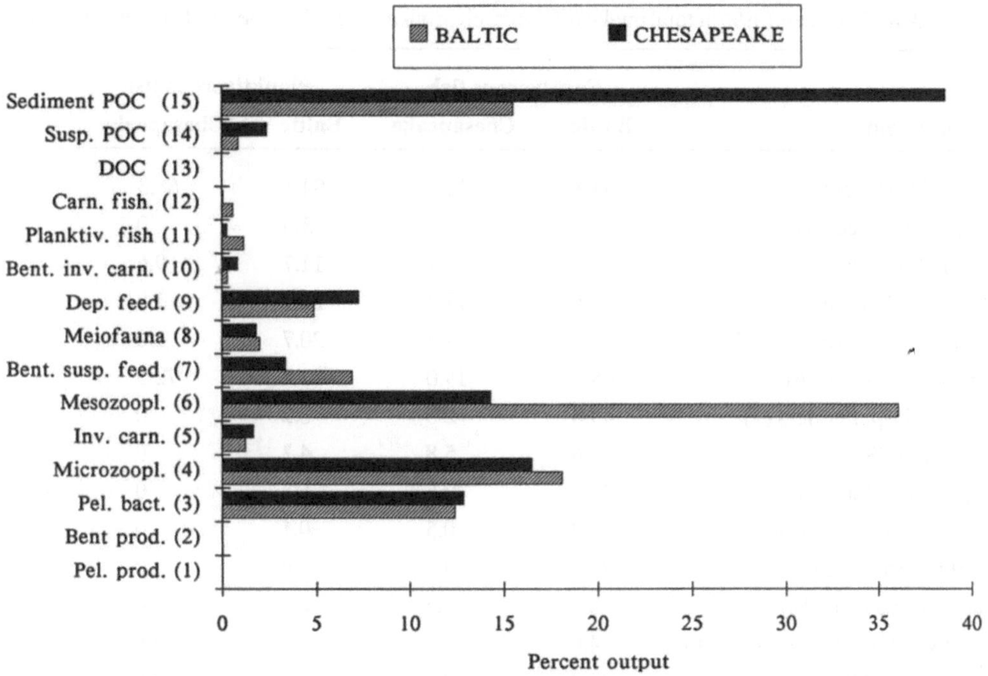

Figure 11.3. Results from the input/output analysis showing where net primary production leave the system.

Table 11.2. The percentage of input from three major sources that reaches the four commercially important compartments.

Source	System	Benthic filterf.	Benthic carn.	Planktiv. fish	Carn. fish
Phytoplankton	Baltic	11.9	0.4	1.7	0.6
production	Chesapeake	2.9	0.6	0.4	0.1
Benthic	Baltic	0	1.2	1.8	0.6
production	Chesapeake	0	1.0	0	0.1
Allochtonous	Baltic	8.9	0.6	2.5	0.8
input	Chesapeake	2.0	1.6	0.4	0.2

Table 11.3. Indirect diets (in % of total intake) of the carnivorous and the filter-feeding fishes in both systems.

Compartment	Carnivorous fish Baltic	Carnivorous fish Chesapeake	Planktivorous fish Baltic	Planktivorous fish Chesapeake
Pelagic Producers (1)	94.8	41.5	94.1	68.3
Benthic Producers (2)	2.9	9.6	3.0	0
Pelagic Bacteria (3)	8.0	9.8	11.7	9.6
Microzooplankton (4)	15.5	13.2	22.6	13.8
Invertebrate Carnivores (5)	16.4	1.6	20.7	1.0
Mesozooplankton (6)	48.2	19.0	83.2	72.5
Benthic Susp. Feeders (7)	49.8	2.6	8.2	0
Meiofauna (8)	4.6	5.8	4.3	0
Deposit Feeders (9)	24.5	79.8	26.3	0
Benthic Inv. Carnivores (10)	2.9	0.5	0.1	0
Planktivorous Fish (11)	32.7	19.5	1.0	0
Carnivorous Fish (12)	0.2	1.5	0.3	0
Dissolved Org. Matter, DOC (13)	4.6	9.8	6.7	9.6
Suspended POC (14)	43.4	79.5	51.7	52.6
Sediment POC (15)	25.0	79.8	26.3	0

Again, the predominance of the Baltic pelagic production (1) is obvious in the indirect diet of the carnivorous fish (95 %), while less than half (42 %) of the diet of Chesapeake carnivorous fishes was once phytoplankton. Other items in the indirect diets underscore this difference. The Baltic fishes are more heavily dependent upon the mesozooplankton (6), benthic suspension feeders (7) and benthic invertebrate carnivores (10), whereas the benthic deposit feeders (9) and detrital material (15) are the primary resources for the Chesapeake carnivores. The surprisingly heavy reliance of Chesapeake carnivores upon the deposit feeders is illuminating. In the Baltic the suspension feeders that are actively removing materials from the water column play a vital link in the food chain to the carnivores. In the Chesapeake the benthic deposit feeders take over the corresponding role. This difference between emphasis on suspension and deposit feeding probably reflects the fact that much of the primary production in the Chesapeake is not grazed and winds up being passively deposited on the sediments and respired or buried.

In the Chesapeake there is a sharp demarcation between the carnivorous fish (12), which depend on detrital food sources, and the plankton-feeding (11) fish which rely mostly on the grazing food chain. In the Baltic this difference is much less pronounced. This is mainly because primary and secondary

Table 11.4. The average path lengths (APL) in the two networks.

Compartment	Baltic	Chesapeake
Pelagic Producers (1)	3.28	3.14
Benthic Producers (2)	2.80	3.52
Pelagic Bacteria (3)	2.55	2.78
Microzooplankton (4)	2.59	3.19
Invertebrate Carnivores (5)	2.45	3.42
Mesozooplankton (6)	2.73	2.59
Benthic Susp. Feeders (7)	2.28	2.50
Meiofauna (8)	2.50	3.36
Deposit feeders (9)	2.53	4.14
Benthic Inv. Carnivores (10)	1.93	1.91
Planktivorous Fish (11)	2.09	3.40
Carnivorous Fish (12)	1.72	3.26
Dissolved Organic Matter, DOC (13)	3.55	3.78
Suspended POC (14)	3.54	4.88
Sediment POC (15)	2.34	4.24

A measure that bears a loose inverse correlation to trophic position is the **average path length (APL)**, which measures the average number of compartments that a particle starting at one particular compartment will pass through before leaving the system (Table 11.4). In contrast to the trophic index, the average path length includes the effects of recycling within the system. In systems without recycling (food chains or tree-like webs) the average path length and the trophic position would be tightly correlated in an inverse fashion.

One does perceive a decrease in the APL for higher members of the Baltic food chain. However, any such correlation in the Chesapeake is confounded by more intensive recycling. The small APL of the benthic invertebrate carnivores (10) reflects both the large metabolic losses and the large exports (i.e., harvesting of blue crab in the Chesapeake or predation by Baltic carnivores [cod], which in turn are heavily exploited.) The big APL for Chesapeake fishes (11, 12) can be attributed to their high production of fecal matter and natural mortality, both of which contribute to the intensive benthic recycling. Exploitation of and respiratory losses by the Baltic fishes give them a reduced APL.

As indicated in Kay *et al.* (1989, Chapter 2, this volume) any arbitrary network can be mapped into a **Lindeman-like trophic chain with accompanying routes of recycling.** The "trophic foundations" of the two ecosystems are presented in Figures 11.5a and b. It should be repeated here that the trophic compartments (as indicated by the roman numerals) are not discrete aggregations of the components shown in Figure 11.2. Any one real compartment may have its activity distributed among several of the trophic compartments in accordance with the trophic structure of its diet. In comparing the two chains in Figure 11.5 one is struck by three significant differences: (a) The food chain in the Baltic possesses one more level than does the Chesapeake chain. (b) The larger chain in the Baltic is probably due to the generally higher trophic efficiencies inherent in that system. (c) Detritivory (the flow from D to II) strongly predominates over herbivory (the flow from I to II) in the Chesapeake (by a ratio of more than 4:1). The corresponding ratio in the Baltic is only 1.5:1.

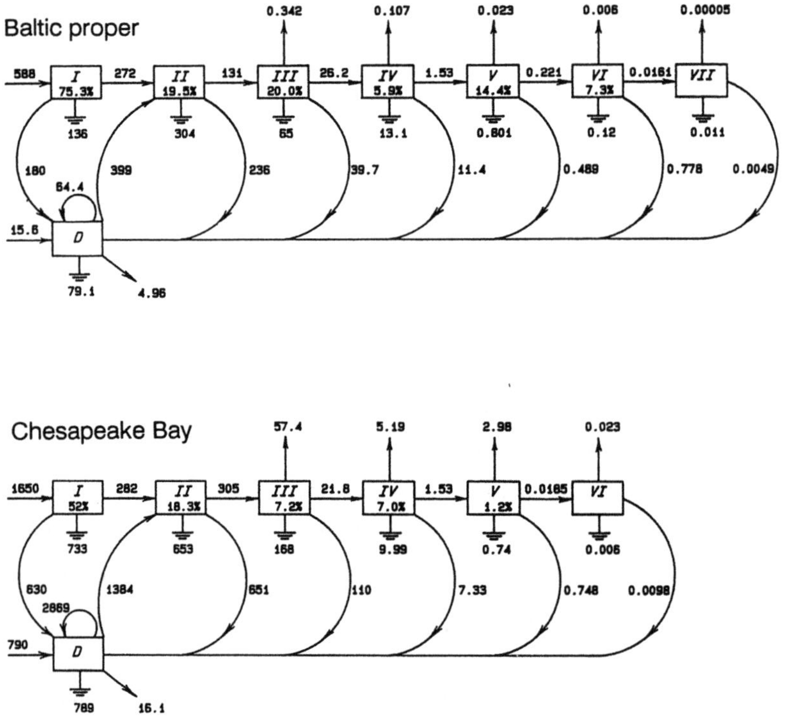

Figure 11.5. The two networks mapped into Lindeman-type trophic levels (= I-VII, D = detritus) with recycling included. Percentages inside boxes refers to overall trophic efficiencies.

The pathways of recycling in an ecosystem may correspond to the system-level means of control in the system (see Kay *et al.* 1989, Chapter 2, this volume). Therefore, much information about the system functioning can be obtained from an **enumeration of the individual cycles in a system** along with some relative weighing of those circuits (Ulanowicz 1983). The two systems possess a comparable number of cycles (22 in the Baltic vs. 20 in the Chesapeake), however, most of the recirculation in the Chesapeake occurs over cycles of very short lengths, whereas the Baltic cycles are generally longer (and hence distributed among more trophic levels) as shown in Figure 11.6. When an ecosystem is stressed, the relative amount of cycling that occurs usually rises. However, the more intense cycling appears to occur in shorter, faster and trophically lower loops (Ulanowicz 1984). Set against this observation, the Chesapeake Bay appears to be more stressed than the Baltic.

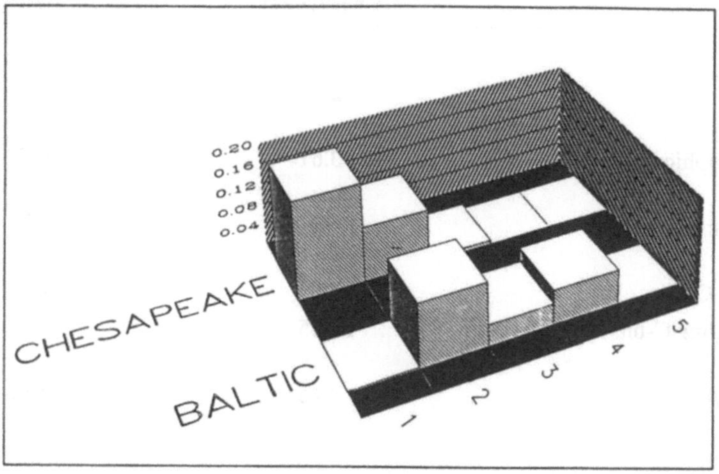

Figure 11.6. Cycle distributions against cycle lengths (1-5) in the two systems, weighted for total activities.

One may actually separate the cycled flow from the network and study its structure, as shown in Fig. 11.7. Visually and topologically the two structures are quite similar. They both possess a bipartite structure, with the cycles among the plankton remaining separated from those occurring in the benthos and nekton. The primary cycling route in the Baltic is in the water column between the zooplankton and the suspended POC (6-14-6). In the Chesapeake the dominant loop lies in the benthos between the deposit feeders and the sediment bacteria-POC (9-15-9). The pelagic bacteria in Chesapeake Bay do not participate in any recycling. They function as a sink for excess productivity instead (Ducklow *et al.* 1986). Filter-feeding fish and benthos are likewise absent from Chesapeake cycling, but their apparent function is to shunt material and energy from the planktonic domain of control into the benthic-nektonic subsystem. The function of the benthic filter feeders in the Baltic appears to parallel its Chesapeake counterpart. The separation of both systems into distinct domains of cycling appears to be due to the absence of any significant route for carbonaceous material deposited on the bottom to return to the plankton and thereafter be consumed by suspension feeders. The importance of resuspension and feeding on previously deposited POC is not taken into account in the networks and is an obvious area for further research. Before drawing any conclusions comparing the two systems as wholes, it is useful to consider the magnitudes of various whole-system indices as listed in Table 11.5.

Table 11.5. Indices of total system behaviour.

Index	**Chesapeake**	**Baltic**
Finn cycling index (%)	29.7	22.8
Average path length (APL, trophic links)	3.61	3.27
Total system throughput (T, mg C-bits)	11224	2577
Development capacity (C, mg C-bits	33000	8007
Ascendancy (A, mg C-bits)	16335	4452
Relative Ascendancy (A/C, %)	49.5	55.6
Overhead on inputs (O_o/C, %)	2.6	0.9
Overhead on exports (O_e/C, %)	0.4	0.0
Overhead on respirations (O_s/C, %)	19.4	21.4
Redundancy (O_r/C, %)	28.1	22.0
Flow diversity (bits)	2.94	3.10

The percentage of total system activity involved in recycling is called the **Finn cycling index** (Finn 1976). It was originally proposed to test Odum's (1969) conjecture that a greater degree of cycling is an indicator of a more mature system. However, as mentioned above, the sheer magnitude of cycling may actually be more indicative of a stressed system. In the Chesapeake considerably more activity (30 %) consists of recycling than in the Baltic (23 %). In the Chesapeake cycling is confined mostly to short and fast cycles. The Baltic, in contrast, possesses longer and slower loops. Hence, it is necessary to qualify a Finn cycling index with a more thorough study of the cycling structure before drawing any conclusions concerning the state of a system's development.

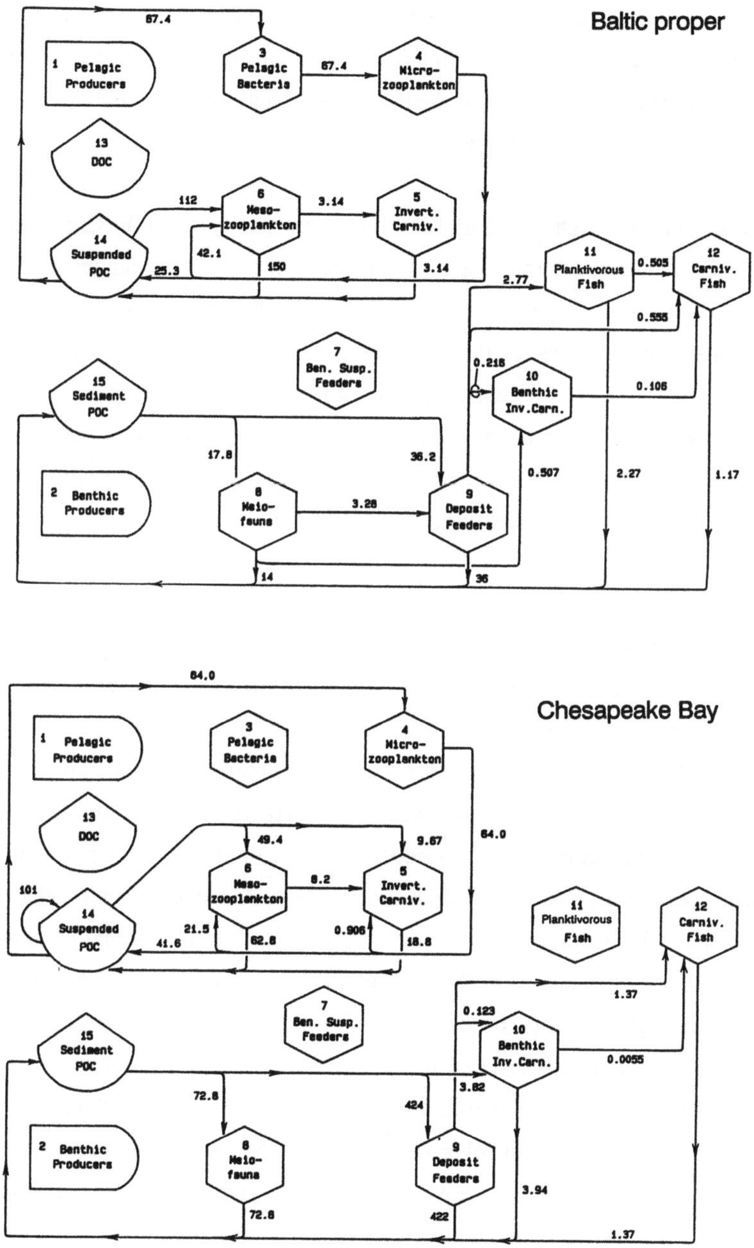

Figure 11.7. The networks for the Baltic and the Chesapeake showing only the cycled flows.

That the Baltic is more highly developed than the Chesapeake is underscored by the differences in the **whole system APL**. As the APL is correlated in inverse fashion to the average trophic positions of a system, the lower value of this index for the Baltic shows that relatively more activity is taking place at higher trophic positions in this estuary.

That the **total system throughput (T)** and **development capacity (C)** of the Chesapeake are both about fourfold higher than in the Baltic is a consequence of the markedly higher mean temperature and productivity of the more temperate estuary. When the development capacities are decomposed into their five components (ascendancy, three overhead terms and the redundancy) one notices only minor differences in their relative magnitudes (as indicated by the percentages of their respective development capacities.) To a certain extent these similarities may be attributed to the procrustean aggregation scheme used to make the systems comparable, but it is probably also due to a true underlying similarity between these systems.

CONCLUSIONS

The classical idea that a higher species diversity imparts stability to the ecosystem (e.g. Hurd *et al.* 1971) has commonly been used as an argument that the Baltic is more sensitive to pollution than are the more diverse marine systems. Jernelöv & Rosenberg (1976) and Rosenberg (1976) have challenged this hypothesis and proposed instead that low species diversity leads to a high genetic diversity, which would make the Baltic communities more adaptive than their marine counterparts. On the other hand, species exposed to the osmotic stress imposed by the low salinities might be more sensitive to additional stresses from pollution (Tendengren & Kautsky 1987). A reduction of species diversity does not necessarily mean that the number of functional units has decreased nor that the major pathways in the food web have been altered (Pimm 1984). Elmgren (1984) has shown that the food webs of the Baltic subareas are strikingly similar to those of other marine areas. Only in the Bothnian Bay, where the low salinity has eliminated the marine bivalves, do the foodwebs decrease in functional diversity. Nixon *et al.* (1986) showed that the yield of fish per unit of primary production for all the Baltic subareas accords well with those observed in most other marine areas. The various techniques used here to analyze networks make it possible to take these comparative studies of the structure and function of ecosystems one step further. It must be remembered that the networks, although fairly complicated, themselves represent gross simplifications of the real systems. Applying network analysis techniques certainly highlighted some real differences between the two estuaries, but some of the differences could be consequences of how the networks were constructed.

According to most of the "classical" theories (Odum 1969), brackish systems should be less developed with fewer feedback control loops and lower transfer efficiencies than their marine and freshwater production in the Chesapeake is sufficiently high during the winter to continue to support the plankton-feeding habits of fish in that category, but the pelagic production in the Baltic drops so low in the cold months that most zooplankton (6) are forced to overwinter in diapause, thereby forcing the planktivorous fish to go to the bottom and utilize the benthic deposit-feeders (9) (Aneer, 1980).

An important index of how a particular component of an ecosystem is performing is its **average trophic level.** This index is the average trophic level at which a species feeds as weighted by the amounts arriving at that species over all actual pathways (see Kay *et al.* 1989, Chapter 2, this volume). The average trophic pathways for the living species in the two systems are plotted as bar graphs in Figure 11.4. In both systems compartments 3, 6, 7, 8 and 9 feed chiefly at the second trophic level, while 4, 5, 10 and 12 function at higher, but similar, trophic positions. Clear differences between the systems exist in the plankton-feeding fish (11), which feed roughly one step higher in the Baltic. There appears to be a real biological difference between the feeding habits of the planktivorous fishes in the two systems. In the Chesapeake the menhaden (*Brevoortia tyrannus*) do act in large part as herbivores consuming phytoplankton, whereas their Baltic counterparts feed exclusively on trophically higher zooplankton during the warmer months and on benthic invertebrates during the winter.

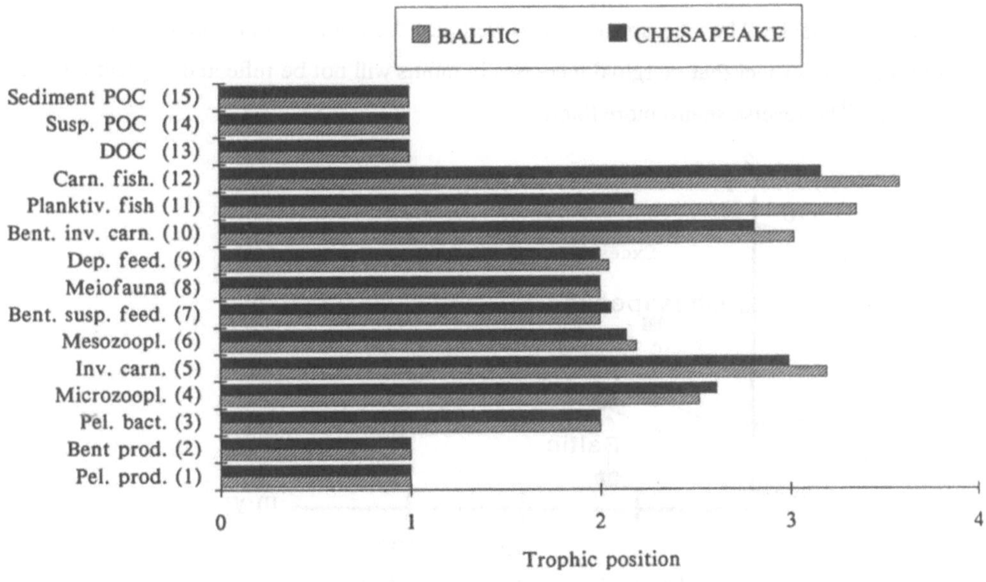

Figure 11.4. Average trophic levels in the two systems.

counterparts. Our analyses show quite the opposite on all counts. The Baltic, in spite of its lower species diversity compared to the Chesapeake Bay, has a higher relative ascendancy and diversity of flows (development capacity divided by T). Furthermore, we showed earlier that the Baltic was trophically more efficient and possesses a more highly structured array of recycling loops than does the Chesapeake. Following the reasoning of Ulanowicz (1986) these characteristics suggests that the Chesapeake is subject to a more intense set of stresses that the Baltic.

This view is also supported by other evidence. Empirical regressions between the areal nutrient input and mean depth/water retention time have been useful in predicting the trophic state in lakes (Vollenweider 1982; Peters 1986). It is interesting to place the present condition of the Baltic and the Chesapeake Bay on such a diagram and compare them with Vollenweider's lines separating permissable from excessive phosphorus loading in lakes (Fig 11.8). The diagram should be used with caution since it was developed for phosphate limited lakes, and the Baltic and Chesapeake are believed to be primarily nitrogen limited.

This is not to say that the Baltic is a pristine body of water. It, too, is the recipient of significant anthropogenic enrichment. However, either such enrichment still serves to increase productivity at all trophic levels (as occurs when oligotrophic waters receive modest increases in nutrient inputs), or else the upper trophic levels have not begun to decline in productivity to anywhere near the extent they have in the Chesapeake. The Chesapeake seems well past the point of diminishing returns, and one has every reason to expect that marginal increases in inputs will not be reflected in greater levels of useful outputs. The reverse seems more likely.

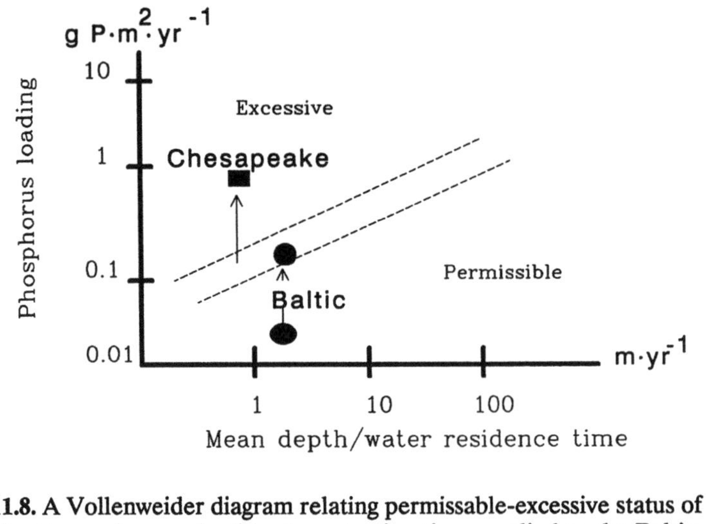

Figure 11.8. A Vollenweider diagram relating permissable-excessive status of lakes to phosphorus loading and mean depth/water retention time, applied to the Baltic and the Chesapeake Bay. Modified from Larsson *et al.* (1985) with data from Smullen *et al.* (1982)

That the Chesapeake Bay ecosystem is in poor shape is hardly news these days. However, that the Baltic appears not to be ecologically as degraded as had been thought is a new viewpoint that might not have appeared were it not for the network analyses performed here. Of course, the semblance of "health" persisting in the Baltic ecosystem should under no circumstances be used as an excuse to relax efforts to protect its biological resources. Rather, the observation should stimulate greater action to preserve the valuable assets that have survived the insults to date. Large mammals, particularity seals were important top predators in the Baltic food web at the beginning of this century (Elmgren 1987). Hunting and biomagnified toxic substances have now reduced the populations to the verge of extinction (Bergman & Olsson 1985). Although the seven Baltic countries have started to take joint action against the increasing pollution, efforts to limit inputs of nutrients and toxic substances have had limited success so far.

Finally, from an academic standpoint, we hope that what we have presented here will stand as a first example of what we hope will become a very fruitful discipline of the "comparative anatomy" of ecosystems.

REFERENCES

Azam, F., T. Fenchel, J. G. Field, J. S. Gray, L. A. Meyer-Reil & F.Thingstad 1983. The ecological role of water-column microbes in the sea, Mar. Ecol. Prog. Ser. 10: 257-263.

Aneer, G. 1980. Estimates of feeding pressure on pelagic and benthic organisms by Baltic herring (*Clupea harengus v: membras* L.), Ophelia, Suppl. 1: 265-275.

Ankar, S. 1977. The soft bottom ecosystem of the Northern Baltic Proper with special reference to the macrofauna, Contrib. Askö Lab., Univ. Stockholm, 19: 1-62.

Baird, D. and R.E. Ulanowicz. 1986. A network analysis of the Chesapeake Bay ecosystem, Ref. No. UMCEES-CBL 87-77, Center for Environmental and Estuarine Sciences, Solomons, MD.

Bergman, A. & M. Olsson 1985. Pathology of Baltic Grey Seal and Ringed Seal females with special reference to adrenocorrtial hyperplasmia: Is environmental pollution the cause of a widely distributed disease? Finnish Game Res. 44: 47-62.

Boynton, W.R., W.M. Kemp & J.M. Barnes 1985. Report to: Maryland Office of Environmental Programs. Chesapeake Bay Water Quality Monitoring Program. Ecosystems Processes Component (EPC). July 1984-June 1985, University of Maryland [UMCEES] CBL 85-21:1-42.

Boynton, W.R., W.M. Kemp, J.H. Garber & J.M. Barnes 1986. Report to: Maryland Office of Environmental Programs. Chesapeake Bay Water Quality Monitoring Program. Ecosystems Processes Component (EPC), July 1985-May 1986, University of Maryland [UMCEES] CBL 86-56: 1-52.

Cederwall, H., & R. Elmgren 1980. Biomass increase of benthic macrofauna demonstrates eutrophication of the Baltic Sea, Ophelia, Suppl. 1: 287-304.

Ducklow, H.W. 1983. Production and fate of bacteria in the oceans, Bioscience 33: 494-501.

Ducklow, H.W., D. A. Purdie, P. J. L. Williams & J. M. Davies 1986. Bacterioplankton: A sink for carbon in a coastal marine plankton community, Science. 232: 865-867.

Elmgren, R. 1978 . Structure and dynamics of Baltic benthos communities, with particular reference to the relationship between macro- and meiofauna, Kieler Meeresforsch., Sonderh. 4: 1-22.

Elmgren, R. 1984. Trophic dynamics in the enclosed, brackish Baltic Sea, - Rapp. P.-v. Réun. Cons. int. Explor. Mer 183: 149-166.

Elmgren, R. 1987. Effects of increased nutrient loading on marine ecosystems: the Baltic, Background paper GESAMP WG "Nutrients and Eutrophication in the Marine Environment", Paris, 22 pp.

Flemer, D.A. & J. Olman. 1971. Daylight incubator estimates of primary production in the mouth of the Patuxent River, Maryland, Chesapeake. Sci.12: 105-110.

Finn, J. 1976. The cycling index, J. Theor. Biol., 56: 363-73.

Goodman, D. 1975. The theory of diversity-stability relationships in ecology, Quarterly Review of Biology, 50: 237-266.

Hällfors, G. & Å. Niemi 1981. Vegetation and primary production, pp. 220-238. In: Voipio, A. (ed.) *The Baltic Sea*, Elsevier, Amsterdam, 418 pp.

Heinle, D.R. 1966. Production of a calanoid copepod, *Acartia tonsa* in the Patuxent River estuary, Chesapeake. Sci. 7: 59-74.

Heinle, D.R. & D.A. Flemer. 1975. Carbon requirements of a population of the estuarine copepod *Eurytemora affinis*, Mar. Biol. 31: 235-247.

Homer, M. & W.R. Boynton 1978. Stomach analysis of fish collected in the Calvert Cliffs region, Chesapeake Bay-1977, University of Maryland. UMCEES 78-154-CBL, 1-363.

Hurd, L.E., M.V. Mellinger, L.L. Wolf & S.J. McNaughton 1971. Stability and diversity at three trophic levels in terrestrial ecosystems, Science 173: 1134-1136.

Jansson, B.-O., G. Aneer & S. Nellbring 1985. Spatial and temporal distribution of the demersal fish fauna in a Baltic archipelago as determined by SCUBA census, Mar. Ecol. Prog. Ser., 23: 31-43.

Jansson, A.-M. & N. Kautsky 1977. Quantitative survey of hard bottom communities in a Baltic archipelago, pp 359-366. In: Keegan, B. F., P. O. O'Ceidigh, and P. S. J. Boaden (eds) *Biology of benthic organisms*, Pergamon Press, Oxford. 630 pp.

Jernelöv, A. & R. Rosenberg, 1976. Stress tolerance of ecosystems, Environ. Cons. 3: 43-46.

Kay, J.J., L. Graham & R. E. Ulanowicz 1989. A detailed guide to network analysis, In: Wulff, F., J.G. Field & K. H. Mann (eds) *Network Analysis in Marine Ecology. Methods and Applications.* Lecture Notes in Coastal and Estuarine Studies, Springer-Verlag, New York.

Kautsky, N. 1981. On the trophic role of the blue mussel (*Mytilus edulis* L.) in a Baltic coastal ecosystem and the fate of the organic matter produced by the mussels, Kieler Meeresforsch., Sonderh. 5: 454-461.

Kremer, P. 1979. Predation by the ctenophore *Mnemiopsis leidyi* in Narragansett Bay, Rhode Island, Estuaries 2: 97-105.

Larsson, U., R. Elmgren & F. Wulff 1985. Eutrophication and the Baltic sea: causes and consequences, Ambio, 14: 9-14.

Larsson, U. & Å. Hagström 1979. Phytoplankton exudate release as an energy source for the growth of pelagic bacteria, Mar. Biol. 52: 199-206.

Leggett, W.C., W.R. Boynton, V. F. Gallucci, L. Johnson, R. Margalef, J. McGlade, D. Mikelecky, R. Rosen, R. O. Ulloa & J. Watson 1985. Hypothesis testing and sampling design in exploited ecosystems, pp 237-240. In: Ulanowicz, R.E. & T. Platt (eds) *Ecosystem theory for biological oceanography*. Can. Bull. Fish. Aquat. Sci. 213

Malone, T.C., W.M. Kemp, H.W. Ducklow, W.R. Boynton, J.H. Tuttle & R.B. Jonas 1986. Lateral variation in the production and fate of phytoplankton in a partially stratified estuary, Mar. Ecol. Prog. Ser. 32: 149-160.

Mann, K. H. 1982. *Ecology of Coastal Waters. A Systems Approach*, Studies in Ecology, Volume 8. Blackwell Scientific Publications, Oxford, 322 pp.

Nehring, D., G. Aertebjerg, P. Alenius, V. Astok, S. Fonselius, M. Hannus, V. Tervo, A. Troszinska, P. Tulkki & A.K. Yurkovskis, 1987. Nutrients, pp 35-81. In: Lassig, J. (ed). *First periodic assessment of the state of the marine environment of the Baltic Sea area, 1980-1985; Background document*. Baltic Sea Environm. Proc. 17B.

Nixon, S. W., C.A. Oviatt, J. Frithsen & B. Sullivan 1986. Nutrients and the productivity of estuarine and coastal marine ecosystems, J. Limnol. Soc. sth. Afr. 12 : 43-71.

Odum, E.P. 1969. The strategy of ecosystem development, Science 164: 262-270.

Odum, H.T. 1971. *Environment, Power and Society*, Wiley-Interscience, New York, 331 pp.

Ojaveer, E. 1981. Marine pelagic fishes , pp. 276-292, In: A. Voipio (ed) *The Baltic Sea*. Elsevier, Amsterdam, 418 pp.

Peters, R. H. 1986. The role of prediction in limnology, Limnol. Oceanogr. 31: 1143-1159.

Pimm, S. L. 1984. The complexity and stability of ecosystems, Nature, Vol. 307: 321-326.

Roman, M.R. 1984a. Utilization of detritus by the copepod, *Acartia tonsa*, Limnol. Oceanogr. 25: 949-959.

Roman, M.R. 1984b. Ingestion of detritus and microheterotrophs by pelagic marine zooplankton, Bulletin of Marine Science 35: 477-494.

Rosenberg, R.1976. The relation of treatment and ecological effects in brackish water regions, Pure and Appl. Chem. 45: 199-203.

Segerstråle, S.G. 1957. Baltic Sea, Mem. Geol. Soc. America 67: 751-800.

Sellner, K. G., R.V. Lacouture, D. C. Brownlee & C. R. Parrish 1985. Report to: Maryland Office of Environmental Programs (OEP), Chesapeake Bay Water Quality Monitoring Program. Phytoplankton and microzooplankton; data summary. The Academy of Natural Sciences, Benedict Estuarine Research Laboratory, Benedict, MD

Sherr, E. B., B. F. Sherr & L . J. Albright 1987. Bacteria: Link or sink? Science. 235: 88-89.

Smullen, J. T., J. Taft. & J. Macknis 1982. Nutrient and sediment loads to the tidal Chesapeake Bay system, pp. 147-262, In: E.G. Macalaster, D., D.A.Barker. and M. Kasper (eds.), Chesapeake Bay Program Technical Studies: a Synthesis. U.S. Environmental Protection Agency, Washington, DC.

Stoecker, D. K. & N. K. Sanders. 1985. Differential grazing by *Acartia tonsa* on a dinoflagellate and a tintinnid, J. Plankton Res. 7: 85-100.

Szymer, J. & R.E. Ulanowicz 1987. Total flows in ecosystems, Ecol. Mod. 35: 123-136.

Tuttle, J.H., T. Malone, R. Jonas, H. Ducklow & D. Cargo. 1985. Nutrient-dissolved oxygen dynamics: Roles of phytoplankton and microheterotrophs under summer conditions, Report to U.S. Environmental Protection Agency. University of Maryland [UMCEES] CBL 85-39:1-50.

Tuttle, J.H., T. Malone, R. Jonas, H. Ducklow & D. Cargo. 1986. Nutrient-dissolved oxygen dynamics: Roles of phytoplankton and microheterotrophs under summer conditions, Report to U.S. Environmental Protection Agency. University of Maryland [UMCEES] CBL 86-125:1-153.

Ulanowicz, R. E. 1983. Identifying the structure of cycling in ecosystems, Mathematical Biosciences 65: 219-237.

Ulanowicz, R. E. 1986. *Growth and Development: Ecosystems Phenology*. Springer-Verlag, New York, 203 pp.

Vollenwieder, R. A. (ed) 1982. *Eutrophication of Waters. Monitoring, Assessment and Control.* OECD, Paris, France, 154 pp.

Wiebe, W. J. & D. F. Smith 1977. Direct measurement of dissolved organic carbon release by phytoplankton and incorporation by micro-heterotrophs, Mar. Biol. 42: 213-223.

Winberg, G. G. 1956. *Rate of Metabolism and Food Requirements of Fish*, Fisheries Research Board, Canada Translation Series, No. 194.

Wulff, F., G. Aertebjerg, G. Nicolaus, Å. Niemi, P. Ciszevski, S. Schulz & W. Kaiser, 1986. The changing pelagic ecosystem of the Baltic Sea, Ophelia, Suppl. 4: 299-319.

PART IV

DISCUSSION AND CONCLUSIONS

CHAPTER 12

NETWORK ANALYSIS IN MARINE ECOLOGY: AN ASSESSMENT.

Kenneth H. Mann
Department of Fisheries & Oceans
Bedford Institute of Oceanography
PO Box 1006, Dartmouth, NS
CANADA B2Y 4A2

John G. Field.
Marine Biology Reseach Institute
University of Cape Town
7700 Rondebosch
SOUTH AFRICA

Fredrik Wulff
Askö Laboratory
University of Stockholm
S-10691 Stockholm
SWEDEN

WHY WORK AT THE ECOSYSTEM LEVEL?

In this chapter we review the accomplishments and the potential of network analysis in marine ecology. As we have seen, the starting point is a network of fluxes of energy or materials in one or more ecosystems. For those not already committed to the collection of data for such a network, it may be useful to consider the advantages and disadvantages of approaching ecological problems at the ecosystem level. One way of addressing the question is through the concept of hierarchies. At the lower level of a hierarchy is some unit that can be considered basic, or relatively simple. Each successive higher level in the hierarchy comprises an interacting group of elements from the lower level. In basic biology the hierarchy:

molecules ⇒ cells ⇒ tissues ⇒ organs ⇒ organisms

is very familiar, and the idea that cells are systems of interacting molecules, tissues are systems of interacting cells, and so on, needs no elaboration. In recent years the recognition that populations are systems of interacting organisms, communities are systems of interacting populations and ecosystems are communities in interaction with their physical and chemical environment has led to the idea that there is a hierarchical continuum from the molecular (or even lower) level to the ecosystem level of organization:

molecules ⇒ cells ⇒ tissues ⇒ organs ⇒ organisms ⇒ populations ⇒ communities ⇒ ecosystems

The emerging world view in which it is perceived that the earth, its atmosphere and its living component, the biosphere, comprise a self-maintaining system ("The Gaia Hypothesis", Lovelock 1979, 1986) can be regarded as the end-point of the series. Early definitions of ecology emphasized the relationship between organisms and their living and non-living environment. A study of a particular species from this point of view used to be called autecology, while studies of the interactions between many species was called synecology. As the term ecosystem, first used by Tansley (1935) came slowly into use, the tendency was to refer to two broad categories, population ecology and ecosystem ecology. Population ecology was particularly useful to those responsible for managing populations of commercial importance, such as fish and wildlife, and the subject of population interactions and population regulation came for a time to dominate the literature of ecology. It was rightly perceived that at the lower levels of a hierarchy it was possible to make more precise, more reproducible experiments. If the subject of study is one species and its physico-chemical environment there are fewer variables to control, and the results are more likely to be reproducible, than if one is studying the interactions between a large number of species with each other and with their non-living environment. These were powerful arguments for directing efforts into population-level studies rather than into ecosystem studies.

On the other hand, even for those whose chief interest was in populations of commercial importance, there came a time when it became necessary to take ecosystem science into consideration. Many of them found that in spite of their very detailed knowledge of the population biology of the species of interest, unforeseen changes were occurring in their populations because radical changes were occurring at the ecosystem level. For example, changes in some of the fish stocks of the Baltic Sea and of Chesapeake Bay area are explicable only in terms of changes that have occurred at the ecosystem level (Wulff and Ulanowicz 1989).

The study of ecosystems shows that they possess properties that are distinctive to that level of organization, and which do not emerge by summing the properties of the component parts. For example, primary production in a system is often limited by the flux of the essential plant nutrient that is most scarce, and in marine environments it is commonly found to be nitrogen. However, the regeneration of nutrients by heterotrophs, and the transport of nutrients into the euphotic zone by water movements are system processes that help determine the total amount of primary production.

This quantity cannot be determined solely from the nutrient requirements of each of the plant species. Hence we arrive at the concept of emergent properties of ecosystems: properties that are found at the ecosystem level of organization but not at lower levels. The study of the ecosystem as a functional whole is referred to as the holistic approach.

The difficulties of this approach are considerable, and progress in searching for general principles and laws has often been slow (Mann, 1988). Nevertheless, the importance of such studies has gradually come to be recognized. In a recent survey of the membership of the British Ecological Society (Cherrett 1988) in which they were asked to list in order the 10 most important concepts in ecology, "The ecosystem" was ranked no. 1.

The early achievements in this field were in the description of the networks of flow of energy or materials. From the pioneering example of the Odum brothers in describing the trophic structure and productivity of Eniwetok atoll (Odum and Odum 1955), we have progressed to a large volume of information of the kind reported in Fasham (1984). The conventional method of representing such results is by a series of boxes which represent the biomasses and connecting arrows which represent the flows between them. The flows are normally averaged over some convenient time period, commonly one year, although for planktonic systems where primary production is measured on a scale of hours, it is not uncommon to present daily mean flows. For practical purposes, as well as for testing the quality of ones understanding of a system, it is desirable to be able to make predictions about the future state of that system. The technique that has been adopted for transforming a static flow model in to a dynamic model capable of forecasting future states is simulation modelling. Borrowed from engineering, the method has been applied with considerable success to ecology, permitting the modeller to see the results of the interactions of a large number of processes in an ecosystem occurring simultaneously. For many years ecologists had high hopes that they would be able to improve the quality of their models to the point where they accurately reproduced the integrated behaviour of ecosystems and could be used to predict the consequences of perturbing those systems. This goal has so far eluded them.

SIMULATION MODELLING, ITS STRENGTHS AND LIMITATIONS.

A simulation model may be regarded as a large number of process models coupled together. Models of single processes are often based on studies conducted under laboratory conditions where all the relevant variables can be controlled, and varied one at a time. For example, the rate of metabolism of a particular species may be measured as a function of body weight, temperature and nutritional status. The resulting model may be a multiple regression of the rate of metabolism on those variables and may be capable of making predictions with acceptable confidence limits. Similar studies may be carried out for the feeding rates of all the species in a system. By linking all the organisms in a network of trophic relations, and coupling all the process models, it is possible to construct a simple simulation model. After setting the initial values of the biomasses, the model can be made to simulate the time course of events in the system.

Numerous models have been constructed which mimic very well the behaviour of the systems on which they are based. They give excellent insights into the probable behaviour of those systems but in most cases are not able, nor are they expected, to make predictions about the future state of that system, especially if the prediction involves ranges of variables which lie outside the range of the observations on the natural system. Examples which are referred to in this volume include : the Baltic Sea (Wulff and Ulanowicz 1989; Jansson *et al.* 1984; Stigebrandt and Wulff 1987); Naragansett Bay, Rhode Island (Kremer and Nixon 1978) ; North Inlet, South Carolina (Summer *et al.* 1980) and a plankton community in an upwelling area (Moloney *et al.* 1986).

The question arises: if simulation models in engineering can predict the behaviour of engineering systems, why can ecological simulation models not give equally good predictions about ecosystems? There are three factors to be considered: (i) the inherent greater variability in the parameters fed into biological models compared with engineering models, (ii) the hierarchical nature of ecosystems and (iii) the capacity for self-organisation which ecosystems possess.

Biological Variability .

When discussing a process model that might be a typical component of an ecological simulation model we took as an example the rate of metabolism of a particular species as a function of body weight, temperature and nutritional status. If, under well-controlled conditions, a plot is made of metabolic rate as a function of body size, and the regression equation is used as a predictor of metabolic rate, the confidence limits on that prediction would typically be of the order of ± 25% of the mean. This variability is for the most part genetically determined. Variability is a fundamental property of organisms in a population. Without variability, natural selection would have nothing on which to operate. When a large number of such process models with their inherent variance are coupled together in a simulation model, the confidence limits on the output are very wide indeed. In fact, very often, a model can only be made to give a sensible output by inserting values that are means rather than representatives of the true range of variability.

Engineering models, by contrast, are constructed of process models with much smaller confidence limits. For example, the coefficient of expansion of a metal with changing temperature is known very precisely within very narrow limits. Hence, biological variability is one very major cause of limitation on the predictive ability of ecosystem simulation models.

Hierarchical Structure.

The hierarchical structure of ecosystems has been discussed very fully by O'Neill *et al.* (1986). One of the points that they make is that the various levels of the hierarchy of an ecosystem are characterized by different time scales. In general, lower levels of a hierarchy have smaller and more rapid scales of fluctuation. For example, during the course of the yearly cycle of phytoplankton the individual species populations show far greater fluctuations in numbers and productivity than the total phytoplankton community.

Weiss (1969) expressed the same idea when he said that the variability of the whole is less than the sum of the variability of the parts. A consequence of this property is that the behaviour of the total system cannot be inferred solely from observations of the behaviour of the parts. Yet the construction of an ecosystem simulation model does just this. It takes the process models appropriate to each species or functional group of species and links them in order to reproduce the behaviour of the total system. In order to make the model conform to the behaviour of the total system it is necessary first to observe the behaviour of that system. When model outputs are compared with data from the real system and are found to be diverging, it is normal for the modeller to modify some aspect to give a better correspondence between the two, but there is no guarantee that the constraints that the modeller imposes are those operating in the natural system.

Allen and Starr (1982) suggested that many of the constraints in natural ecosystems result from the formation of interacting groups of species, such as guilds, trophic levels etc (known in systems theory as holons). They suggested that if the models were constructed to account for these aspects of the organization, more of the constrained behaviour of the system would be accounted for and "the model could explain what actually happens as well as what can happen."

In discussing the empirical evidence for the occurence of hierarchical structure in ecosystems, O'Neill *et al.* (1986) suggested that it is useful to distinguish between hierarchies of species and hierarchies of processes. The arrangement of species into hierarchical groups with strong interactions within the groups and weaker interactions between the groups is of great value in understanding phenomena such as keystone species (whose removal leads to major changes in ecosystem structure) and differences in stability in different ecosystems. The concept of hierarchies in relation to processes provides a framework for discussing the topic mentioned earlier, that the variability of the whole is less than the sum of the variability of the parts. O'Neill *et al.* (1986) use a forest as their example of an ecosystem at a high level in the hierarchy. The lower-level units the trees, may be detroyed by small-scale fires, or localized pest outbreaks, but the forest as a unit persists and its processes proceed with only minor variations.

Seen in this light, the dilemma of population versus ecosystem ecology, mentioned earlier, is a false dichotomy. An ecosystem is an entity with at least two dimensions. In one it is structured according to constraints related to organismal interactions, in another according to constraints related to fluxes and mass balances. Much of the inadequacy of contemporary simulation models reflects their inability to take these two hierarchies into account.

The capacity for self-organization.

The concept of self-organization in ecosystems has its roots in the thermodynamic theory of dissipative structures (see Prigogine 1945, 1947) and in Bronowski's (1973) concept of stratified stability. The subject has recently been reviewed by Allen (1985) and by O'Neill *et al.* (1986). Briefly , a dissipative structure is one that maintains its structure by the dissipation of energy supplied from outside itself. An organism is a dissipative structure, maintaining a high state of internal organisation (entropy production at a local minimum) by the flow of energy usually derived either from the sun (in the case of plants) or from food (in the case of animals).

The evolution of life from non-living systems and the evolution of organisms with increasing degrees of complexity are seen as the result of dissipative structures interacting with one another to form new and more complex combinations within the framework of the total system. At each stage of the process the structures, or subsystems, are required for further stages of development. The total system is seen to be undergoing self-organization. It is now suggested that this process applies not only to hierarchies up to the level of organisms, but also to populations, communities and ecosystems.

Parallel with the development of these ideas there had been put forward mathematical models to account for the sudden changes of state that can occur in populations and ecosystems, the existence of bifurcations and multiple stable states (e.g. May and Oster 1976) . These two lines of thought come together in the idea that the sudden changes of state are manifestations of an ecosystem's ability to create new forms of internal organization, in keeping with its function as a dissipative structure.

While these ideas are still very speculative, lacking anything but circumstantial evidence as to their validity, they are sufficient to make us realize that mechanistic simulation models which do not permit the existence of multiple stable states are unlikely to be good predictors of the future behaviour of ecosystems. The small number of models that do try to take into account the capacity for self- organiz ation of an ecosystem (e.g. Allen and McGlade 1986) are of necessity forced to deal with only one small subset of the great range of possibilities for self-organization.

In the light of these limitations on the use of simulation models, it seemed worthwhile to explore alternative approaches to the understanding of the properties of ecosystems. This was one of the tasks undertaken by SCOR Working Group 73 on Ecological Theory and Biological Oceanography (Ulanowicz and Platt 1985).

THE EMERGENCE OF A COMPARATIVE SCIENCE OF ECOSYSTEMS

In their search for alternatives to simulation modelling as an approach to the dynamics of ecosystems, members of SCOR Working Group 73 identified five areas which showed potential for supporting alternative approaches: *Thermodynamics, Statistical Mechanics, Ataxonomic aggregations, Information theory* and *Input-output analysis* . Each of these was explored in a symposium "Ecosystem Theory for Biological Oceanography" (Ulanowicz and Platt 1985).

Thermodynamics.

Thermodynamics, especially the thermodynamics of open systems far from equilibrium and the concept of dissipative structures, was discussed in the previous section. The idea that the flow-through of energy can be used to create and maintain local zones of reduced entropy production and hence more complex structure is proving a fertile ground for understanding the self-organizing aspect of ecosystems.

Statistical Mechanics.

The idea that statistical mechanics might prove helpful in ecology goes back at least as far as Lotka (1925) who wrote " ...what is needed is an analysis... that shall envisage the units of a biological population as the established statistical mechanics envisages molecules, atoms and electrons; that shall deal with such average effects as population density, population pressure, and the like, after the manner thermodynamics deals with the average effects of gas concentration, gas pressure....".

The approach was explored in Platt *et al.* (1981) and in Trainor (1985) but it was concluded that more work would be needed before useful applications could be found.

Ataxonomic Aggregations.

An approach that offers interesting possibilities for the generalized understanding of ecosystem function is one in which organisms in a planktonic community are classified strictly on the basis of particle size, and not at all on the basis of taxonomic position. From the observations that most heterotrophic organisms feed on organisms smaller than themselves and that in general smaller particles multiply faster than larger particles, it has been possible to develop a body of ecosystem theory that forms an interesting complement to the more usual taxonomically based work.

The basic ideas were inherent in Elton's (1927) concept of the pyramid of numbers, but in the 60 years since the publication of that book we have learned how to quantitatively relate feeding, metabolism, growth and life expectancy to body size, (reviewed by Banse 1982, Peters, 1983, Calder 1985) and have thus paved the way for quite sophisticated size-dependent ecosystem models (e.g. Platt and Denman 1978, Cousins 1985, Moloney and Field 1989).

Information Theory.

The best-known and perhaps earliest use of information theory in ecology was the use of the Shannon-Wiener information theory index (set out in Shannon and Weaver 1949) as a measure of species diversity in a community. The original formulation was devised to measure the information content per symbol of a code composed of a fixed number of symbols whose probabilities of occurrence are known. Species diversity is a function of both species richness and the evenness with which the various species are represented in the population, and the Shannon-Wiener index seemed to express this efficiently. However, it has since been shown to have some difficulties of ecological application (primarily since species diversity is not neccesarily a measure of information or organization) and is less used than in earlier years.

More recently information theory has been used to develop a formal theory of self-organization (Atlan 1985). It has come to be known as the "complexity from noise principle", and works on the basis that while noise has a negative effect on the transmission of messages at a particular level in a system, when viewed from a higher level in the hierarchy it is seen as generating additional variability , some of which may prove to be a useful increase in complexity. The interested reader is referred to Atlan's original papers and also to the work on adaptability by Conrad (1983, 1985). The concluding paragraph of Atlan (1985) reads "...information theory has proven to be a useful tool in quantitative ecology. However, in its classical form, where the only quantities are static measures of probabilistic distributions having no functional meaning, it has severe limitations. The extension of information theory into a theory of self-organization that incorporates the inter-relationships between different levels and the creation of meaning , may be particularly relevant to the analysis of evolving ecosystems."

Network analysis.

It is a truism to state that in marine ecology it is easier to measure the biomasses of organisms than to measure the fluxes between them. One of the early conclusions of SCOR Working Group 59 (Platt *et al.* 1981) was that it was important to pay more attention to physiological rates for the computation of ecological fluxes. The 1982 symposium "Flows of energy and materials in marine ecosystems" (Fasham 1984) was a direct result of this recommendation. Focussing attention on fluxes rather than biomasses leads to a different concept of the functioning of an ecosystem, one that is much more dynamic, and which is more in keeping with contemporary ideas about dissipative structures and the capacity for self-organization.

Once a significant number of ecological flow diagrams have been obtained, it is an obvious next step to try to compare them. At the organismal level comparative anatomy and comparative physiology have, in their time, been fields in which rapid advances in understanding were made. It appears to us that the time is now ripe for a comparative science of ecosystems. The subject is not without its difficulties.

The flow network of an ecosystem can be made to look very different according to the conventions used in putting it together. One very striking example concerns the method of representing uptake of CO_2 in photosynthesis and its release in respiration. It is usual to regard the supply of CO_2 as effectively infinite, so that the gas is represented as flowing through the ecosystem from an infinite source to an infinite sink. However, for some purposes (a rock pool or a laboratory microcosm, for example) it is convenient to represent a finite pool of CO_2 with removals and additions. In that case, the amount of carbon recycling in the system will change dramatically. In the real world it is probable that a significant proportion of carbon atoms released by heterotrophs are taken up and rapidly recycled by autotrophs, but this question is seldom addressed explicitly. This is clearly the case with symbiotic algae and their hosts, such as tropical coral reefs and *Tridacna* clams, and many planktonic protozoa.

Nevertheless, a comparative study of ecosystem flows has been found to be well worth developing. Hannon (1973) was the first to borrow from economics the techniques of input-output analysis (Leontief 1951) and apply them to questions of ecological flow analysis. Finn (1976) used input-output analysis to develop the concept of average path length in ecological flow networks, and Ulanowicz (1980) developed indices of the development of natural communities based entirely on the structure of the flow networks.

It was a decision of SCOR Working Group 73 (the successor of Working Group 59 mentioned earlier) to pursue network analysis as a first step towards a comparative science of marine ecosystems. As mentioned in Chapter 1 (Field *et al.* 1989, this volume), a workshop was organized to bring together data from a number of marine systems and experiment with the techniques of network analysis. The results are displayed in this volume. It is now time to try to assess progress and evaluate the usefulness of the techniques.

NETWORK ANALYSES - HOW HELPFUL ARE THEY?

Ascendency, growth and development

It has been argued that ecosystems evolve in succession towards maturity, and Odum (1969) presented a summary of 24 attributes believed to characterize mature ecosystems. These attributes may be grouped under four headings as the tendencies: a) to internalize flows, b) to increase cybernetic feedback, c) to increase specialization of the components, and d) to increase the number of components in the system. Ulanowicz (1980, 1986) has suggested that both the size and organization (both growth and development) of an ecosystem can be characterized by a single index, Ascendency which should tend to increase with growth and development in a succession. Ascendency has the units of total system throughput of the currency of the ecosystem model (e.g. g C. $m^{-2}\ yr^{-1}$) and information content (bits). Ascendency and Total System Throughput are system-level properties.

Ulanowicz' hypothesis of changes in Ascendency during ecosystem succession are difficult to test because flows of matter among ecosystem components are difficult to measure and very difficult to monitor over a period of time. The closest practical tests appear to be models of ecosystem flows. Simulation models of ecosystem succession (e.g. Kremer 1989 ; Field *et al.*, 1989; Warwick & Radford, 1989) confirm the strong influence of the growth component (Total System Throughput, T) on Ascendency (A) and on its ceiling, Development Capacity (C) which are all strongly correlated. The organization component of system development (information content) appears to some extent to be dominated by the growth component (T), although both Kremer (1989) and Warwick & Radford (1989) note that Development Capacity (C) increases slightly faster than throughput (T) as their systems developed in summer, indicating that flows became more diverse as well as larger.

Kay (1984) and Kay *et al.* (1989) argue that it may be preferable to use two indices to describe system development, one to describe the diversity of flows (D) among all components in the ecosystem, and another to describe the average breadth of resource niches (S) used by all the living components in the community. These indices have the growth component (T) omitted from their calculation and thus they only describe the *proportions* of resources flowing and utilized in the system, not their magnitudes.

Both of these indices peak a month later than T, A and C in the seasonal benthic model of Warwick & Radford, indicating a greater diversity of flows (D) and broadest resource niches (S) a month after greatest throughput. A comparable result was obtained in the model of a plankton system after upwelling (Field *et al.*, 1989) with greatest flow diversity and broadest resource niches following after peak throughput and Ascendency. S and D contain different information from Ascendency, and are independent of the magnitude of flows through the system. It would appear, therefore, that if one is following the development of same system through time, one is extracting maximum value from the data by separating out the magnitude (T), flow diversity (D) and resource niche breadth of living components (S) by using all three indices. This may also be done by using Ulanowicz's unscaled measures i.e. Overheads/ Throughput (O/T), A/T, and C/T (see Fig. 7.9 for a graphic comparison of the measures).

However, all the above indices are sensitive to the model structure and its degree of aggregation and will therefore vary from investigator to investigator according to his/her individual approach to studying the system. For example, a system might be modelled by breaking it down to all its component species, in which case the network would be much larger and its flow diversity greater than the same system modelled by pooling species into trophic guilds. Similarly, even total system throughput, which is calculated by summing all inputs or outputs from all components (see Kay *et al.* 1989) will be larger if one separates all the predator-prey interactions in a food web, as opposed to pooling together several predators and their prey (such as all the zooplankton, for example) into a single compartment.

Thus Ascendency, Total System Throughput, Flow Diversity, Resource Niche Breadth and Development Capacity are all sensitive to the structure of the model network, and it may be argued that they are therefore not suitable for comparing different ecosystems, unless the systems have been modelled in the same way, probably by the same investigator(s), as in the comparison of the Baltic and Chesapeake systems (Wulff & Ulanowicz, 1989).

Eutrophication

Ulanowicz (1986b) hypothesizes that eutrophication may be defined at the ecosystem level as an increase in system Ascendency (A) due to a rise in Total System Throughput (T) that more than compensates for a concommitant fall in the mutual information of the flow network. The model of an upwelling event (Field *et al.* 1989) depicts the consequences of a rapid influx of nitrogen into a plankton system which may be viewed as a natural eutrophication event. In this case, the model plankton system takes 5-6 days before it shows signs of responding to eutrophication at the system level, and indeed all the increases in Ascendency and Capacity to day 7 are due entirely to increased nitrogen throughput (Fig. 7.10b). Thereafter, Ascendency and Capacity increase faster than Throughput, indicating that flow diversity is increasing, as is confirmed by the sudden rises in the indices D and S eight days after upwelling (Fig. 7.10c). It therefore appears that the initial system effect of eutrophication may indeed be to increase nitrogen throughput, in this upwelling example, while flow diversity increases slightly later in the succession.

So far there have been very few studies of flows in whole ecosystems as they change during and after eutrophication, and it remains an interesting hypothesis as to whether whole system effects of eutrophication can be indexed by using a combination of Ascendency and Total System Throughput.

Comparison of ecosystems

The indices of ecosystem flow networks mentioned above all have the dimensions and units of information content (bits), and some (e.g. Ascendency) also have the additional units of the flow medium (e.g. gN m^{-2} y^{-1}). As discussed above, these measures are all sensitive to the flow network structure or topology, and this depends upon the investigator's conceptual model of the ecosystem. These indices are thus unlikely to be good comparative measures for comparisons among ecosystems. Relative Ascendency (A/C) on the other hand, is a dimensionless ratio of the realised growth and development of the network relative to the potential of that particular network. In a sense, the magnitude of flow and the topology of the ecosystem network cancel out between the numerator and denominator of the ratio. Thus Relative Ascendency is a good candidate as an index for comparing the states of ecosystem flow networks.

Relative Ascendency has been calculated for a number of the flow networks described in this book, and values are summarised in Table 12.1. As noted by several contributors to this volume, A/C appears to be fairly stable and usually has values in the range 45%-65%. It is likely that the currency of the flows will influence the value of Relative Ascendency, since energy, for example, has an open flow through ecosystems, whereas nutrients such as nitrogen may be conserved and tightly recycled. The composite currency of "embodied energy" used to describe the flows of different commodities in the biosphere (Costanza and Hannon 1989), gives A/C values of 48.1%-49.5% for the Biosphere network. In the model of Narragansett Bay, the carbon based A/C values remain constant at 60% over all four seasons, whereas the nitrogen-based values fluctuate seasonally from 50% in winter to 70% in summer, suggesting that nitrogen flows are more organized, with fewer parallel paths, in summer. The nitrogen-flow upwelling plankton model of Field *et al.* (1989) however, has much higher values of A/C ranging from 95% soon after upwelling, settling down to about 70% later. The large values probably occur because the model is essentially one of a closed system, with no new nitrogen entering it after the single upwelling event. Thus both the Narragansett Bay and Benguela upwelling model analyses suggest that nitrogen flows give relative ascendency values which vary considerably with time in the same ecosystem, whereas the carbon version shows much less variation (in Narragansett Bay).

Table 12.1. Comparison of dimensionless whole system indices: Relative Ascendency (A/C) and Finn Cycling Index (CI). The model currency, type of model and number of compartments are also given (see chapters for details).

SYSTEM	SOURCE	CURRENCY	MODEL	No of Comp.	A/C	FINN CI
Biosphere	Chap 5	Embodied Energy	Commod.	9	0.481	-
	-		Process	9	0.495	-
Narrag.	Chap 6	Carbon	Jan-Sept	6	0.6	-
Bay	-	Nitrogen	Jan. (winter)	10	0.5	-
-	-	-	Mar.	10	0.59	-
-	-	-	June (summer)	10	0.7	-
-	-	-	Sept.	10	0.6	-
Benguela	Chap 7	Nitrogen	Day 1	7	0.87	0.2
Plankton	-	-	Day 2	7	0.85	0.5
after	-	-	Day 6	7	0.95	0.3
upwelling	-	-	Days 9-40	7	0.67-0.73	0.7
Baltic Sea	Chap 11	Carbon	Annual	14	0.56	.23
Chesapeake Bay	-	-	-	14	0.50	.30

Carbon flows have been compared in an 11- compartment estuarine mudflat system by Warwick and Radford (1989), with A/C values ranging from 52% in winter to 68% in summer, a range comparable to the nitrogen A/C range in Narragansett Bay given above. They also analyzed different versions of the model using annually averaged flows of carbon. The results show that Relative Ascendency is sensitive to the degree of aggregation of the model, ranging from 60.5% for the 19-compartment unaggregated model to 47.6% for the 8-compartment model aggregated on the basis of size, with intermediate values for models of intermediate numbers of compartments, aggregated in different ways.

Thus it appears that even Relative Ascendency (A/C), a dimensionless ratio, is sensitive to model structure and to the currency of the model.

While this may be disquieting for those who would compare different ecosystems on the basis of a single index, it is at the same time encouraging that the index is sensitive to the structure of foodwebs and to the currency of the flows that are being used to study the systems.

In comparing carbon flows of two coastal systems, Chesapeake Bay and the Baltic Sea, Wulff and Ulanowicz (1989) found that Chesapeake Bay had a lower A/C compared to the Baltic. The difference is used to deduce the hypothesis that the Baltic is less stressed than Chesapeake Bay. Both systems were modelled in the same way with the same type of aggregation and the same number of compartments. It would appear that this is thus a valid comparison and a reasonable conclusion.

Cycling

Odum (1969) hypothesized that one of the characteristics of mature ecosystems was that a large amount of their nutrient flow was recycled. This hypothesis can be tested in several of the models analyzed in this volume using the Finn Cycling Index which is the proportion of Total System Throughput that recycles in the system. The Finn Cycling Index is a system-level property.

Detailed models of nitrogen flows in plankton systems (Ducklow *et al.* 1989) are based on observations of warm core rings at different seasons in the North Atlantic. The inverse and a-priori models of nitrogen flow networks give different values of the Cycling Index (Table 12.1), but in all cases the values increase from April - June - October, as one might expect from Odum's (1969) hypothesis in a maturing system. Values range from 42% to 89%. It is demonstrated that this is inversely related to the f-ratio, commonly used to describe nitrogen relations in plankton ecology. Nitrogen cycling has been modeled in a much simpler way at different stages of succession after upwelling in the Benguela current (Field *et al.* 1989) and shows fluctuating values of the Finn Cycling Index after the upwelling event as the phytoplankton bloom develops, new nitrogen is used up and recycling commences in a period of dynamic succesional change (Table 12.1; see also Fig. 7.10a). Once the bloom declines, the system gradually runs down and the Cycling Index stabilizes at about 70%. On this limited evidence, the Finn Cycling Index for nitrogen flow in plankton systems appears to corroborate Odum's (1969) hypothesis when the system is approximately in steady state, as modeled in the warm core rings and after collapse of the upwelling bloom. During the early stages after model upwelling, however, there was an erratic response by the Finn Index, suggesting that it is not a simple or smooth relationship.

Energy and carbon flows have also been subjected to Finn Cycling analysis. The South Carolina North Inlet saltmarsh system (Asmus and McKellar 1989) has an index of 69% based on steady state energy flow. Carbon budgets of Chesapeake Bay and the Baltic Sea (Wulff & Ulanowicz 1989) have Finn Indices of 30% and 23% respectively. Thus a wide range of Finn Cycling Index values have been obtained in the various models and currencies studied, and the index appears to be sensitive to model structure and difficult to use on its own as a comparative index between different ecosystem studies.

Cycle distributions

Network analysis also permits more detailed studies of material cycling in ecosystems. Flows can be split into cycled and straight-through (acyclic) flows, and the cycled flows in turn into flows that cycle through arcs of one, two,... or more nodes or compartments. One may analyze which arcs have the smallest flows and those which, if disrupted, might greatly reduce nutrient recycling in the whole system. Ducklow *et al.* (1989) found that most nitrogen cycling through their 7- component warm core rings models, cycled through arcs of 3 components (Table 8.12), and very little cycled through long arcs of 5, 6 or 7 components. There is little pattern of shifts from short to long paths of cycling through the seasons in the a priori analyses, but June and October inverse analyses have a larger proportion of cycling through longer arcs. They were also able to show that most recycling in their models was through the microzooplankton, which nearly always had the highest cycling efficiency of the individual components (Table 8.13).

When cycle distributions were analysed in the salt-marsh energy flow model (Asmus & McKellar 1989), it was found that most energy cycled in loops of 2 components (69% of the cycled flows), followed by loops of 5, 3 and then 4 components, with no clear pattern to their distribution. The comparative carbon flow study of the Baltic Sea and Chesapeake Bay (Wulff & Ulanowicz 1989) found that most carbon cycled in short, fast loops of one or two components in Chesapeake Bay, whereas in the Baltic most cycles were through slower longer cycles of 2, 4 or 3 components. They conclude from this that cycle distribution is a much better indicator of ecosystem stress (through nutrient loading or pollution) than the Finn whole system Cycling Index. Systems stressed through nutrient loading are hypothesized to have a higher Finn Cycling Index (in contrast to the hypotheses of Odum 1969) but this cycling is through shorter, faster arcs. There is a measure of support for this hypothesis in the inverse analysis of warm core rings, but not in the a priori results (see paragraph above).

Diversity

Network Analysis allows analysis of foodwebs at three different hierarchical levels: the level of the whole ecosystem, the level of inter-component (bilateral) comparisons, and the level of the individual component (Kay *et al.* 1989). Flow diversity is a system-level property, like Total System Throughput, Finn's Cycling Index and Ascendency.

The diversity of flows in an ecological network has been measured in this volume using two information theory measures, Kay's index D and Ulanowicz's Development Capacity (C) divided by Total System

Throughput (T). The diversity of flows is specified by both the number of interactions in the foodweb and the evenness of those flows (if they are they all of equal magnitude, evenness is maximal). Kay's index D is based upon outflows only (the evenness and number of flows from sources) whereas Ulanowicz's C/T may be based upon both inflows and outflows (see Kay *et al.* 1989 for details). For systems in steady state, inflows balance outflows and the two indices should give similar results.

Flow diversity has been calculated in four chapters of this volume. In the Narragansett Bay estuarine study, Kremer (1989) compares the values of Development Capacity and Throughput graphically (Fig. 6.6) and finds that diversity is static through winter-spring, but increases in summer and yet more in the fall. This applies to both the carbon and nitrogen currency analyses. Flow diversity appears to increase as the seasons progress, as might be expected.

The Lynher estuary benthic ecosystem has also been modelled on a seasonal scale (Warwick & Radford 1989). Diversity (D) varies considerably from month to month, reaching a peak in July-August, with troughs in April and November and a minor peak in February. There is no obvious explanation for all these changes in flow diversity which are described as counter-intuitive, and were not obvious before computation.

The Benguela upwelling study (Field *et al.* 1989) includes both indices of flow diversity. Both measures decline slightly over the first seven days after the upwelling event as a phytoplankton bloom develops (Fig. 7.11), the indices then rise rapidly as the bloom crashes and the heterotrophic foodweb comes into play. C/T remains constant at this high level from day 8 onwards, whereas Kay's D, based on outflows only, appears to be more sensitive and reaches a gentle peak at day 12 (1.2 bits) before declining gradually to about 0.6 bits as the system ages and runs low on nitrogen. The flow diversity changes were not obvious in advance, but once calculated the general trends can be explained and make sense.

Flow diversities (C/T) have also been calculated for the comparable steady-state models of the Baltic Sea (3.10 bits) and Chesapeake Bay (2.94 bits) (Wulff and Ulanowicz 1989). The higher flow diversity, together with other attributes such as Relative Ascendency, Total System Throughput, and the number and length of cycling loops, is interpreted as evidence that the Baltic is a less stressed system than Chesapeake Bay.

The fates and sources of flows

At the level of comparisons between components in a foodweb, network analysis allows tracing the fate of flows forward to their destinations, or conversely to trace flows backwards from a component of interest to their sources ("extended diets"). Tracing flows forwards, it may be pertinent to enquire about where most of the primary production leaves a system. For example, Wulff & Ulanowicz (1989) note that most of the primary production in Chesapeake Bay leaves via respiration of the sediment bacteria, whereas in the Baltic most carbon is lost via meso-zooplankton (Fig. 11.3). They also examine the amount of primary production reaching products of special interest such as commercial fish, and find that the Baltic is four times more efficient in producing planktivorous fish from phytoplankton than Chesapeake Bay (Table 11.2). Tracing flows backwards, one can readily see how much products of special interest, such as fish, depend upon other components in the system. These dependency coefficients or "extended diets" show that in the Baltic 95% of the diet of carnivorous fish originates as phytoplankton production, compared to a figure of 42% for Chesapeake Bay where most of the indirect diet is via suspended and sediment particulate organic carbon and deposit feeders (Table 11.3). These are important and fundamental differences between the systems which are readily revealed using techniques of network analysis.

In a salt-marsh system, Asmus & McKellar (1989) use dependency coefficients to show that their oyster reef depends upon the intertidal marsh for 35% of its energy and on the water column for 69% (Table 9.4). The two sum to more than 100% because of recycling and the fact that the same energy may pass through both components. Similarly, the water column sub-system depends upon the intertidal marsh for 50% of its energy input. The subtidal benthos have an indirect diet of 52% from the intertidal marsh and 94% from the water column.

Ducklow *et al.* (1989) have examined the dependency of phytoplankton on various components of North Atlantic warm-core ring ecosystems during different seasons. Tracing the fate of flows forwards, fractional outflows are tabulated from phytoplankton to protozoa, macrozooplankton, dissolved organic nitrogen, detritus etc. (Table 8.8), showing that very little of the nitrogen outflow from phytoplankton flows directly to macrozooplankton and a considerable proportion flows to protozoa in most seasons. By backtracking, they have analysed how much phytoplankton depend upon ammonium/urea, protozoa, bacteria, and macrozooplankton for their nitrogen supply, as well as looking at the recycling of nitrogen from phytoplankton production to phytoplankton uptake (Table 8.9). They also note that macro-zooplankton depend surprisingly heavily upon protozoa in all seasons. As one might expect, most of these dependencies tend to increase in the summer and autumn, showing greater inter-dependence of components as the proportion of recycled nitrogen increases and the f-ratio ("new": total- nitrogen uptake) decreases in maturing plankton systems. The large dependencies are much greater than would be imagined from examination of the direct flows alone.

In another plankton study, nitrogen flows were simulated in the Benguela ecosystem after an upwelling event (Field *et al.* 1989). They found varying dependence of bacteria and flagellates on other components as the food web evolves with water column stability after an upwelling disturbance. Similarly, mesozooplankton depend largely upon net-phytoplankton during the phytoplankton bloom, but switch to micro-zooplankton after collapse of the bloom.

Michaels & Silver (1988) have examined different hypotheses of microbial foodweb structures using network analysis to estimate the flux of nitrogen out of the photic zone under different conditions. They used extended diets to compare the grazer utilization of nitrogen originally incorporated by each of the algal groups and found that small size classes of producer, such as picoplankton and bacteria, although constituting 78 % of production, only contributed less than half the sinking flux of nitrogen, unless generalist grazers such as salps were incorporated in the model. Using total contribution coefficients, Michaels & Silver were able to calculate the probability that incorporated inorganic nitrogen would be exported as sinking material within a unit of time for each group of producers. Although most nitrogen flow is through the pico-plankton and bacteria, most export is via net-phytoplankton (which contribute less than 10% of primary production, but over 35% of export) because of the few trophic steps between large algae and meso-zooplankton that generate sinking material.

Even in oligotrophic oceanic environments, sporadic patches of net phytoplankton can contribute substantially to particle flux, as can occasional swarms of generalist grazers such as salps which can consume small particles. The results are described as surprising and counter-intuitive because the network analyses show that most of the sinking flux is via rare, large organisms, although most of the activity is in small organisms.

Thus, both for academic and practical purposes, those researchers who have used dependency coefficients and fractional outflows have provided new insights into the dependency of species or groups of interest upon their indirect sources of food, or into the fates of flows in the system.

Trophic positions and path lengths

Network analysis allows calculation of a number of measures at the level of ecosystem components and bilateral measures for comparing components.

Among these are the Average Path Length (APL) which is the average number of components a unit of matter must pass through before leaving the system (via export, emigration, or respiration), and the Effective Trophic Position or the weighted average of the number of steps between non-living detritus or inorganic nutrients and the component of interest. The two are roughly inversely correlated, except that Average Path Length is based on calculations including recycled matter, and Effective Trophic Position is calculated omitting recycling. It is well-known to ecologists that few species feed only at one trophic level, and the Effective Trophic Position allows for that.

In the Benguela upwelling system (Field *et al.* 1989,), Effective Trophic Positions have been calculated for the various size-components at different times after upwelling. Thus flagellates are always near position 3.0, being about two steps removed from the nitrogen pool. The Effective position of meso-zooplankton, on the other hand, changes from about 3.0 to around 4.0 after the phytoplankton bloom collapses and the heterotrophic food web comes into action (Fig. 7.6).

Lindeman Spines are diagrams depicting a complex food web as a simpler food chain in which the compartments are abstract discrete steps. They permit rapid visualisation of the main grazing pathways and grazing efficiencies, and show the relative amounts of re-cycling and export at different steps along the abstract chain. They give a good overview of the whole foodweb. In the Benguela system, one can readily observe the increases in grazing efficiency, the lengthening of the food web, and the increased length of nitrogen recycling loops with time as the community evolves in a stable water column after upwelling (Figs 7.7, 7.8).

The warm-core rings study of Ducklow *et al.* (1989) shows that the Effective Trophic Position of most components in the plankton community increases with season as the food web gets more complex, there is more nitrogen recycling and trophic efficiency increases. The increased trophic efficiency in later seasons is due to a larger proportion of phytoplankton production being grazed rather than decomposed. They also note that trophic efficiency decreases with increased trophic position, i.e. further along the Lindeman Spine increasing proportions of nitrogen are recycled or exported (Table 8.11).

Asmus & McKellar (1989) analyse the effective trophic positions of components of the salt-marsh system in terms of energy flow. They find that meiofauna feed entirely at position 2, macrofauna at position 2.2, nekton at 2.85, and birds at position 3.44 (ranging from positions 2 - 5). There is an order of magnitude decrease in energy flow at each step along the Lindeman Spine, accompanied by a decrease of similar magnitude in the detritus recycled at each step.

In the Baltic Sea, planktivorous fish feed at Effective Trophic Position 3.4, as compared to 2.2 in Chesapeake Bay (Wulff & Ulanowicz 1989). This is because Baltic herring feed largely on zooplankton in summer and benthic invertebrates in winter when zooplankton are in diapause, whereas Chesapeake menhaden feed to a considerable extent on phytoplankton. The Average Path Length is reduced in Baltic carnivorous fish (1.72) which are heavily exploited compared to their Chesapeake counterparts (3.26) which suffer heavy natural mortality, resulting in intensive recycling. Comparison of Lindeman Spines between the Baltic and Chesapeake systems shows that the Baltic spine is one link longer, and this is associated with greater trophic efficiencies along the grazing chain, and that detritivory is much more important than herbivory in Chesapeake Bay than in the Baltic.

CONCLUSIONS

This volume reports the results of studies stimulated by a workshop held in at the University of California, Santa Cruz in December 1986. Most of the contributors to this volume had little or no prior experience with network analysis, but they did have many years of experience of studying the ecosystems they have analysed here. The general conclusion of the contributors is that they all gained new insights into the structure and functioning of the systems they had been studying.

Network analysis of ecosystems facilitates study of one or more of the many feedbacks and controls that operate to and fro between the ecosystem level of organization on the one hand, and community, population and even physiological levels on the other. The analyses used in this book help us understand the feedbacks and controls between different hierarchical levels, but we are very aware that flows of energy and matter are not the only controls that operate in ecosystems. For example, keystone predators may influence the community structure, visual cues may affect animal behaviour, and flows of energy or matter (like trace amounts of hormones or pheromones) may have an effect vastly greater than their volumes. Therefore these methods do not allow prediction of future states or conditions in an ecosystem, but they do give insights which may lead toward that goal.

Network analysis offers formal methods for analyzing ecosystem structure. The networks drawn by ecologists can be viewed as hypotheses of ecosystem structure which can be analyzed formally to see which hypotheses are most consistent with observations and measurements. Even more important though, is the use of network analyses to suggest key variables (such as particular flows of energy or matter, and internal variables like assimilation efficiency) which need to be measured accurately to test the hypothesis. We suggest that, when research programs are being designed in order to gather critical field or laboratory data to test hypotheses, network analyses be carried out on the best available flow diagrams as a guide to the selection of key variables to be measured.

Foodweb networks can be quantified using several means described in this volume: a) with a priori methods, used in most chapters of this volume, but formally defined by Ducklow *et al.* (1989) b), by a combination of a priori methods and simulation modelling (Ulanowicz 1989; Kremer 1989; Field *et al.* 1989; Warwick and Radford 1989), or c) by inverse methods (Vezina 1989; Ducklow *et al.* 1989).

We have seen that component-level and bilateral-level measures show clear patterns that can be used to compare different ecosystems if they are carefully studied in the same way. We have also seen that general trends become apparent when the succession of ecosystems is analysed by compressing the foodwebs into Lindeman Spines and their associated measures of Average Path Length and Effective Trophic Position. These patterns include a tendency to decreased flows and decreased trophic efficiency with greater distance along the Lindeman Spine. As ecosystems mature in succession there is a trend to lengthening of the Lindeman Spine, increased Effective Trophic Positions and more recycling along longer loops.

At the level of the whole ecosystem there are also a number of information theory measures that have been found helpful in this volume. It is suggested that Flow Diversity and Resource Niche Breadth are useful indices for following the development of ecosystem structure in the succession of a single system after disturbance, whereas Relative Ascendency, a dimensionless ratio, looks like a promising candidate for comparing the growth and development status of different ecosystems, provided that they have been carefully studied and modelled in the same way and at the same level of aggregation.

The indices are really useful for comparing the same ecosystem at different seasons or stages of succession and are also very helpful for comparing two ecosystems that have been meticulously modelled to have the same degree of aggregation. The comparison of the Baltic Sea with Chesapeake Bay is an important first. We believe that the techniques have much potential to give new insights into ecosystem structure and functioning, and they reveal patterns in ecosystem data that are not readily obtainable in another way. We trust that this book will enable other ecologists to gain as much insight into the ecosystems they study, as we have gained from using its techniques.

Ecosystem ecology has been troubled by a lack of consistency in the approach used by different investigators, and lack of an ability to make meaningful comparisons between different systems. We hope that this volume may indicate a small beginning in the new science of comparative ecosystem ecology

REFERENCES

Allen, P. M. 1985. Ecology, thermodynamics and self-organization: towards a new understanding of complexity. Can. Bull. Fish. Aquat. Sci. 213: 3-26.

Allen, P. M. and J. M. McGlade 1986. The dynamics of discovery and exploitation: The case of the Scotian Shelf groundfish fisheries. Can. J. Fish. Aquat. Sci. 43: 1187-1200.

Allen, T. F. H. and T. B. Starr 1982. *Hierarchy: Perspectives for ecological complexity*. Univ. of Chicago Press, Chicago. 353 p.

Asmus, M. L. and H. N. McKellar, Jr 1989. Network analysis of the North Inlet salt marsh ecosystem. Chapter 9 in Wulff, F., J. G. Field & K. H. Mann (eds), *Network Analysis in Marine Ecology: Methods and Applications*, Lecture Notes on Coastal and Estuarine Studies, Springer-Verlag, New York.

Atlan, H. 1985. Information theory and self-organization in ecosystems. Can. Bull. Fish. Aquat.Sci.213: 187-199.

Banse, K.1982. Mass-scaled rates of respiration and insintric growth in very small invertebrates. Mar. Ecol. Prog. Ser. 9: 281-297.

Bronowski, J. 1973. *The Ascent of Man*. Little, Brown and Co, Boston.

Calder, W. A.1985. Size and metabolism in natural systems. pp. 65-75 in: Ulanowicz, R. E., and T. Platt (eds.) *Ecosystem theory for biological oceanography.* Can. Bull. Fish. Aquat. Sci. 213, 260 p.

Cherrett, M. 1988. Ecological Concepts - the results of the survey of members views. Bull. Brit. Ecol. Soc. 19: 80-82.

Conrad M.1983. *Adaptability: the significance of variability from molecule to ecosystem.* Plenum Press, New York.

Conrad, M.1985. The statistical basic of ecological potentiality.pp.179-186 in: Ulanowicz, R. E., and T. Platt (eds.) *Ecosystem theory for biological oceanography.* Can. Bull. Fish. Aquat. Sci. 213, 260 p.

Costanza, R. and B. Hannon 1989. Dealing with the "mixed units" problem in ecosystem network analysis. Chapter 5 in Wulff, F., J. G. Field & K. H. Mann (eds), *Network Analysis in Marine Ecology: Methods and Applications*, Lecture Notes on Coastal and Estuarine Studies, Springer-Verlag, New York.

Cousins, S. H.1985. The trophic continuum in marine ecosystems: structure and equations for a predictive model..pp. 76-93 in: Ulanowicz, R. E., and T. Platt (Eds.) *Ecosystem theory for biological oceanography.* Can. Bull. Fish. Aquat. Sci. 213, 260 p.

Ducklow, H. W., M. J. R. Fasham and A. F. Vézina 1989. Derivation and analysis of flow networks for open ocean plankton systems. Chapter 8 in Wulff, F., J. G. Field & K. H. Mann (eds), *Network Analysis in Marine Ecology: Methods and Applications*, Lecture Notes on Coastal and Estuarine Studies, Springer-Verlag, New York.

Elton, C. 1927. *Animal Ecology*. Macmillan, New York. 207 p.

Fasham, M. J. R. (Ed.). 1984. *Flows of energy and materials in marine ecosystems: Theory and practice.* Plenum, New York. 733 p.

Field, J. G., C. L. Moloney and C. G. Attwood 1989. Network analysis of simulated succession after an upwelling event. Chapter 7 in Wulff, F., J. G. Field & K. H. Mann (eds), *Network Analysis in Marine Ecology: Methods and Applications*, Lecture Notes on Coastal and Estuarine Studies, Springer-Verlag, New York.

Field, J. G. , F. Wulff and K. H. Mann 1989. The need to analyze ecological networks. Chapter 1 in Wulff, F., J. G. Field & K. H. Mann (eds), *Network Analysis in Marine Ecology: Methods and Applications*, Lecture Notes on Coastal and Estuarine Studies, Springer-Verlag, New York.

Finn, J. T. 1976. Measures of ecosystem structure and function derived from analysis of flows. J. Theor. Biol. 56: 363-380.

Hannon B. 1973. The structure of ecosystems. J. Theor. Biol. 41: 535-546.

Jansson, B.-O., W. Wilmot and F. Wulff. 1984. Coupling the sub-systems - The Baltic Sea as a case study. pp 549-595 In: M.J. Fasham (Ed.) *Flows of Energy and Materials in Marine Ecosystems.* Plenum, New York. 733 p.

Kay, J. J., L. A. Graham and R. E. Ulanowicz 1989. A detailed guide to network analysis. Chapter 2 in Wulff, F., J. G. Field & K. H. Mann (eds), *Network Analysis in Marine Ecology: Methods and Applications*, Lecture Notes on Coastal and Estuarine Studies, Springer-Verlag, New York.

Kremer, J. N. 1989. Network information indices with an estuarine model. Chapter 6 in Wulff, F., J. G. Field & K. H. Mann (eds), *Network Analysis in Marine Ecology: Methods and Applications*, Lecture Notes on Coastal and Estuarine Studies, Springer-Verlag, New York

Kremer, J.N. and S.W. Nixon. 1978. *A Coastal Marine Ecosystem - Simulation and Analysis.* Springer-Verlag, New York. 217 p.

Leontief, W. 1951. *The structure of the American economy, 1919-1939*. Oxford University Press, New York. 2nd ed. 264p.

Lotka, A.J. 1925. *Elements of Physical Biology*. Williams and Wilkins, Baltimore. 460 p.

Lovelock, J.E. 1979. *Gaia, a new look at life on earth.* Oxford University Press. Oxford, 157 p.

Lovelock, J. E.1986. Geophysiology: A new look at earth science. Bull. Am. Met. Soc. 67: 392-397.

Mann, K. H. 1988. Towards predictive models for coastal marine ecosystems. pp. 283-308 In: L. R. Pomeroy and J. J. Alberts (eds.) *Concepts of Ecosystem Ecology*. Springer-Verlag, New York.

May, R. M. and G. F. Oster. 1976. Bifurcations and dynamic complexity in simple ecological models. Am. Nat. 110: 573-599.

Michaels, A. F. and M. W. Silver.1988. Primary production, sinking fluxes and the microbial food web. Deep-Sea Res. 35: 473-490

Moloney, C. M. and J. G. Field 1989.Modelling carbon and nitrogen flows in a microbial plankton community. In Reid, P. C., P. H. Burkhill and C. M. Turley (eds), *Protozoa and their role in marine processes*, NATO ASI series, Springer-Verlag, New York (In press).

Moloney, C. L., M. O. Bergh, J.G. Field and R.C. Newell 1986. The effect of sedimentation and microbial nitrogen regeneration in a plankton community: A simulation investigation. J. Plankton Res. 8: 427-445.

O'Neill, R.V., D. L. DeAngelis, J. B. Waide and T. F. H. Allen. 1986. *A hierarchical concept of ecosystems.* Princeton Univ. Press, Princeton. 253 p.

Odum, E. P.1969. The strategy of ecosystem development. Science 164: 262-270.

Odum, H. T. and E. P. Odum. 1955. Trophic structure and productivity of a windward coral reef community on Eniwetok atoll. Ecol. Monogr. 25: 291-320.

Peters, R.1983. *Ecological implication of body size relationships.* Cambridge University Press, London.

Platt, T., K. H. Mann and R. E. Ulanowicz. 1981. *Mathematical models in biological oceanography.* The UNESCO Press, Paris. 156 p.

Platt. T., and K.L. Denman. 1978. The structure of pelagic marine ecosystems. Rapp. P. -V. Reun. Cons. Int. Explor. Mer.173: 60-65.

Prigogine, I. 1945. Modération et transformations irréversibles des systemes ouverts. Bull. Acad. Roy. Belg. Cl. Sci. 31:600.

Prigogine, I.1947. *Etude thermodynamique des processus irrversibles.* Desoer. Liege.

Shannon,C.E. and W. Weaver. 1949. *The mathematical theory of communication.* Uiversity of Illinois Press, Urbana.

Stigebrandt, A. and F. Wulff 1987. A model for the dynamics of nutrients and oxygen in the Baltic proper. J. Mar. Res., 45: 729-759.

Summer, J. K., H. N. McKellar, R. F. Dame and W. M. Kitchens. 1980. A simulation model of an estuarine subsystem coupling and carbon exchange with the sea. II. North Inlet model structure, output and validation. Ecol. Modelling, 11: 101-140.

Tansley, A. G. 1935. The use and abuse of vegetational concepts and terms. Ecology 16: 284-307.

Trainor, L. E. H. 1985. Statistical mechanics in biology - applications to ecology. Can. Bull. Fish. Aquat. Sci. 213: 41-51.

Ulanowicz, R. E. 1980. An hypothesis on the development of natural communities. J. theor. Biol. 85: 223-245.

Ulanowicz, R. E. 1986. *Growth and Development: Ecosystem Phenology.* Springer-Verlag, New York, 203 p.

Ulanowicz, R. E. 1989. A generic simulation model for treating incomplete sets of data. Chapter 4, this volume.

Ulanowicz, R. E. and T. Platt. (eds.) 1985. *Ecosystem Theory for Biological Oceanography.* Can. Bull. Fish. Aquat. Sci. 213. 260 p.

Warwick, R. M. and P. J. Radford 1989. Analysis of the flow network in an estuarine benthic community. Chapter 10 in Wulff, F., J. G. Field & K. H. Mann (eds), *Network Analysis in Marine Ecology: Methods and Applications*, Lecture Notes on Coastal and Estuarine Studies, Springer-Verlag, New York.

Weiss, P. 1969. The living system: determination stratified. p. 3-55 in: A. Koestler and J. R. Smythies (eds.). *Beyond Reductionism.* Hutchinson, London.

Wulff, F. and R. E. Ulanowicz 1989. A comparative anatomy of the Baltic Sea and Chesapeake ecosystems. Chapter 11 in Wulff, F., J. G. Field & K. H. Mann (eds), *Network Analysis in Marine Ecology: Methods and Applications*, Lecture Notes on Coastal and Estuarine Studies, Springer-Verlag, New York.

INDEX

Acyclic flow 186
Aggregation 38, 46, 52, 56, 129, 177, 181, 207, 221, 225, 251, 269, 271, 279
Agulhas 133
Ammonium 73, 101
Amphipods 239
Anchovies 6, 133, 240
Anoxia 233
Arcs 16, 43,182
A priori method 9, 71, 161
Arthropods 223
Ascendency 7, 53, 55, 103, 119, 127, 150, 225, 249
Aschelminthes 223
Assimilation 74, 136, 167
Ataxonomic aggregations 4, 265
Average Path Length 6,47, 246, 276, 279
Backward flow analysis 21
Bacteria 73, 88, 133, 162, 223
Bacterial production 162, 220, 238
Baltic Sea 232, 260
Benguela Upwelling System 133
Benthos 207, 220, 233
Birds 213
Bivalves 239
Bluegreen algae 236
Capacity (see development capacity)
Carbon 73, 120, 162, 207, 221
Carnivores 88, 223, 239
Chesapeake Bay 29, 232, 269
Chlorophyll 240
Ciliates 199, 238
Coastal ecosystem 120
Coelenterate 223
Commensurability 91
Commodity 92
Compartmental analysis 4, 90, 94
Computer programs 15
Cone Spring 17, 86
Contribution coefficients 8, 29, 210
Copepods 223, 238
Coral reef 6, 267
Crystal River
Ctenophores 239
Cycle analysis 41, 181, 212, 247, 272
Cycled flow, 160, 211
Cycling efficiency 186
Cycling index 5, 25, 46, 103, 151, 213, 249
Decapods 240
Dependency coefficients (indirect diets) 8, 29, 141, 178, 211
Deposit feeders 210, 221, 240
Destination probabilities 34
Detritivores 88
Detritus 73, 164, 213
Development capacity (C) 7, 54, 103, 119, 127, 140, 151, 225,249, 268, 273
Diatoms 6, 124, 236
Dinoflagellates 234
Dissolved inorganic nitrogen /DIN 120, 134, 161
Dissolved organic carbon /DOC 136, 236
Dissolved organic matter /DOM 73, 134
Dissolved organic nitrogen /DON 134, 161
Dissolution 74
Echinoderms 240
Edges 16
Effective trophic position 8
Embodied energy 102, 214, 270
Energy intensities 10, 102
English Channel 63, 72
Eniwetok atoll 261
Entropy 54, 59, 264, 265
Euphausids 139
Euphotic zone 134, 163
Euryhaline 233
Eutrophication 54, 233, 269
Extended diets 29, 177, 275
Faeces 135, 236
Filter feeders 210, 239
Finfish 206, 240
Finn cycling index (see cycling index)
Fish 6, 8, 29, 126, 194, 233, 240, 275
Fish larvae 7, 133
Flagellates 7, 134, 236
Flow analysis 4,19, 90
Flow diversity (D) 6, 7, 120, 127, 140, 225, 249, 273, 274
Flow speciation (S) 7
Food chains 6, 8, 180, 246
Food webs 6, 7, 62, 73, 155, 160
Forward outflow analysis 21
Grazing chain 40
Halocline 233
Herbivore 239
Heterotroph 85
Hierarchy 259, 263
Indirect diets 144, 242
Information theory 4, 7, 15, 52, 59, 120, 266
Ingestion 65, 74, 121, 136, 163, 192

Input environs 24, 28, 33
Input-output analysis 4, 15, 19, 90, 160, 241
Inverse method 9, 63, 161, 203
James River 30
Kinorhyncha 223
Keystone species 263
Lindeman (spine) 8, 38, 144, 155, 180, 188, 213, 246
Lynher estuary 220
Macrobenthos 221
Macrofauna 125, 213, 233
Macrozooplankton 6, 163
Marine snow 169
Markovian process 30
Meiobenthos 213, 221, 223
Mesohaline 234
Mesozooplankton 73, 134, 201, 238
Metazoans 73
Mirobial loop 78, 159, 119, 201, 238
Microheterotrophs 134
Microzooplankton 73, 134, 163
Mixed units 9
Molluscs 223
Mudflat 220
Nanofytoplankton 134
Narragansett Bay 120, 229
Nekton 213, 240
Nematoda 221
Network analysis 4, 5, 90, 267
Network thermodynamics 5
Nexus 44
Niche 149
Nitrate 73, 101, 160
Nitrogen 6, 73, 126, 134, 162, 207
Nitrogen fixation 236
Nitrogen, new 135, 160
Nitrogen, regenerated 135, 155
Nodes 16
North Inlet 207
North Sea 6
Oligochaetes 223
Oligohaline 234
Oligotrophic 233
Open weighted networks 16
Origin probabilities 34
Ostracods 223
Overhead 54, 150, 228, 249
Oyster 207
Path length 25, 212, 246
Particulate organic carbon /POC 223, 239
Particulate organic matter /POM 76
Particulate organic nitrogen /PON 76
Phosphate 233
Phosphorus 6, 233
Phytobenthos 223, 239
Phytoplankton 120,133,163,223
Picoplankton 6, 78, 201, 276
Plankton 62, 120, 132, 159, 262
Pollution 251
Polychaetes 240
Potomac River 234
Predator 30, 37, 45, 84, 121, 137, 159, 210, 239
Prey 45, 84, 121, 137
Primary production 6, 74, 162, 233
Production efficiency 74
Production, new 159
Production, regenerated 199
Protozoa 73, 163
Redundancy 61, 225, 249
Regeneration 160
Relative Ascendency 155
Residence time 34
Resource niche (S) 47, 49, 150
Respiration 18, 74, 136
Rhode Island 120
Rotifers 238
Salt marches 206
Sardines 6, 133
Seals 253
Sedimentation (sinking) 135, 155, 160, 204, 233
Shellfish 206
Simulation modelling 9, 82, 120, 132, 159, 207, 220, 261
Spawning ground 133, 239
Species diversity 6, 120, 130, 251
Spring bloom 7
Stability 130, 251, 263
Statistical mechanics 4, 265
Storage partition 34
Stress 46, 53, 251
Structural organization (D) 49, 55, 150,177
Succession 7, 133, 155, 170
Suspension feeders 240, 249
Susquehana River 234
Synecology 260
System properties 21, 45, 47
Thailand 6
Thermocline 233
Thermodynamics 4, 45, 97, 265
Throughflow 20 25
Tidal creek 207
Total System Throughflow (T) 7, 21, 45, 127, 150, 225, 249
Toxicity 253
Trophic efficiency 9, 37, 144, 180
Trophic level (position) 8, 37, 155, 180, 207, 245
Tuna 6
Turnover 25, 31
Upwelling 6, 51, 132
Warm core ring 50, 162
Zooplankton 7, 121

Coastal and Estuarine Studies

(formerly Lecture Notes on Coastal and Estuarine Studies)

Vol. 1: J. Sündermann, K.-P. Holz (Eds.), Mathematical Modelling of Estuarine Physics. Proceedings, 1978. 265 pages. 1980.

Vol. 2: D.P. Finn, Managing the Ocean Resources of the United States: The Role of the Federal Marine Sanctuaries Program. 193 pages. 1982.

Vol. 3: M. Tomczak Jr., W. Cuff (Eds.), Synthesis and Modelling of Intermittent Estuaries. 302 pages. 1983.

Vol. 4: H.R. Gordon, A.Y. Morel, Remote Assessment of Ocean Color for Interpretation of Satellite Visible Imagery. 114 pages. 1983.

Vol. 5: D.C.L. Lam, C.R. Murthy, R.B. Simpson, Effluent Transport and Diffusion Models for the Coastal Zone. 168 pages. 1984.

Vol. 6: M.J. Kennish, R.A. Lutz (Eds.), Ecology of Barnegat Bay, New Jersey. 396 pages. 1984.

Vol. 7: W.R. Edeson, J.-F. Pulvenis, The Legal Regime of Fisheries in the Caribbean Region. 204 pages. 1983.

Vol. 8: O. Holm-Hansen, L. Bolis, R. Gilles (Eds.), Marine Phytoplankton and Productivity. 175 pages. 1984.

Vol. 9: A. Pequeux, R. Gilles, L. Bolis (Eds.), Osmoregulation in Estuarine and Marine Animals. 221 pages. 1984.

Vol. 10: J.L. McHugh, Fishery Management. 207 pages. 1984.

Vol. 11: J.D. Davis, D. Merriman (Eds.), Observations on the Ecology and Biology of Western Cape Cod Bay, Massachusetts. 289 pages. 1984.

Vol. 12: P.P.G. Dyke, A.O. Moscardini, E.H. Robson (Eds.), Offshore and Coastal Modelling. 399 pages. 1985.

Vol. 13: J. Rumohr, E. Walger, B. Zeitzschel (Eds.), Seawater-Sediment Interactions in Coastal Waters. An Interdisciplinary Approach. 338 pages. 1987.

Vol. 14: A.J. Mehta (Ed.), Estuarine Cohesive Sediment Dynamics. 473 pages. 1986.

Vol. 15: R.W. Eppley (Ed.), Plankton Dynamics of the Southern California Bight. 373 pages. 1986.

Vol. 16: J. van de Kreeke (Ed.), Physics of Shallow Estuaries and Bays. 280 pages. 1986.

Vol. 17: M.J. Bowman, C.M. Yentsch, W.T. Peterson (Eds.), Tidal Mixing and Plankton Dynamics. 502 pages. 1986.

Vol. 18: F. Bo Pedersen, Environmental Hydraulics: Stratified Flows. 278 pages. 1986.

Vol. 19: K.N. Fedorov, The Physical Nature and Structure of Oceanic Fronts. 333 pages. 1986.

Vol. 20: A. Rieser, J. Spiller, D. VanderZwaag (Eds.), Environmental Decisionmaking in a Transboundary Region. 209 pages. 1986.

Vol. 21: Th. Stocker, K. Hutter, Topographic Waves in Channels and Lakes on the f-Plane. 176 pages. 1987.

Vol. 22: B.-O. Jansson (Ed.), Coastal Offshore Ecosystem Interactions. 367 pages. 1988.

Vol. 23: K. Heck, Jr. (Ed.), Ecological Studies in the Middle Reach of Chesapeake Bay. 287 pages. 1987.

Vol. 24: D.G. Shaw, M.J. Hameedi (Eds.), Environmental Studies in Port Valdez, Alaska. 423 pages. 1988.

Vol. 25: C.M. Yentsch, F.C. Mague, P.K. Horan (Eds.), Immunochemical Approaches to Coastal, Estuarine and Oceanographic Questions. 399 pages. 1988.

Vol. 26: E.H. Schumann (Ed.), Coastal Ocean Studies off Natal, South Africa. 271 pages. 1988.

Vol. 27: E. Gold (Ed.), A Law of the Sea for the Caribbean: An Examination of Marine Law and Policy Issues in the Lesser Antilles. 507 pages. 1988.

Vol. 28: W.S. Wooster (Ed.), Fishery Science and Management. 339 pages. 1988.

Vol. 29: D.G. Aubrey, L. Weishar (Eds.), Hydrodynamics and Sediment Dynamics of Tidal Inlets. 456 pages. 1988.

Vol. 30: P.B. Crean, T.S. Murty, J.A. Stronach, Mathematical Modelling of Tides and Estuarine Circulation. 471 pages. 1988.

Vol. 31: G. Lopez, G. Taghon, J. Levinton (Eds.), Ecology of Marine Deposit Feeders. 322 pages. 1989.

Vol. 32: F. Wulff, J.G. Field, K.H. Mann (Eds.), Network Analysis in Marine Ecology. 284 pages. 1989.

MIX
Papier aus verantwortungsvollen Quellen
Paper from responsible sources
FSC® C105338

If you have any concerns about our products,
you can contact us on
ProductSafety@springernature.com

In case Publisher is established outside the EU,
the EU authorized representative is:
Springer Nature Customer Service Center GmbH
Europaplatz 3, 69115 Heidelberg, Germany

Printed by Libri Plureos GmbH
in Hamburg, Germany